Biosocial Bases of Violence

NATO ASI Series

Advanced Science Institutes Series

A series presenting the results of activities sponsored by the NATO Science Committee, which aims at the dissemination of advanced scientific and technological knowledge, with a view to strengthening links between scientific communities.

The series is published by an international board of publishers in conjunction with the NATO Scientific Affairs Division

A	Life Sciences	Plenum Publishing Corporation
B	Physics	New York and London
C	Mathematical	Kluwer Academic Publishers
	and Physical Sciences	Dordrecht, Boston, and London
D	Behavioral and Social Sciences	
E	Applied Sciences	
F	Computer and Systems Sciences	Springer-Verlag
G	Ecological Sciences	Berlin, Heidelberg, New York, London,
H	Cell Biology	Paris, Tokyo, Hong Kong, and Barcelona
I	Global Environmental Change	

PARTNERSHIP SUB-SERIES

1. Disarmament Technologies	Kluwer Academic Publishers
2. Environment	Springer-Verlag
3. High Technology	Kluwer Academic Publishers
4. Science and Technology Policy	Kluwer Academic Publishers
5. Computer Networking	Kluwer Academic Publishers

The Partnership Sub-Series incorporates activities undertaken in collaboration with NATO's Cooperation Partners, the countries of the CIS and Central and Eastern Europe, in Priority Areas of concern to those countries.

Recent Volumes in this Series:

Volume 289 — Neurobiology: Ionic Channels, Neurons, and the Brain
edited by Vincent Torre and Franco Conti

Volume 290 — Targeting of Drugs 5: Strategies for Oligonucleotide and Gene Delivery in Therapy
edited by Gregory Gregoriadis and Brenda McCormack

Volume 291 — Recollections of Trauma: Scientific Evidence and Clinical Practice
edited by J. Don Read and D. Stephen Lindsay

Volume 292 — Biosocial Bases of Violence
edited by Adrian Raine, Patricia A. Brennan, David P. Farrington, and Sarnoff A. Mednick

Series A: Life Sciences

Biosocial Bases of Violence

Edited by

Adrian Raine
University of Southern California
Los Angeles, California

Patricia A. Brennan
Emory University
Atlanta, Georgia

David P. Farrington
Cambridge University
Cambridge, United Kingdom

and

Sarnoff A. Mednick
University of Southern California
Los Angeles, California

Plenum Press
New York and London
Published in cooperation with NATO Scientific Affairs Division

Proceedings of a NATO Advanced Study Institute on
the Biosocial Bases of Violence,
held May 12 -- 21, 1996,
in Rhodes, Greece

NATO-PCO-DATA BASE

The electronic index to the NATO ASI Series provides full bibliographical references (with keywords and/or abstracts) to about 50,000 contributions from international scientists published in all sections of the NATO ASI Series. Access to the NATO-PCO-DATA BASE is possible in two ways:

—via online FILE 128 (NATO-PCO-DATA BASE) hosted by ESRIN, Via Galileo Galilei, I-00044 Frascati, Italy

—via CD-ROM "NATO Science and Technology Disk" with user-friendly retrieval software in English, French, and German (©WTV GmbH and DATAWARE Technologies, Inc. 1989). The CD-ROM also contains the AGARD Aerospace Database.

The CD-ROM can be ordered through any member of the Board of Publishers or through NATO-PCO, Overijse, Belgium.

Library of Congress Cataloging-in-Publication Data

Biosocial bases of violence / edited by Adrian Raine ... [et al.].
        p.   cm. -- (NATO ASI series. Series A, Life sciences ; v.
    292)
      Includes bibliographical references and index.
      ISBN 0-306-45601-X
      1. Violence--Physiological aspects. 2. Violence--Psychological
    aspects. 3. Violence--Social aspects.   I. Raine, Adrian.
    II. Series.
    RC569.5.V55B56   1997
    616.85'82--dc21   97-13920
                                                                  CIP

ISBN 0-306-45601-X

© 1997 Plenum Press, New York
A Division of Plenum Publishing Corporation
233 Spring Street, New York, N. Y. 10013

http://www.plenum.com

10 9 8 7 6 5 4 3 2 1

Printed in the United States of America

PREFACE

There are notable features about violence research. First, research over the past 50 years has identified many important familial, psychosocial, and community influences on violence. Furthermore, within the past decade there has been a sudden growth in interest in biological contributions to violent behavior. Second, it is clear that violence is a complex form of behavior that defies a simple explanation. Consequently, any successful approach to understanding such behavior must take into account the multiple social, psychological, and biological processes that conspire together to create the violent individual. Despite these two obvious facts, most research on violence is not conducted within a multidisciplinary, integrative biosocial framework.

The goal of this book is to begin to lay the groundwork for a genuine integration between psychosocial and biological processes in attempting to explain violence. While there is cursory acknowledgment of the biosocial approach in the literature, this is largely superficial and does not tackle the specific manner in which biological factors interact with social factors in the development of violent behavior in humans. Indeed, there are only a handful of empirical studies which have tackled this issue head-on. The reason for this gap in the field is that such integration is extremely difficult for both practical and conceptual reasons. The contributions in this book cannot hope to suddenly fill this void overnight. Nevertheless, it is hoped that they can at last begin to lay the foundations to the evolution of biosocial violence research in the forthcoming years by both adding to the very few empirical demonstrations of the biosocial interaction perspective, and also by clarifying conceptual and theoretical issues.

This book is aimed at academics who study crime and violence. It focuses on violence because this is a critical problem in society, but it also contains perspectives on aggressive and antisocial behavior in children because such behaviors are critical contributions to adult violence. As such, this book should be of interest to all research scientists, graduate students, and advanced undergraduates in psychology, sociology, criminology, psychiatry, social work, law, and medicine who have an interest in the causes of violent and antisocial behavior in both children and adults.

The first chapter provides an overview of previous biosocial research and discusses conceptual and theoretical questions and issues which future biosocial research must address. The next section on theory includes four chapters which offer biosocial theories of violence hinging on the concepts of personality, temperament, and cognitive processes, and one chapter which deals with secular trends. The section on psychophysiology contains four contributions which focus largely on autonomic and emotional processes and their interactions with family and demographic factors in understanding aggression in children, official and self-report violence in adulthood, psychopathic behavior, and wife batterers. Perinatal factors (birth complications) and constitutional factors (physique) are dealt with in the next section which is in turn is followed by a section on neurotransmitters. The three contributions making up this latter section analyze the ways in which serotonin interacts with environmental influences (social attachment, cohesion, and dominance) in predisposing to aggression and violence in both man and animals. This is followed by two chapters on hormones in conjunction with social and individual influences such as social dominance and depression. The last chapter attempts to highlight some of the key issues raised in these chapters and provide directions for future research. Finally, a series of short papers, largely from junior scientists, is included which cover topics of brain imaging, neuropsychology, obstetrics, psychophysiology, hormones, and neurotransmitters, and their interaction with psychosocial variables.

This book is the product of a NATO Advanced Study Institute on the Biosocial Bases of Violence which was held at the Rodos Palace Hotel on the island of Rhodes, Greece from May 12–21, 1996. This meeting was directed by Adrian Raine, and co-directed by Patricia Brennan, David Farrington, and Sarnoff Mednick. This interdisciplinary conference brought together junior and senior scientists from Europe and North America with the two key aims of stimulating new findings on the interaction between biological and psychosocial factors in predisposing to violence, and transmitting this knowledge base to the next generation of young scientists throughout the world. In addition to the scientists listed as authors in this book, this conference would not have been a success without the participation of many other participating scientists, including John Archer, Laura Baker, Ernest Barratt, Robert Cancro, Avshalom Caspi, Michael Dawson, Marian Junger, Malcolm Klein, Joan McCord, Peter Venables, Frank Verhulst, Jan Volavka, and Per-Olof Wikstrom.

Great debts of thanks are due to Susan Bihrle (conference assistance), Lori Lacasse (subject and author indexes), Dyanne van Peter (manuscript organization), Susan Stack (conference organization), and Pauline Yaralian (conference assistance). Finally, the book and conference would not have been possible without the generous grant support from NATO, and a Research Scientist Development Award (5 KO2 MH)1114-04) from the National Institute of Mental Health.

Adrian Raine

CONTENTS

This book is dedicated to the memory of Hans Eysenck, a pioneer in biosocial theories of crime. He made an extraordinary contribution to the NATO meeting, which was one of the last major conferences he attended.

BIOSOCIAL BASES OF VIOLENCE

Conceptual and Theoretical Issues

Adrian Raine,[1] Patricia Brennan,[2] and David P. Farrington[3]

[1]Department of Psychology
University of Southern California
Los Angeles, California 90089-1061
[2]Department of Psychology
Emory University
Atlanta, Georgia 30322
[3]Institute of Criminology
Cambridge University
Cambridge, United Kingdom

One of the most surprising features of research on violence is the discrepancy between the belief that a biosocial approach to violence is important and noteworthy (Mednick & Christiansen, 1977), and the reality that there are few good examples of ongoing biosocial research into the origins of antisocial and violent behavior (Brennan & Raine, 1995). One reason for the dearth of empirical biosocial research may be due to the fact that this approach has not previously been clearly enunciated, with numerous conceptual and theoretical issues requiring clarification.

The overarching goal of this chapter is to provide an initial attempt to address conceptual and theoretical issues that must be resolved before true advances can be made in understanding the causes of violence. Because of the overlap between violence and other types of criminal and antisocial behavior (since individuals who commit one type have an elevated risk of committing other types), we will focus not only on violence but also on these other types of antisocial behavior.

This chapter has six specific aims:

1. To define what we mean by "social" and "biological".
2. To define more precisely what we mean by the term "biosocial", and differentiate it from non-biosocial approaches.
3. To outline briefly the few biosocial theories developed to date.
4. To outline studies which have shown an interaction between social and biological factors in predisposing to violence, illustrating the different approaches and designs that have been (and can be) used.

Biosocial Bases of Violence, edited by Raine *et al.*
Plenum Press, New York, 1997

    5. To develop a list of key empirical and conceptual questions that need to be addressed and which are *specific* to a biosocial approach (e.g. not questions that are applicable to violence research in general).

    6. To outline an initial heuristic biosocial model which may be of use in generating hypotheses that can be empirically addressed.

It must be acknowledged at the outset that our clarification of these issues is initial and exploratory, and that there is room for considerable disagreement about definitions. Nevertheless, it is hoped that this chapter will provide a background and common context within which the other chapters in this book may be viewed, and that it will provoke questions and debate that will help to advance the biosocial approach to violence research.

THE BIOSOCIAL APPROACH TO VIOLENCE: DEFINITION AND CONCEPTUAL CLARIFICATIONS

Definitions of "Biological" and "Social" Variables

The definition of a variable depends partly on the theoretical nature of the construct being measured and partly on the operational definition (method of measurement) of that construct. We define a variable as "biological" when it reflects a biological measure of a biological construct; for example, when physiological arousal is measured by the basal level of skin conductance. Similarly, we define a variable as "social" when it reflects a social measure of a social construct; for example, when socioeconomic status is measured by the occupational prestige of the main family breadwinner.

For ease of exposition, we will include "psychological" variables (individual difference variables such as intelligence and impulsivity) in the "social" category, along with family, peer, school, neighborhood, situational and other environmental influences. If we did not do this, we would have define "biopsychological" or "biopsychosocial" research, which is a degree of complexity that we wish to avoid. Nevertheless, the fact that IQ has a substantial genetic component (and hence biological basis) underlines the difficulty in inferring a process (biological or social) simply from the nature of the measure.

Most variables can be classified relatively unambiguously as either "biological" or "social". However, at the boundary, it can be difficult to decide which category a variable belongs in. For example, if a neuropsychological deficit (a biological construct) is measured using a psychological test such as Verbal Fluency (a social measure) it is unclear whether this variable should be defined as biological or social. While our definition gives no clear guidance on this issue, we would tend to view neuropsychological measures as biological in nature. This issue nevertheless illustrates the potential for a false dichotomies in attempting to distinguish biological from social processes.

It is also noted that a social variable may influence violence through a biological mechanism, and vice versa. For example, a broken home may cause a low heart rate (Wadsworth, 1976), which in turn may cause violence; or physical child abuse may cause brain dysfunction (Milner & McCann, 1991), which in turn may cause violence. Conversely, head injury may cause low socioeconomic status which in turn may cause violence.

Definitions of "Risk" and "Protective" Factors

"Risk factors" are factors associated with a relatively high risk of a negative out-come such as violence. Such factors may or may not have a causal influence. For example, the link between poor parental supervision and violence may be causal, but the link be-tween childhood aggression and adult violence may reflect continuity in the underlying behavioral phenomenon.

A "protective factor" could be defined as a factor associated with a relatively low risk of a negative outcome. In some cases, a protective factor could merely be the opposite side of the coin to a risk factor. For example, just as low intelligence may be a risk factor, high intelligence may be a protective factor. However, a variable could have a "protec-tive" effect in the absence of any "risk" effect, if it is nonlinearly related to the negative outcome (Farrington, 1994). For example, having few friends may be associated with a low risk of violence (compared with having an average number of friends), but having many friends may not be associated with a high risk of violence.

Risk and protective factors imply categorical variables, and they are often investi-gated using dichotomized variables. When Stouthamer-Loeber et al. (1993) trichotomized variables, they found that most risk factors had correspondingly opposite protective fac-tors, and that no protective factor existed without a correspondingly opposite risk factor. Because of the link between risk and protective factors, Rutter (1985, 1990) argued that protective factors should be defined on the basis of interaction effects, to the extent that they buffer or nullify the effect of a risk factor. Similarly, Garmezy et al. (1984) defined a protective factor as one that helps an individual to compensate in the face of risk factors or high levels of stress.

Definition of Biosocial

A necessary condition for biosocial research is that a researcher measures both bio-logical and social variables. However, in our view it is not sufficient merely to investigate the links between biological variables and violence separately from the links between so-cial variables and violence. We would define research as "biosocial" only if:

    a. the researcher investigates if biological variables are related to violence inde-pendently of (after controlling for) social variables, and vice-versa; or if the re-searcher investigates whether biological and social factors have additive main effects; or

    b. the researcher investigates biological and social variables in causal sequences (e.g., a social variable influencing a biological variable which in turn influences violence); or

    c. the researcher investigates whether biological and social variables interact in their influence on violence.

We will focus especially on such interaction effects, as we believe they have been sorely neglected and are particularly important.

We would again stress that "biosocial" means different things to different re-searchers and that ours is an initial working definition. In the next section we expand on our definition with examples. However, we should note that we do not include the socio-biological perspective (e.g., Daly & Wilson, 1988) in our "biosocial" framework. Ac-cording to the sociobiological perspective, antisocial behavior results from genetic selection processes in response to changing social conditions. We have excluded this perspective because we do not feel that it fits well with our focus of providing a conceptual basis for hypothesis testing and empirical research. Nevertheless, we believe a sociobiological approach to crime is important in its own right (see Raine 1993 for a more detailed discussion).

Variations on a Biosocial Approach to Violence. There are numerous "biosocial" approaches used in research on violence. Several are listed below. The simplest forms are given first, progressively increasing in sophistication.

1. Correlated Biosocial Effect. In perhaps the simplest example of what may be termed the correlated biosocial effect, a biological risk factor (e.g. prefrontal dysfunction) and a social risk factor (e.g. child abuse) are both related to violence, but the high correlation between the two measures means that there is a nonsignificant increment in explaining violence when both sets of variable are used. Unless there are strong correlations between social and biological risk factors such findings may not be frequently expected.

2. Additive Biosocial Effect. A more interesting example of a biosocial effect is where the researcher shows that when subjects have *both* biological and social deficits, they are more likely to become violent than if they have only one kind of deficit. For example, Raine et al. (1996) found that individuals with both biological and social deficits (neurological problems and adverse family environments) were twice as likely to become violent as those with either biological deficits alone (obstetric problems) or social problems alone (poverty). In this case, however, the effects were additive and not interactive (see Figure 1).

3. Sequential Biosocial Effect. The sequential biosocial effect refers to cases in which it can be demonstrated, for example, that a social factor causes changes in biologi-

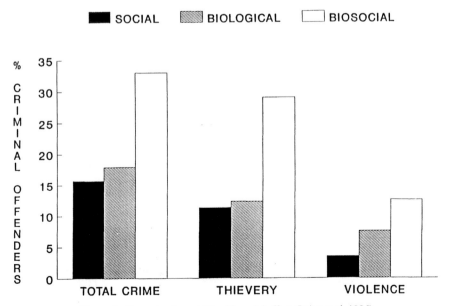

Figure 1. Example of an additive biosocial effect (Raine et al. 1996).

cal functioning which in turn cause violence. Alternatively, it is conceivable that a biological factor could cause a change in social circumstances which in turn predispose towards violence.

One example of the first scenario would be where physical child abuse (e.g. severe shaking of the child by the parent or blows to the child's head) results in damage to the prefrontal cortex which in turn predisposes to violence by virtue of releasing inhibitory control over the limbic system. Another example would be where a loss of social status results in a reduction in serotonin which in turn results in violent behavior as a means of increasing the individual's social ranking.

Sequential biosocial effects require prospective longitudinal research to delineate the pathways towards later violence and the temporal ordering of variables. While both of the examples given above are plausible, they have not been systematically tested in humans.

4. Experimental Biosocial Effect. Experimental biosocial effects can be demonstrated most convincingly in experiments which actively manipulate the social context or a biological variable. Such studies are especially important to the biosocial perspective.

The first type of experimental study involves manipulating the social context and looking for changes in a biological measure. For example, Pitts (1993) measured tonic heart rate in both a resting state and also in a state in which male children were "challenged" by having them listen to a tape designed to instill aggression. Reactively aggressive boys, but not controls, showed an increase in heart rate to a reactive aggression challenge, suggesting that aggressive situations might act as a source of stimulation in this group to increase arousal back to normal levels (since reactive aggressive boys had lower heart rates in a resting state than controls).

The second type of experimental study is almost a reversal of this design. In this scenario, a biological variable is manipulated, and its effects on a social variable are stud-

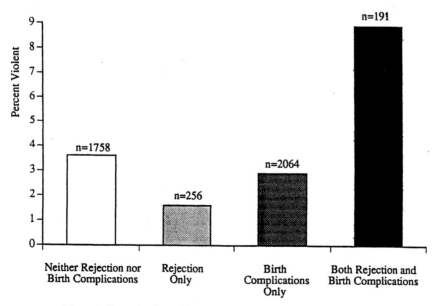

Figure 2. Example of a multiplicative biosocial effect (Raine et al. 1994).

ied. For example, physiological arousal level may be manipulated using biofeedback and its effects on reducing sensation-seeking behavior and eventual aggression tested.

5. Multiplicative Biosocial Effect. Interactive effects may be essentially multiplicative. In the example in Figure 2, violence is disproportionately higher in those with both birth complications and early maternal rejection, relative to those who only possess one of these risk factors (Raine et al. 1994). Critically, there is a multiplicative effect as opposed to the simple additive effect that was illustrated in Figure 1. If a researcher tests for an interaction and fails to find one, this would still fall within our definition of biosocial research.

6. Protective Biosocial Effect. The second example of an interactive biosocial effect is conceptually different. In this, the strength of the relationship between one variable (e.g., biological) and violence is different at different levels of the other variable (e.g., social). If one variable (a risk factor) had no effect on violence at one level of the other variable, the latter variable may be regarded as a protective factor in Rutter's (1990) sense. For example, in Figure 2, birth complications do not predict violence among those whose mothers were nonrejecting, and conversely maternal rejection does not predict violence among those who experienced normal births.

Similarly, Figure 3 illustrates an example of a biosocial study which demonstrates the compensatory protective factor model (Brennan et al. 1996). In this study, children with criminal fathers (high-risk) who did not become criminal themselves, evidence heightened levels of autonomic responsiveness in comparison to high-risk individuals who became criminal and also in comparison to low-risk controls. This study suggests that enhanced autonomic responsiveness may compensate for (be protective in) environments characterized as high risk for criminal outcome.

7. Crossover Biosocial Effect. Finally, the influence of one variable may be opposite at different levels of the other variable. For example in the research of Raine and Venables (1981), high SES antisocial boys showed poor classical conditioning, while low SES antisocial boys showed good conditioning (see Figure 4). Conversely, high SES prosocial boys showed good classical conditioning, whereas low SES prosocial boys showed poor conditioning, viz. a crossover effect. One interpretation of this result is that boys who condition well will tend to become antisocial in an antisocial environment and prosocial in a prosocial environment, whereas boys who condition poorly in a prosocial environment will tend to become antisocial, and boys who condition poorly in an antisocial environment will tend to become prosocial. However, it is difficult to draw firm conclusions about the direction of causal influence from a non-longitudinal study where the measure of antisocial - prosocial environment (SES) is relatively indirect.

In studying biosocial interactive effects, an important theoretical issue is whether the effects of a biological variable will be stronger in favorable social conditions (where the "social push" toward violence is weaker) or whether it has stronger effects in adverse social conditions (because of multiplicative risk). More research is needed on this key issue.

DATA ANALYTIC TECHNIQUES

It is not difficult to establish that a biological variable is a risk factor for violence. Since this merely requires showing that violence is significantly more likely when the bio-

Heart Rate Orienting Response and Criminal Status

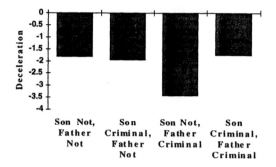

Skin Conductance Responders and Criminal Status

Skin Conductance Orienting Response and Criminal Status

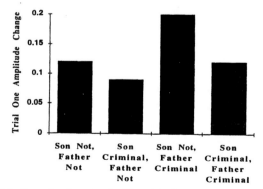

Figure 3. Example of a protective biosocial effect (Brennan et al. 1996).

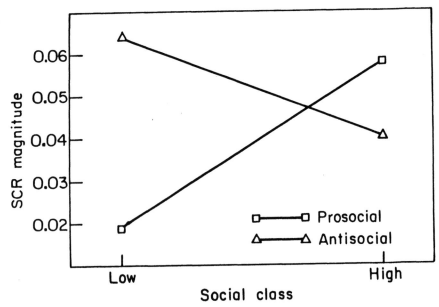

Figure 4. Example of a crossover biosocial effect (Raine and Venables, 1981).

logical variable is present than when it is absent (or that changes in the biological variable are followed by changes in violence). Similarly, it is not difficult to establish that a social variable is a risk factor for violence. It is also straightforward to establish whether a bio-logical variable predicts violence independently of a social variable or vice-versa; stand-ard regression techniques can be used.

It is more difficult to establish biosocial pathways or sequential causal models, sug-gesting for example that changes in a social variable produce changes in a biological vari-able with in turn produce changes in violence. It is possible to contrast simple models (e.g. the above versus a more complex model that also includes a direct causal effect of the so-cial variable on violence). However, real life analyses commonly (and indeed should) in-volve several biological and several social variables which greatly increases the number of possible models that could be tested. It is difficult to prove conclusively that one model is the best model, although it is easy to test whether a particular model can fit the data.

In deciding on the most plausible causal model, it may be useful to carry out a number of regression analyses with different variables as the dependent variable, or to deal with the variables in blocks. For example, assume that the best predictor of violence is an individual difference variable such as impulsivity. If so, the best predictor of impul-sivity out of other types of variables could be established. If the best predictor of impul-sivity is found to be poor parental supervision, then the best predictor of poor supervision out of other types of variables could be established. If the best predictor of poor supervi-sion is found to be a poor neighborhood, for example, these successive regressions might suggest that a poor neighborhood causes poor parental supervision, which in turn causes impulsivity, which in turn causes violence.

Establishing interaction effects is also difficult. Traditionally, interaction effects have been entered in equations as multiplicative variables (A x B). If this multiplicative variable predicted violence independently of the main effects (A and B in this example), this would show either that the combination of high A and high B was associated with a disproportionately high level of violence and/or that the combination of low A and low B was associated with a disproportionately low level of violence. However, in this strategy, the effects of a combination of high A and low B cannot be disentangled from the effects of the reverse combination of low A and high B, since both lead to the same value of the interaction term. Note also that it is incorrect to enter the main effects and interaction effects simultaneously in regression analyses; to establish interaction effects, the interaction term must be entered *after* the main effects.

Interaction effects can be demonstrated most clearly with dichotomous variables. The levels of violence in each of the four categories high biological/high social, high biological/low social, low biological/high social, and low biological/low social can be investigated and different types of interaction effects can be distinguished. Interaction effects could also be studied with slightly more complex variables, such as trichotomous variables. An advantage of the dichotomous variables is that they encourage a focus on types of individuals (e.g., individuals in the high biological/high social category), making it possible to assess in what proportion of violent persons the violence might be attributable to some particular interaction. In contrast, the use of continuous variables assumes implicitly that everyone is homogeneous; difference in violence between one person and another are quantitative and not qualitative differences, and are attributable to some persons having a greater amount of violence-producing variables. This method encourages a focus on explaining variance rather than explaining groups (types) of persons.

These problems are not specific to biosocial research on violence, but they need to be solved in this as well as other research areas. Further in-depth discussion of interaction effects may be found in Rutter (1983) and Rutter and Pickles (1991).

BIOSOCIAL STUDIES OF ANTISOCIAL BEHAVIOR

There have been surprisingly few studies of biosocial interactions and violence. One reason for this is that of the few crime researchers who assess both social and biological variables, most analyze for main effects only, and do not take the necessary next step of assessing interaction effects in their regression or ANOVA models. A major problem in studying interaction effects is that they are rarely predicted by theories and that the number of possible interactions is often very large.

Perhaps even more surprisingly, criminologists who have focused exclusively on social variables have almost always ignored interaction effects between these social risk factors. The exception to this is the first large-scale, systematic assessment of interaction effects between social risk factors for offending conducted by Farrington (1994). Although not concerned with biosocial effects per se, two important findings from this analysis should be highlighted. First, Farrington (1994) found that the likelihood of obtaining interaction effects decreased with age of offending. That is, more interactions occurred for the prediction of early offending (age 10–14 years) than later offending (age 27–32 years). This was interpreted as indicating that antisocial behavior becomes more ingrained with time and is less susceptible to moderating effects of other variables. Second, the best replicated interaction effect was one which showed that having a convicted parent predicted offending only in the offspring from low SES families, suggesting that high SES acts as a protective factor against the risk factor of a convicted parent.

The aim of this section is to describe briefly the few studies that we have found which illustrate a biosocial interaction effect, in the hope that they will give pointers to other researchers who are beginning to analyze their data from a biosocial perspective. We focus on interaction effects because they have been rarely tested and are viewed as potentially important.

1. *Wadsworth (1976).* In a study of 1,813 British male children who were assessed on resting heart rate at age 11 years, a low pulse rate was found to be associated with criminal convictions between the ages of 8 and 21 years. However, when the sample was divided into those from intact homes and those from homes broken by divorce or separation in the first four years of life, the heat rate - crime relationship was found only in those from intact homes. This interaction was interpreted by Raine (1988) as indicating that where the "social push" towards crime is minimized (intact homes), biological variables such as low arousal may have greater explanatory power.

2. *Christiansen (1977).* An impressive study of 3,586 unselected twin pairs from Denmark illustrated significant heritability for crime (Christiansen, 1997). What tends to have been largely overlooked in Christiansen's work, however, is the fact that social and demographic factors mediate heritability for crime. For example, Christiansen (1977) found that heritability for crime was higher in those from high social classes, and also in those who were rural born. Such findings are consistent with the study of Wadsworth (1976) in suggesting that where the social push to crime is minimized, heritable and biological factors may play a greater role in explaining crime.

3. *Raine and Venables (1981).* One of the few empirical studies to examine a biosocial *interaction* was an analysis of the interaction between psychophysiological functioning and social class (Raine & Venables, 1981). This study examined 101 15-year old male school children in England. Social class was assessed according to parental occupation. Antisocial conduct was measured by self-report and teacher ratings. Conditionability was measured by skin conductance responses in a laboratory-based classical conditioning paradigm. Raine and Venables (1981) hypothesized that the biological variable under study (individual differences in conditioning) might have stronger explanatory power in the higher social class where the "social push" toward crime is low. The results of the study supported their hypothesis. When subjects from the high social classes were examined, antisocial individuals were found to evidence poor conditionability. However, the opposite effect was noted in the low social classes; antisocial individuals from low social class backgrounds were found to evidence good conditionability. This broadly fits with the theory that antisocial behavior can result from good learning in an antisocial context or from poor learning in a prosocial context (Eysenck, 1977).

4. *Raine and Venables (1984).* In the same sample as that described above, Raine and Venables (1984) found that low resting heart rate characterized antisocial behavior. Following the previous work of Wadsworth (1976), the sample was then divided into those from high social classes and those from low social classes. The heart rate - antisocial behavior relationship was found in those from high, but not low, social classes. This interaction effect thus supports the notion that biological variables may have greater explanatory in power in benign (e.g. high social class) home backgrounds.

5. *Cloninger and Gottesman (1987).* Strong evidence for a biosocial interaction between genetics and environment in accounting for crime is reported by Cloninger and Gottesman (1987) in a cross-fostering analysis of "petty" criminality.

Swedish adoptees (862 males and 913 females) were divided into four groups depending upon the presence or absence of (1) biological parent crime and (2) a negative rearing environment. Cross-fostering analysis revealed a multiplicative interaction of heredity and environment. The highest rates of crime were found for the male and female groups in which *both* biological and social risk factors were present. Similar interaction effects have also been reported by Cadoret et al. (1983) and Crowe (1975).

6. *Moffitt (1990)*. Moffitt (1990). used data gathered in the context of the longitudinal Dunedin (New Zealand) Multidisciplinary Health and Development study to examine the relationship between neuropsychological functioning and self-reported aggressive acts during adolescence. The sample consisted of 1037 children examined every two years from birth to age 13 years. The neuropsychological and self-reported delinquency measures were gathered at age 13, and measures of family functioning were gathered at all stages of assessment. Moffitt noted a significant interaction between family adversity and verbal neuropsychological deficits for self-reported delinquent acts involving aggressive confrontation. The group who evidenced *both* family adversity and neuropsychological deficits was found to have a mean aggressive score four times higher than the other groups in the sample.

7. *Raine, Brennan and Mednick (1994)*. Raine et al. (1994) tested the biosocial interaction hypothesis that delivery complications when combined with maternal rejection of the child would predispose to adult violent crime. This hypothesis was tested within the context of a birth cohort of 4269 Danish males. Delivery complications were recorded at birth by a Danish obstetrician, and were interpreted as a reflection of perinatal neurological damage. Maternal rejection was operationalized as unwanted pregnancy, attempt to abort the fetus, and public institutional care of the infant. Violent criminal status was assessed at age 18 through a check of the Danish official police register. In this study, a significant interaction was noted between maternal rejection and delivery complications in predicting violent crime. The combination of both maternal rejection and delivery complications led to a disproportionate increase in violence.

8. *Mednick and Kandel (1988)*. This study focused on neurological deficits and used the biological measure of minor physical anomalies (MPAs) as a reflection of neurological damage. Mednick & Kandel (1988) found that MPAs in combination with an unstable family environment predicted a violent criminal outcome. MPAs were measured in 129 males at age 12. Family stability was operationalized as living consistently with both parents throughout childhood. Violent outcome was assessed through a check of official police records for arrest. Results of the study suggested that children who had high numbers of MPAs and who were raised in an unstable family environment had a disproportionately high probability of violence.

9. *Lewis et al. (1988)*. One study which suggests that the interaction between social and biological factors may be more complex than first imagined is a follow-up of 95 violent delinquents. Evidence for a biosocial interaction was observed in that delinquents with two or more biological vulnerabilities (neurological deficits, cognitive impairment, psychotic symptoms) and who also had a social deficit (history of abuse and/or family violence) were more likely to become violent criminals in adulthood. However, those with only one biological deficit coupled with the social deficit were no more likely to become serious criminals

than those with social deficits only or those with biological deficits only. The important implication of this finding is that it suggests that studies which fail to find an interaction between a single biological deficit and a single social deficit do not necessarily negate a biosocial perspective on violence, and that more complex interaction effects may be worth pursuing.

Biosocial Theories of Antisocial Behavior

The above studies, while having individual limitations, are of value in providing an initial empirical base upon which to build. Nevertheless, we believe that major advances in knowledge are most likely to emerge when studies are set within the context of a well-developed hypothetic-deductive framework. As such, theoretical models of the biosocial bases of violence need to be developed and tested. This section will outline the very few theories which have been articulated to date. Again we will focus on biosocial theories highlighting interactions between biological and social variables. There are four major interactive biosocial theories of delinquency and crime:

1. Eysenck's Theory of Criminal Behavior. Eysenck theorized that certain biologically-based personality features increase the risk for antisocial outcomes, given a particular social upbringing (Eysenck, 1964). Individuals are assumed to inherit particular personality features, along with associated autonomic and central nervous system characteristics. These biological characteristics affect their responsiveness to punishment and their propensity for antisocial outcomes. For example, extraverted personality types are considered to have chronic cortical underarousal and to be unresponsive to punishment (poor conditionability). Eysenck states that antisocial behavior is inhibited through a classical conditioning process in which a child associates antisocial acts with punishment after the two have been repeatedly paired in the child's environment. This classical conditioning process will be effective if punishment (or reprimand) is delivered consistently and quickly, *and* if the child evidences good conditionability.

Eysenck's theory suggests an apparently paradoxical biosocial interaction for antisocial outcomes. Under normal environmental circumstances, poor conditioners will become antisocial. However, given a criminogenic social environment, individuals with deficits in biological functioning (poor conditioners) would be less likely than individuals with well-functioning biological systems (good conditioners) to evidence antisocial outcomes (Eysenck, 1977). In this process of "antisocialization", children who are highly conditionable and who have antisocial parents will become "socialized" into their parents' antisocial habits, whereas children who condition poorly would avoid becoming antisocial. This cross-over prediction met with some support in the study by Raine and Venables (1981).

2. Mednick's Biosocial Theory of Criminal Behavior. According to Mednick (1977), children learn "civilized" behavior through passive avoidance conditioning. In other words, a child will avoid committing an act that has resulted in punishment in the past. This avoidance occurs because of the child's fear of punishment. When the child avoids the act, a reduction in fear occurs, and this fear reduction acts as a reinforcement. The more quickly the fear dissipates, the stronger the reinforcement is to the child. Mednick theorized that a child with an autonomic nervous system that recovers very quickly from fear will easily learn civilized behavior, whereas a child with an autonomic nervous system that recovers slowly will have difficulty learning to inhibit aggressive behavior.

There are necessary social and biological components to Mednick's biosocial theory. In terms of the social environment, a child must experience consistent and adequate punishment for aggressive acts (or fear will not be induced). In terms of the biological component, a child must have immediate and complete dissipation of fear following the inhibited act (or reinforcement will not occur). If either the biological or the social components are absent for a child, antisocial behavior is a likely outcome. The most antisocial children are those lacking both social and biological factors.

3. Buikhuisen's Biosocial Theory of Chronic Juvenile Delinquency. Like Mednick, Buikhuisen (1988) also characterized an antisocial outcome as a failure of the socialization process. He stated that children who cannot learn avoidance behavior will be at a higher risk for a delinquent outcome. There are several necessary components for avoidance learning. First, an individual must have intact information processing capabilities. Second, an individual must have an adequately responsive autonomic nervous system. Third, an individual must have personality traits that facilitate avoidance learning.

Buikhuisen discussed in detail the negative results of a failure of avoidance learning. At home, the parent rejects the child which impedes the development of a conscience, decreases feelings of empathy, and lowers the child's self esteem. At school, failed avoidance learning results in poor academic performance and eventual drop-out. If the child cannot find a positive social identity through achievements in sports, school or music, then normative social integration cannot occur. This transactional, developmental process typically results in chronic delinquent behavior. However, interventions that strengthen the social regulatory mechanisms for antisocial behavior can work to offset this process.

4. Moffitt's Life-Course-Persistent Offender Theory. Moffitt's biosocial theory (1993) states that the biological roots of antisocial outcomes are present before or soon after birth. Moffitt theorizes that congenital factors—heredity and perinatal complications—produce neuropsychological deficits in the infant's nervous system. These neuropsychological deficits manifest themselves as temperament difficulties, cognitive deficits, and motor delays in the child. Moffitt demonstrates that children with these biological deficits often find themselves in deficient social environments as well. If the biological deficits were inherited, this implies that at least one parent also suffers from neuropsychological, temperamental or cognitive deficits. These inherent deficits in the parents then have a direct negative impact on the child's social environment. Alternatively, a deficient social environment might actually cause perinatal complications, poor nutrition, or early child abuse that, in turn, result in biological deficits in the child. Children who are unfortunate enough to have both biological and social deficits are theorized to be at the highest risk for persistent antisocial behavior.

A BIOSOCIAL MODEL OF VIOLENCE

These theories are important first steps towards the specification of a biosocial approach to violence. However, they each have some limitations. First, none of them have violence explicitly as the outcome variable. Second, some of them are very broad. For example, Buikhuisen invokes a number of quite broad "biological" processes (information-processing, autonomic nervous system, avoidance learning, personality) and a number of complex social processes (parental rejection, empathy, self-esteem, academic ability, social identity) in a complex transactional model. The advantage of this model is that it can incorporate a wide variety of findings, but this also produces the disadvantage that it is hard to disprove the model. Good models need to be testable and make explicit some key hypotheses which stem from the model. Few studies are wide-ranging enough to test all the constructs in Buikhuisen's model, and none that we know of are able to test any part of the complex developmental, transactional process predicted by the model.

At the other end of the spectrum is Mednick's biosocial theory. Mednick delineates a very specific physiological process (skin conductance half-recovery time) in conjunction with a specific social process (consistency of punishment) in producing antisocial and criminal behavior. The advantage lies in generating clear, testable hypotheses within the scope of the experimenter. The downside is that such experimental specificity will inevitably mean that the theory can never account for all criminal behavior which probably involves many constructs. The models of Moffitt and Eysenck lie somewhere in between those of Buikhuisen and Mednick.

Rather than attempt a new biosocial *theory*, we propose below an initial, heuristic biosocial *model* of violence. Our goal is to provide a broad framework which some contributors may use to place in context their research findings. In addition, such research findings could help answer certain questions which the model poses. Consequently, our goals have been to propose quite broad processes in order to accommodate the many constructs that contributors will use, while at the same time allowing for some degree of specificity. Thus, the model strives to generate several specific conceptual questions and to avoid being so complex that it is incomprehensible.

The downside of this approach is that the model is inevitably overly simplistic. For example, as noted earlier in this chapter, separating biological from social factors may be a false dichotomy. Furthermore, we have not broken down biological and social processes into subclasses of processes (e.g. hormonal, psychophysiological, family, school) or specified cognitive processes explicitly. These specific processes are clearly important, but in order to make this model relatively comprehensible, we have only made explicit the broader biological and social processes.

The model is outlined in Figure 5. Roughly running from top to bottom, the key processes are as follows:

1. *Basic processes.* The core of the model is based on biological and social risk and protective factors.

2. *Genes and environment as determinants of risk and protective factors.* Both genetic and environmental forces are assumed to be the building blocks for later processes, and each may directly influence both risk factors (solid lines from "genetics" and "environment" to "biological risks" and "social risks"), and protective factors.

   The model also suggests however that these forces may also interact (dotted lines connecting "genetics" and "environment"); environmental forces can give rise to the expression of a latent genetic trait (e.g. poor environment magnifying genetically predisposed low IQ), while genetic factors can alter the environment (e.g. a low IQ individual drifts into more criminogenic environments). A comprehensive discussion of the interplay between genetic and environmental factors may be found in Rutter et al. (in press).

   While genetic factors may largely give rise to biological risk factors, and while environmental factors give rise to social risk factors, environmental factors may produce biological deficits (e.g. accidents causing brain damage), while genetic factors may bring about social risk factors (e.g. child abuse poten-

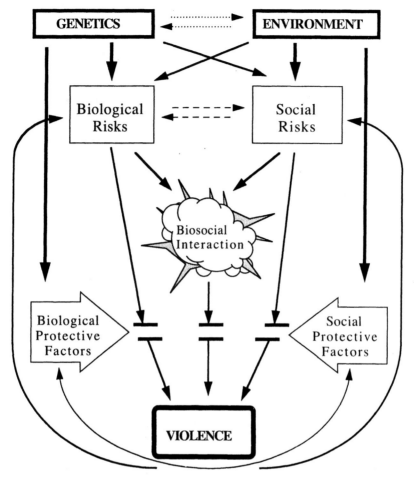

Figure 5. Heuristic biosocial model of violence.

tial). These processes are indicated in the model by crossed lines connecting "genetics" to "social risks" and "environment" to "biological risks".

3. *Reciprocal relationships between biological and social risk factors.* Just as genetic and environmental factors may have reciprocal relationships, so may have biological and social risk factors (indicated by broken lines connecting "biological risks" and "social risks"). For example, the biological risk factor of low arousal may push the stimulation-seeking child into the social risk factor of delinquent gangs. The social risk factor of lack of parental supervision may result in the child suffering head injury (e.g. falling out of trees, car accident), a biological risk factor.

4. *Biological and social risk factors as direct determinants of violence.* Biological risk factors may directly result in adult crime independently of social risk factors, and social risk factors may similarly have direct pathways to crime (see solid lines from "biological risks" and "social risks" to "adult violence"). The vast majority of research to date assumes these direct effect pathways.

5. *Risk factors interacting to produce violence.* The more interesting but much less tested proposition is that biological and social risk factors interact together to produce violence. This is indicated by solid lines from "biological risks" and "social risks" to "biosocial interaction".

6. *Protective factors.* An important feature of the model is that it suggests that biological and social protective factors can break all three pathways (biological, social, and biosocial) to violence. This is symbolized in Figure 5 by the interruption of lines from "biosocial interaction" to "adult violence" by "biological protective factors" and "social protective factors".

The model also allows for the possibility that biological protective factors may produce social protective factors (curved line from "biological protective factors" to "social protective factors"), and vice-versa. For example, high arousal and good information-processing may have a direct effect in nullifying the risk factor of inconsistent parental punishment because those with high arousal condition well despite erratic reinforcement schedules imposed by the parents. However, high arousal may have a protective effect instead by producing the social protective factor of academic achievement (assuming high arousal results in more sustained attention) which is then the key process that prevents violence outcome.

7. *Violence affecting risk / protection status.* The model suggests that the outcome variable of violence could produce new risk factors for violence. For example, imprisonment may result in new social risk factors (divorce, unemployment) or new biological risk factors (being subjected to assaults in prison resulting in head injury) which increase the risk of later violence. This process is symbolized in the model by curved lines from "adult violence" to both "biological risks" and "social risks".

QUESTIONS POSED BY THE MODEL

The model gives rise to a number of questions which some participants may be able to address with their data. The most important of these are as follows:

1. *Are social risk factors more likely to interact with biological than social risk factors?* Due to its nature, the model focusses on interactions across social and biological variables. However, interactions *within* a class of variables may be very important. There is very little research on interactions amongst social risk factors (Farrington, 1994), and to our knowledge no research on interactions between biological risk factors. It would be of use to know whether social factors are more likely to interact with biological risk factors than social risk factors, and vice-versa for biological risk factors. It may be that the relatively high covariation been social risk factors (e.g. family size, socio-economic status) may reduce the likelihood of significant interactions between them, whereas the lower correlations between social and biological variables may be more conducive to such interaction effects. This speculation needs to be tested empirically.

2. *Do biological protective factors exist?* The model assumes the presence of biological as well as social protective factors. Although there is a literature on social protective factors against crime development (e.g., Farrington et al., 1988), there have only been three studies to date on biological protective factors

(Raine, Venables, and Williams et al., 1995, Raine, Venables and Williams, 1996, Brennan, Raine, and Mednick, 1996). Do biological protective factors really exist, and how important is their influence?

3. *Do protective factors operate at their own level?* To what extent do social protective factors operate at the level of nullifying a social risk for violence? Alternatively, do social protective factors best operate in nullifying biological risk factors, and vice-versa? To date there are no data which have contrasted these two alternative views.

4. *Does nullifying a biosocial risk pathway to violence require more protective factors than nullifying a social risk pathway?* The model specifies that protective factors can operate both in nullifying the biosocial pathway to violence as well in nullifying the more direct social and biological pathways. The model also suggests that the biosocial pathway is a more potent force (three lines connecting "biosocial interaction" to "adult violence" as opposed to one line for "biological risks" and one line for "social risks"). Therefore, the question is raised as to whether an individual with both social and biological risk factors for violence, but who nevertheless desists from violence, possesses more protective factors against violence than a desistor who has only biological or only social risk factors.

5. *Does violence affect risk factors?* One question raised by the model which appears to have received little attention in the literature concerns whether violence and its sequelae alter risk and protective factors. It was earlier suggested that imprisonment could create new risk factors for violence, but can it also create new protective factors? For example, imprisonment could disrupt the negative community influences of the offender, and relocation into a new community away from negative peer groups might result in desistance from future offending (Buikhuisen and Hoekstra, 1974). Alternatively, imprisonment may cause the social risk factors of divorce and increased difficulty in obtaining employment. Yet again, leading a violent criminal career may increase risk for head injury through fighting or increased drug abuse, and therefore increased brain dysfunction.

6. *Different biosocial models for different developmental stages towards violence?* The model in Figure 5 has adult violence as its outcome. Yet understanding the early developmental trajectories towards such an outcome may require different models for each stage. For example, intermediate outcomes might include disinhibited temperament in infancy, aggressive behavior in early childhood, conduct disorder in later childhood, delinquency in early adolescence, early adult crime in later adolescence, and then the final process of sustained violence in adulthood. It would be envisioned that different processes may underpin different outcomes. For example, very different processes or biosocial interactions may underlie adolescent delinquency relative to sustained adult violence.

Although the model in Figure 5 may be a common template for all outcomes, contributors may wish to consider the extent to which the biosocial processes underlying their outcome variables require a somewhat different model. Furthermore, different aspects of antisocial development may be aligned more closely to the biological side of the model (left side of Figure 5) than to the social (right) side. For example, disinhibited temperament may be determined more by biological than social factors, while juvenile delinquency may be determined more by social factors.

7. *Is the biosocial pathway more important for certain subgroups of violent offenders ?* An important challenge lies in answering the question of whether different forms of violence (e.g. rape vs robbery; proactive aggression vs reactive aggression) are more likely to be developed through one biosocial pathway rather than another. Alternatively, are some forms of violence mediated through biosocial pathways, whereas other forms are mediated by social or biological pathways? There are almost no data to answer this question. One study showed that birth complications interact with maternal rejection at age 1 year in predisposing to violence at age 34 years, but only with respect to index violence (murder, robbery, rape, and assault), and not with respect to threats of violence or illegal possession of weapons (Raine et al. 1996). To what extent is life-course persistent vs adolescent-limited antisocial behavior (Moffitt, 1993) differentially explained by the biosocial perspective?

Similar questions can be asked with respect to gender and ethnicity. For example, is the integrated biosocial perspective particularly important for understanding female violence? Biological causal factors may be needed in addition to social influences for violent females to overcome the relatively powerful socialization forces against violence in women. Similarly, are biosocial interactions less important in explaining violence in ethnic minority groups (where violence might be better explained by discrimination and adverse social conditions), but relatively more important in explaining violence in Caucasians ?

8. *Individuals vs. groups.* The model focuses on processes which move forward (down the page in Figure 5) to explain violence outcome. Violence is the dependent variable, and the usual question is: "given x biological process interacting with y social process, how much variance in violence is explained?". It may however be important to ask questions about individuals as opposed to groups. For example, reversing the model's flow (starting at the bottom with the adult violent offender and moving up the page) can pose some interesting questions. For example, taking 100 violent offenders, what percent only possess social risk factors (relative for example to a nonviolent control group)? What percent only possess biological risk factors? What percent possess both risk factors? How far can we attribute the violence of groups of offenders to specific combinations of biological and social risk factors?

SUMMARY

This chapter makes an initial attempt to both outline and address some of the conceptual and theoretical issues surrounding biosocial research on violence. Three conditions are outlined for a study to be construed as biosocial. Seven different types of biosocial effects are then illustrated: (1) correlated biosocial effect (2) additive biosocial effect (3) sequential biosocial effect (4) experimental biosocial effect (5) multiplicative biosocial effect (6) protective biosocial effect and (7) crossover biosocial effect. Some data analytic difficulties specific to biosocial research are discussed. Several previous studies which establish biosocial effects for antisocial behavior are reviewed, together with four current biosocial theories of antisocial behavior. An initial, heuristic biosocial model of violence is then presented which may provide a framework within which findings in later chapters may be set. Finally, eight questions which are raised by the model are posed.

ACKNOWLEDGMENTS

This paper was written while the first author was supported by an NIMH grant (RO1 MH46435–02), and also by an NIMH Research Scientist Development Award (1 KO2 MH01114–01).

REFERENCES

Brennan, P., Raine, A., and Mednick. S.A. *Psychophysiological protective factors for males at high risk for crime.* (under review)

Brennan, P. and Raine, A. (in press). Biosocial bases of antisocial behavior: Psychophysiological, neurological, and cognitive factors. *Clinical Psychology Review*

Buikhuisen, W. and Hoekstra, H.A. (1974). Factors related to recidivism. *British Journal of Criminology 11* 185–187.

Buikhuisen, W. (1988). Chronic juvenile delinquency: a theory. In W. Buikhuisen & S. A. Mednick (Eds.) *Explaining criminal behavior* (pp. 27–50). Leaden, The Netherlands: E. J. Brill.

Cadoret, R.J., Cain, C.A. and Crowe, R.R. (1983). Evidence for gene-environment interaction in the development of adolescent antisocial behavior. *Behavior Genetics 13* 301–310.

Christiansen, K.O. (1977). A preliminary study of criminality among twins. In S.A. Mednick and K.O. Christiansen (Eds). *Biosocial bases of criminal behavior*. New York: Gardner.

Cloninger, C. R. & Gottesman, I. I. (1987). Genetic and environmental factors in antisocial behavior disorders. In S. A. Mednick, T. E. Moffitt, and S. Stack (Eds.). The causes of crime: New biological approaches (pp. 92–109). Cambridge: Cambridge University Press.

Crowe, R. (1974). An adoption study of antisocial personality. *Archives of General Psychiatry 31* 785–791.

Daly, M. and Wilson, M. (1988b). *Homicide* New York: Aldine de Gruyter.

Eysenck, H. J. (1964). *Crime and personality* (1st ed.). London: Methuen.

Eysenck, H. J. (1977). *Crime and personality* (3rd ed.). St. Albans, England: Paladin.

Farrington, D.P. (1994). Interactions between individual and contextual factors in the development of offending. In R.K. Silbereisen and E. Todt (Eds.). *Adolescence in context: the interplay of family, school, peers, and work in adjustment* (pp. 366–389). New York: Springer-Verlag.

Farrington, D.P., Gallagher, B., Morley, L., Stedger, R.J., and West, D.J. (1988). Are there any successful men from criminogenic backgrounds? *Psychiatry 51* 116–130.

Garmezy, N., Masten, A.S., & Tellegen, A. (1984). The study of stress and competence in children: A building block for developmental psychopathology. *Child Development, 55*, 97–111.

Lewis, D.O., Lovely, R., Yeager, C. And Femina, D.D. (1989). Toward a theory of the genesis of violence: A follow-up study of delinquents. *Journal of the American Academy of Child and Adolescent Psychiatry 28* 431–436.

Mednick, S. A. (1977). A biosocial theory of the learning of law-abiding behavior. In S. A. Mednick and K. O. Christiansen (Eds.) *Biosocial bases of criminal behavior* (pp. 1–8). New York: Gardner.

Mednick, S.A. and Christiansen, K.O. (1977) *Biosocial bases of criminal behavior* New York: Gardner.

Mednick, S. A. & Kandel, E. (1988). Genetic and perinatal factors in violence. In S. A. Mednick and T. Moffitt (Eds.) *Biological contributions to crime causation* (pp. 121–134). Dordrecht, Holland: Martinus Nijhoff.

Milner, J.S. and McCann, T.R. Neuropsychological correlates of physical child abuse. In J.S. Milner (Ed.). *Neuropsychology of aggression* (pp. 131–146). Boston: Kluwer, 1991.

Moffitt, T. E. (1990). The neuropsychology of juvenile delinquency. In M. Tonry and N. Morris (Eds.) *Crime and justice: A review of research* (Volume 12). Chicago: University of Chicago Press.

Moffitt, T. E. (1993). Adolescence-limited and life-course-persistent antisocial behavior: A developmental taxonomy. *Psychological Review, 100*, 674–701.

Pitts, T. (1993). *Cognitive and psychophysiological differences in proactive and reactive aggressive boys.* Doctoral dissertation, Department of Psychology, University of Southern California.

Raine, A. (1993). *The psychopathology of crime: Criminal behavior as a clinical disorder.* San Diego: Academic Press.

Raine, A., Brennan, P. A., & Mednick, S. A. (1994). Birth complications combined with early maternal rejection at age 1 year predispose to violent crime at age 18 years. *Archives of General Psychiatry, 51*, 984–988.

Raine, A., Brennan, P. and Mednick, S.A. (1996) _The biosocial interaction between birth complications and early maternal rejection in predisposing to adult violence: Specificity to serious, early onset violence_ (under review).

Raine, A., Brennan, P., Mednick, B. and Mednick, S.A. (1996). High rates of violence, crime, academic problems, and behavioral problems in males with both early neuromotor deficits and unstable family environments. _Archives of General Psychiatry 53_ 544–549.

Raine, A. and Venables, P.H. (1984). Tonic heart rate level, social class, and antisocial behavior. _Biological Psychology 18_ 123–132.

Raine, A., Venables, P.H. and Williams, M. (1995). High autonomic arousal and electrodermal orienting at age 15 years as protective factors against criminal behavior at age 29 years. _American Journal of Psychiatry 152_ 1595–1600. (findings also reported and discussed in _Science_ 1995 _270_ 1123–1125).

Raine, A., Venables, P.H. and Williams, M. (1996). Better autonomic conditioning and faster electrodermal half-recovery time at age 15 years as possible protective factors against crime at age 29 years. _Developmental Psychology 32_ 624–630.

Raine, A. & Venables, P. H. (1981). Classical conditioning and socialization—A biosocial interaction? _Personality and Individual Differences, 2,_ 273–283.

Rutter, M. (1983). Statistical and personal interactions: Facets and perspectives. In D. Magnusson and V. Allen (Eds). _Human development: An interactional perspective_ (pp. 295–319). London: Academic.

Rutter, M. (1985). Resilience in the face of adversity: Protective factors and resistance to psychiatric disorder. _British Journal of Psychiatry 147_ 598–611.

Rutter, M. (1990). Psychosocial resilience and protective mechanisms. In J. Rolf, A.S. Masten, D. Cicchetti, K.H. Nuechterlein, and S. Weintraub (Eds). _Risk and protective factors in the development of psychopathology_ (pp. 181–214). Cambridge: Cambridge University Press.

Rutter, M., Dunn, J., Plomin, R., Simonoff, E., Pickles, A., Maughan, B., Ormel, J., Meyer, J. and Eaves, L. (in press). Integrating nature and nurture: Implications of person-environment correlations and interactions for developmental psychopathology. _Development and Psychopathology_

Rutter, M. and Pickles, A. (1991). Person-environment interactions: Concepts, mechanisms, and implications for data analysis. In T.D. Wachs and R. Plomin (Eds). _Conceptualization and measurement of organism-environment interaction_ (pp. 105–144). Washington, D.C.: American psychological Association.

Stouthamer-Loeber, M. and Loeber, R. (1993). The double edge of protective and risk factors for delinquency: Interrelations and developmental patterns. _Development and Psychopathology, 5,_ 683–701.

Wadsworth, M.E.J. (1976). Delinquency, pulse rate and early emotional deprivation. _British Journal of Criminology 16_ 245–256.

2

PERSONALITY AND THE BIOSOCIAL MODEL OF ANTI-SOCIAL AND CRIMINAL BEHAVIOUR

H. J. Eysenck

Institute of Psychiatry
Denmark Hill
London, United Kingdom SE5 8AF

1. INTRODUCTION

In considering the causes of criminal and violent behaviour, we are dealing with two separate but interrelated factors. A crime is committed by a **person** in a certain **situation**; individual differences are responsible for the fact that in similar situations one person will commit the crime, another will not. Situations define not only the narrow circumstances of a particular crime, but the whole attitude of a given society to anti-social conduct, to the child's upbringing, discipline in school, judicial procedures, existing levels of punishment, certainty of detection, religious beliefs, prevalence of TV violence, and many more. We cannot explain the huge differences in anti-social conduct between Singapore and Washington, Switzerland and South Africa, Egypt and England in terms of general differences, or personality factors. Communist countries like Russia used to be relatively crime-free; after the overthrow of communism Russia is one of the most crime-ridden countries in the world.

Sudden changes like this cannot be explained in terms of genetic changes or criminal predisposition; the time factor makes any such explanation impossible. Social constitutions thus remain the major explanatory principles, but clearly such often-adduced causes as unemployment, poverty or income inequality have little evidential support. Periods of unemployment in the USA are characterized by **decreasing** crime rates (Lester, 1994). When unemployment was huge in Germany during the 1920s period of inflation, crime was minimal. Income inequality in the USA, using the Gini coefficient of income distribution, remained steady from 1961 to 1981, while crime rates increased linearly (Rutter, 1995). Poverty has decreased tremendously from the 1920s to the present day in England, but crime has increased geometrically. Clearly the constant reiteration of these shibboleths by politicians do not deserve scientific credence. It should be obvious that **social causes** (whatever they might be) can only act through their influence on people's minds, and hence produce psychological conditions favourable to antisocial conduct; purely sociological theories of crime are a **contradictio in adjecto**.

Fundamentally we may bring together the two sides (individual differences and social causes) in terms of what economists call the **marginal** customers. The price of a com-

Biosocial Bases of Violence, edited by Raine *et al.*
Plenum Press, New York, 1997

modity is not determined by those who would buy it whatever the price (within reason), not by those who could not afford it in any case, but by the marginal customer who would buy it at price X, but not at price X + 1. Let us postulate the concept of **criminal disposition**, determined partly by a person's heredity, partly by social and biological factors encountered during his upbringing, and their interaction. The predisposition is closely related to his personality (to be discussed, and documented presently), and determines the **probability** of his committing a crime, from very low (left of continuum) to very high (right of continuum); this probability is indicated by P in Fig. 1.

Consider now the application of this general theory to our marginal criminal, i.e. a person with a certain position on the predisposition continuum. The social ethos, as defined previously, will not affect people to the right of him; they are so strongly predisposed to crime that they will commit their crime regardless. It will not affect people to the left of him, they are not sufficiently predisposed to crime to commit a crime in any case. It is the marginal criminal who will be influenced by relatively slight changes in the social ethos. Large changes in this ethos, as in post-communist Russia, will of course shift the position of the marginal criminal to the left, rendering even people with rather low predisposition liable to commit crimes. Smaller but still noticeable shifts in diminishing permissiveness, such as happened recently in New York, will shift the position of the marginal criminal to the right, i.e. a higher degree of predisposition is required to indulge in criminal activity.

2. THE NATURE OF PERSONALITY

There is an enormous literature on personality, but for reasons given elsewhere (Eysenck & Eysenck, 1985; Eysenck, 1991) I shall concentrate on a system of personality descriptions anchored in a large nomological network (Garber & Strassberg, 1991) and presented in diagrammatic form in Fig. 2. The central place is given over to the ma-

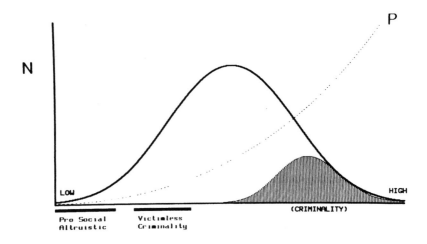

CRIMINAL PREDISPOSITION
(DIATHESIS)

Figure 1. Diagrammatic representation of criminal predisposition. P = probability of becoming a criminal.

jor trait constellations, empirically established; these major dimensions of personality have been designated P (psychoticism), E (extraversion), and N (neuroticism), although other terms have been used by other authors. The causal sequence begins with **distal antecedents,** namely genetic determinants (DNA), and there is a large body of evidence supporting the view that a large part of the total phenotypic variance in personality is genetic (Eaves, Eysenck & Martin, 1989). But this information, although important, only leads to further questions. Hereditary information in the DNA is copied onto RNA by a complementation process; RNA in turn participates with certain intracellular structures to produce polypeptides, which compose proteins, which may be structural, transport, and catalytic (enzymes). Enzymes are of particular interest because they facilitate the chemical reaction of life. Thus clearly we must seek knowledge about the **proximal antecedents** of personality.

To have a scientifically worth-while system we must formulate **theories** concerning the links between personality and such proximal antecedents as may be relevant. (Eysenck, 1967); such theories can be tested directly, by means of psychophysiological measures (Eysenck, 1994). They may also be tested by means of **proximal consequences,** i.e. laboratory studies of behaviour, such as conditioning, vigilance, memory, perception, etc. (Eysenck, 1981; Eysenck & Eysenck, 1985). It is the large body of experimental studies under this heading that presents the best evidence for the validity of the systems as presented.

Finally, and going well beyond the possibility of direct experimental laboratory testing, we have **distal consequences,** such as criminality (Eysenck & Gudjonsson, 1989), creativity (Eysenck, 1995), sexual behaviour (Eysenck, 1976), psychopathology (Eysenck, 1992a), marriage (Eysenck & Wakefield, 1981), etc. It is in making testable predictions mapping distal and proximal antecedents, as well as proximal consequences onto this field

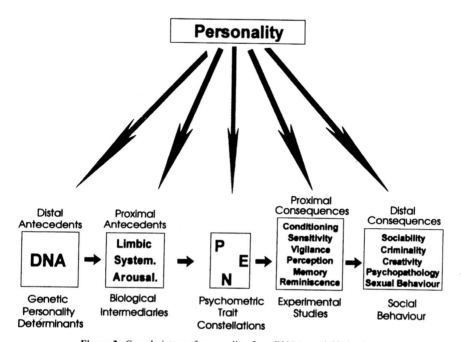

Figure 2. Causal picture of personality, from DNA to social behaviour.

of distal consequences, that the general personality theory proves its scientific and social usefulness, and in this chapter I shall try and document its applicability to the field of criminality and violence (Eysenck, 1977; Eysenck & Eysenck, 1971; Eysenck, Rust & Eysenck, 1977).

Leading into the documentation demonstrating the relationship between personality and criminal predisposition, I shall very briefly set down the general theory I put forward originally 35 years ago (Eysenck, 1980). I argued then that it was not meaningful to ask why people commit crimes; the real problem is that most people most of the time do not commit crimes or other antisocial acts although to do so would be to their immediate advantage. I suggested that they were prevented from doing so by their **conscience**, and I suggested that conscience was a **conditioned response** produced along Pavlovian lines, through innumerable positive and negative reinforcements of pro-social and anti-social acts respectively. This theory was amplified later on (Eysenck, 1977) and has found much empirical support (Eysenck & Gudjonsson, 1989; Raine, 1993). I linked it with personality through the postulation of a cortical **arousal** factor that promoted classical conditioning when high, and slowed it down when low. This was linked with the postulation that extraverts (and people high on psychoticism) had habitually **low** cortical arousal, and would hence condition poorly, and thus have trouble developing a proper conscience. This leads to the prediction that **P and E would be linked with antisocial and criminal conduct**. Emotional instability (N) seems likely to cause difficulties in making sensible and socially acceptable adjustments and through its strong autonomic reactions might lead to impulsive behaviour. This is a rather weak prediction, lacking the strong empirical support for the arousal-conditioning hypothesis.

Constitutionally low arousability may affect criminality **directly** as well as through making for poor conditionability. Essentially, a person with a low-arousal level seeks to increase this level by a variety of means, such as risk-taking, sensation-seeking, impulsive actions, socializing with many other people, drug abuse, multiplicity of sexual partners, etc. These activities are likely to lead such a person towards criminal activity, but not inevitably; risky sports activities may take the place of criminality in middle-and upper-class persons.

In considering this theory, it is important to disregard criticisms suggesting that conditioning theories neglect **cognitive** factors. While this would be true of fundamentalist behaviouristic theories, such as those of Watson and Skinner, it totally misrepresents modern theories which inevitably include cognitive factors as vitally important elements (e.g. Mackintosh, 1974, 1984; Davey, 1983). This consideration goes back to Pavlov and his concerns with language as "the second signalling system", with Platenov (1959) as its main exponent. Cognitive changes are produced as easily, if not more so, by behavioural methods (e.g. exposure) as by purely cognitive manipulations.

3. PERSONALITY AND CRIMINAL PREDISPOSITION

I have surveyed a large body of evidence testing the prediction outlined in the previous section (Eysenck & Gudjonsson, 1989); many of these studies were carried out in cross-cultural experiments in many different countries. The general finding has been that P correlates positively in about every study with criminality and antisocial conduct; for children, youths and adults in pretty equal measure, and both for actual law-breaking leading to incarceration, and for self-confessed semi-criminal and anti-social activity. Neuroticism is more strongly associated with criminality and antisocial conduct in **adults**, while extraversion appears to be more closely involved in **young** samples than incarcerated adults. This latter finding may be due in part to a tendency for prisoners to live a life that makes extraverted behaviour impossible, in part to prisoners understating their degree of extraversion (Eysenck & Gudjonsson, 1989). These correlations are all highly significant, but do they have a high enough effect size to have social significance? Correlations tend to be quite high on the whole. Looking at large-scale studies, giving reasonable stable co-efficients, we may cite results reported by Jamison (1980) for 781 boys and 500 girls, correlating personality traits with the Allsop and Feldman (1976) ASB (antisocial behaviour scale). Correlates for the two sexes were .58 and .59 for P, .31 and .40 for E, and .10 and .09 for N. For L, the so-called Lie Scale, which is essentially a measure of conformity when conditions of testing do not encourage people to give an overly good account of themselves, correlations were -.56 and -.60. (L usually correlates negatively with P, which is not surprising!) (It is of interest, particularly in view of the stress on **violence** in this book, that Choynowski (1995) found a very strong association between L reversed (non-conformity) and aggression.)

Powell (1977) studied 381 boys and 427 girls, and found a similar correlation with personality, using the ASB scale. For senior boys and girls, the correlations were: P: .47 and 44; E: .26 and .17; N: .18 and .30; L: -.64 and -.56. For junior children, correlations were: P: .42 and .48; E: .04 and .10; N .09 and .17; L: -.48 and -.50. These of course are raw correlations; when corrected for attenuation, a correlate of .50 would become over .60. Combining the correlation into a multiple R, we get a value of over .70 or so (uncorrected). Thus personality predicts something like 50% of self-reported antisocial conduct in children, clearly indicating that personality is an important causal factor in antisocial conduct.

For adults in various studies we found similar results, with P always giving the highest correlations, or the largest difference between criminal and control groups. It might be expected that P would correlate more with violent than with non-violent crime, and Chico & Ferrando (1995) have put the matter to the test. The P score for 181 violent delinquents was 10.42 ± 3.79, while for non-violent delinquents it was 6.70 ± 3.01. This is a very significant difference, and indicates, taken together with the Mitchell et al. (1980) finding, that violent offenders are high P-low N as compared with non-violent offenders. This agrees with the usual finding that **maleness** may be related to violence; males as compared to females tend to show high P-low N levels. (Eysenck, 1995b). Given these relations, we would expect certain sex-related hormonal factors (e.g. testosterone to differentiate between violent and non-violent subjects. Ellis & Coontz (1990) have reviewed the evidence, which generally supports this view.

Bernson & Fairey (1984) added to the evidence linking violent crime with P in a study comparing 30 juvenile assaultive offenders with 30 juvenile property offenders. "Juveniles convicted of assaultive offences exhibited significantly higher psychoticism, extraversion, and neuroticism scores, and lower lie scores than those convicted of property offences" (p.527). In addition, Zuckerman's sensation-seeking scores were higher for assaultive offenders. (Sensation-seeking correlates quite highly with P and E - Eysenck, 1983).

I will not here enter further into this field, but will discuss briefly results achieved with the Eysenck Criminality Scale (C Scale)(Eysenck & Eysenck, 1975). The scale brought together items from all three personality scales (mainly P), and the resulting 34-item scale discriminated well between 934 criminals and 189 non-criminals, of similar age and social class. Scores were 9.01 ± 4.54 for non-criminals, and 15.57 ± 5.18 for criminals, with alpha reliability of .75 and .75, respectively. (Test-retest correlations were slightly higher). The difference is over one S.D., and of course a random non-criminal sample will contain a fair number of unascertained criminals, which would lower the discrimination. A similar scale for children was also constructed, with an alpha reliability of .74. The scale correlated .71 with the ASB scale, suggesting good validity (Eysenck & Eysenck, 1975). Follow-up studies have not yet been done, but in view of a large literature showing considerable consistency of behaviour from child to adult with respect to antisocial behaviour (e.g. Olweus, 1984; Farrington , 1986; Staltin & Magnusson, 1989; Wolfgang, Thornberry & Figlio, 1987), it seems likely that the Junior C scale would correlate well with adult criminality. A study by Putsins (1982) shows that this expectation is not unreasonable.

Of some special interest are data concerning the **dual threshold** hypothesis. According to this quite general hypothesis, if you have two groups, say, men and women, who differ in their conduct with respect to a certain type of conduct, say criminality, which in turn is associated with a given trait, say P, then the difference between delinquents and non-delinquents on P should be **larger** for women than men (Eysenck & Gudjonsson, 1989). The reason of course is that men clearly have a stronger propensity (P) than women to indulge in this type of behaviour, so that women require a higher degree of P in order to indulge in the behaviours in question. Gudjonsson et al. (1991) showed that as usual prisoners had higher scores than non-prisoners, but that women prisoners had even **higher** P scores than male prisoners, while women non-prisoners had much lower P scores than men. This finding is fairly general (Eysenck & Gudjonsson, 1989).

The C-scale clearly differentiates criminal and non-criminal populations in England; does the effect obtain in cross-cultural studies? A large-scale study of this kind was undertaken in Zagreb (Croatia) by Sakic, Zuzul, Knezovic, Kulenovic & Zarevsky; unfortunately the war prevented publication, but they communicated the major results to me. (Some of these results have been published in a rather inaccessible form - Sakic, Knezovic & Zuzul, 1987). In the first study, a control group of 128 male subjects was compared with 101 prisoners convicted of violent crimes. C-scale scores were 10.1 ± 4.20 for the controls, 15.0 ± 5.66 for the property offenders, and 15.8 ± 5.13 for the violent offenders. There is no significant difference between the two kinds of offenders. In the second experiment, a control group of 128 males (the same as before) was compared with 205 prisoners who had committed physical assault, either mild (n=62), moderate (n=46), serious (n = 55), or very serious (n = 42); the last group was a recidivist group who had at least twice, on separate occasions, committed a serious physical assault. Ascending C scores were found in the four violent groups, in order 13.3 ± 5.5; 14.3 ± 5.3; 15.8 ± 5.6; and 18.3 ± 5.3 (p <0.0001). More serious violent offences are linked with higher C scores.

In a third study, 976 male offenders were studies, with the total sample graded into 5 groups according to seriousness of crime as indexed by duration of sentence. Fig. 3 shows that all groups exceed the control score of 10.1 ± 4.20, with a linear increase in score according to seriousness of crime (p <0.0001). Clearly C increases with seriousness of offence. Length of time spent in prison did not increase a prisoner's C score.

The C scale is of course not the only criminality scale; the Gough Socialization scale has also been used very widely (Gough, 1994). Unlike the C-scale, the So scale has no underlying theory and is purely inductive, choice of items being dictated entirely by successive administration to criminal and non-criminal samples. Like the C-scale, the So scale differentiates these groups at slightly above the 1 S.D. level, and has done so in a number of cross-cultural studies. The So scale correlates quite well with various stability-neuroticism scales, and negatively with scales related to psychoticism traits (egocentricity, aggression, impulsivity, negativism, sensation-seeking, but positively with agreeableness and conscientiousness). Backorowski & Newman (1985) found a correlation of -.54 be-

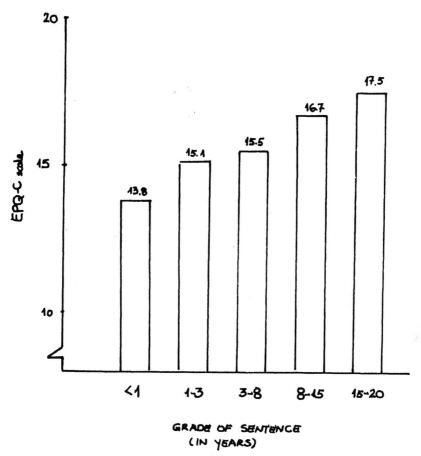

Figure 3. Scores on the EPQ-Criminality Scale for criminals graded according to severity of crime (length of sentence). (After Sakic et al., 1987).

tween So and P. Social status, as with the Eysenck scales, correlates negligibly with So. The So has heritabilities of around .40, with a possible epistatis component (Eysenck, 1995). There is a wealth of material in the almost 200 references cited by Gough (1994) that are relevant to our conclusions, and that should be considered for any final verdict.

4. TRAITS AS INDICATORS OF CRIMINAL PROPENSITY, AND DIFFERENCES BETWEEN VIOLENT AND NON-VIOLENT CRIMINALS

Personality appears to be organized in a hierarchical fashion, with intercorrelations between primary traits resulting in such higher-order factors as P, E and N (Eysenck, 1947). I have reviewed the relation between criminality and these higher-order factors, but much work has been done on primary traits, and while these behave pretty well as might be predicted from the findings related to higher-order factors, a rapid run through may be useful, particularly when a given trait differentiates between violent and property crimes.

One such study (Mitchell, Rogers, Cavanagh & Wasyliw (1980) was concerned with the Cattell **anxiety** factor, in a sample of 2,509 male adolescent offenders of 15.75 to 17.25 years. The more violent delinquents (both black and white) were significantly **less** anxious than the non-violent delinquents (p <.006). Whites were over-represented in the high anxiety group, blacks in the low anxiety group (p <.006). Whites ere over-represented in the high anxiety group, blacks in the low anxiety group (p <.0005). Violent crime was largely composed of murder, rape and battery. Wardell & Yendell (1980) also found violent recidivists less anxious than non-violent recidivists.

A variable correlating with impulsivity and sensation-seeking, and also differentiating between criminal and non-criminal groups is risk-taking, as indicated by Schwenkmetzger (1983), who also cites a number of German studies. Impulsiveness has also been found so to differentiate in Israeli delinquents (Rotenberg & Nachshon, 1979). Perception of risk was found, as expected, negatively related to P and antisocial behaviour by Jamison (1980) in normal children, and Stewart & Hemsley (1984) found risk-taking correlated with high P and criminality. Stimulation-seeking has been found in antisocial pre-adolescent children by Whitehill, Scott and De Myer-Gapin (1976). Aggressiveness was found correlated with extraversion and locus of control (reversed) by De Man and Green (1988).

A study by Hormuth et al. (1973) is of particular interest, because not only did the authors replicate the personality correlates on the FPI discussed below, but they demonstrated a significant difference between criminals and non-criminals on the factor of "impulse control", using a measured ability to control motor behaviour, thus getting away from purely verbal self-description.

A particularly interesting aspect of personality related particularly to violent crime is extent of personal space, i.e. the distance from other people that subjects find comfortable. Kinzel (1970); Booraem, Flowers, Bordner & Satterfield (1977) and McGurk, Davis & Greham (1981) have found evidence that this distance is greatest for violent criminals, less for property criminals, and least for "victimless" criminals, e.g. drug-takers. Eastwood (1985) failed to find any difference on a small sample of criminals; he gives a detailed discussion of previous work. Large personal space requirements are typical of high P scorers, and of persons of low intelligence; they are also found in psychopathological subjects who are not criminal. The precise relationships involved, and the possibility of interaction between causal factors, remain to be discovered.

A questionnaire widely used in Germany, the FPI (Fahrenberg Selg & Hampel, 1978) has been applied several times on criminals and controls (Steller & Hunze, 1984). Unfortunately the inventory does not contain a P scale, or a scale relevant to some of the traits it summarizes, but clear differences are observed for some 3,400 delinquents, as compared with controls, in nervousness, depression and emotional instability. "Increases on the secondary scale for Extraversion are also frequently ascertained, they are essentially due to increases on the primary scales for Aggression and Sociability." (p.87). Lossel & Westendorfer (1976) have reported data with the FPI which are essentially identical with the above, isolating aggressiveness, depression, excitement, dominance and neuroticism.

The "Big Five" have also been considered in relation to criminality (Heaven, 1996). Neuroticism was a significant correlate of criminality, as were agreeableness (reversed) and conscientiousness (reversed); these two scales can be regarded as primary factors in the psychoticism higher-order factor (Eysenck, 1991, 1992). Altogether, there is much support in these various studies for the P-E-N based C scale as a good measure of criminal propensity.

From the large literature a few further studies may be mentioned. Comparisons between criminals and controls have implicated delay of gratification (e.g. Pena and Luengo,

1993), anger/hostility and venturesomeness (e.g. Heaven, 1993). assertiveness and low conventionality, (e.g. Sirder, 1988), low self-control (e.g. Feldman, 1977), low self-esteem (e.g. Rice, 1992), lack of certainty (e.g. Rigby, Mak & Slee, 1989), locus of control (e.g. Shaw & Scott, 1991) and many others. It would go beyond the limits of this chapter to review all published studies, and I have endeavoured to concentrate on studies not dealt with in a previous survey (Eysenck & Gudjonsson, 1989). However, it may be useful to add to our account three indirect deductions from the main theory that P, E and N, and individual traits subsumed under their higher-order factor, are related to criminal activity, more strongly to violent than non-violent behaviour.

1. It is well known that antisocial and criminal activity increases, linearly with age from 12 or thereabouts to 17 or 18, and then steeply declines in a negatively accelerated curve to almost zero at the age of 50 (Moffitt, 1993). It would seem to follow that if personality factors P, E and N are **causally** related to antisocial conduct, such conduct should be associated with an **increase** in P, E and N scores from 10 to 17, and with a **decrease** from 18 onwards. For L scores the relationship should be inverted, decreasing with age up to 17, increasing from 18 onwards. Eysenck & Eysenck (1975) and Eysenck (1983) have shown that this is indeed so.

2. As psychoticism denotes a continuum from Altruism and conformity at one end to schizophrenia at the other, we would expect the observed correlation of P with criminality, especially violent criminality, to be manifested in a higher level of such criminality in schizophrenia. Of course the regression may be non-linear; actually psychotic individuals may be (a) segregated, (b) under drugs, or (c) incapacitated by their disease, and this may prevent them from carrying out assaultive attacks or other crimes. Bocker & Haefner (1973) found that schizophrenics were more likely, depressives less likely than normals to commit assaultive crimes, which agrees with the finding that N is lower in assaultive criminals than in criminals guilty of property crimes.

3. As already mentioned, the sex ratio of male offenders to female offenders is roughly 10 to 1, with female offenders often guilty of "victimless" crimes, such as prostitution; it is particularly in violent crimes that the disproportion is greatest. If violent crime is positively related to P, and negatively to N, this finding, as already noted, supports the general theory linking personality, on the one hand, and crime and violence, on the other, males scoring higher on P, females on N (Eysenck, 1995c). Direct proof, of this kind, constituting deductions from the general theory are always weaker in evidential value because of the weaknesses in relying on published figures and statistics; these are inevitably often questionable. But further experimentation along these lines may produce interesting and important evidence.

5. DISTANT AND PROXIMAL ANTECEDENTS

In terms of our model, as pictured in Fig. 2, we have so far dealt with the central place, i.e. the psychometric criminal predisposition factor diagrammed in Fig. 1. and have established (a) that such a factor exists, and (b) that it is related to higher-order factors P, E and N, and to primary traits associated with these superfactors. To be certain that the argument is rigorous, it needs to be shown that (a) criminality is stable over time, (b) that criminality has a strong genetic basis, and (c) that we can identify proximal (biological) antecedents linking the genetic components with criminal predispositions. An attempt to do so will be made in this section.

(a) As the theme of this book is violent behaviour, I will start with a review of 16 studies by Olweus (1979). concentrating on aggressive reaction patterns in males. "The degree of stability that exists in the area of aggressiveness was found to be quite substantial; it was, in fact, not much lower than the stability typically found in the domain of intelligence testing marked by individual differences in habitual aggression level that manifest themselves early in life, certainly by the age of 3" (p.852). Olweus concluded that "(a) the degree of longitudinal consistency in aggressive behaviour patterns is much greater than has been maintained by proponents of a behavioural specificity position, and (b) important determinants of the observed longitudinal consistency are to be found in relatively stable individuals in differentiating reaction tendencies or motive systems (personality variables) within individuals" (p. 852).

The Philadelphia Cohort Study (Wolfgang, Thornberry & Figlio, 1987) similarly produced evidence of consistency of criminal conduct, mostly based on impulsivity. Statlin & Magnusson (1989) showed that early aggression led to later violent crimes and damage to public property. Farrington (1986) reviewed a number of studies also indicative of consistency. Olweus relates conduct problems to a combination of E and N, again noting consistency of behaviour. In a follow-up study, Klinteberg, Humble & Schalling (1992) also noted consistency. Taylor and Watt (1977) in a follow-up study of more than 6000 school children noted that future criminal activity was predictable in terms of more than 3 deviant items in their early behaviour on the frequency, the seriousness, and the types of later criminal offences.

Consistency in conduct implies predictability, and several studies have shown such predictability. In an early review, Loeber & Dishion (1983) showed that a child's conduct problems were among the most predictive variables for later criminality. Thus prediction in general is certainly possible and useful as far as criminal activity is concerned (Monahan, 1981; Morrison 1994).

Clearly there is **consistency of behaviour**, not mirrored in odd "dynamic" formulations so frequent in predictive circles. Mossman (1994), finds in his review that "past behaviour alone appears to be a better long term predictor of future behaviour than clinical judgments" (p.783).

(b) It would be inappropriate here to discuss in detail the evidence in favour of the great importance of genetic factors in criminality. The study of MZ and DZ concordance for criminality. The study of MZ and DZ concordance for criminality, and work on children, adopted at birth from criminal and non-criminal violent parents has left little doubt about the proposition that something like 50%-60% of total phenotypic variance in criminality is genetic, and this estimate is made without any correction for attenuation which would raise it significantly (Eysenck and Gudjonsson, 1989; Raine, 1993). Our model of criminal predisposition would lead us to expect genetic determination of **altruism**, as the opposite to criminality, and Rushton, Fulzer, Neale, Blizard and Eysenck (1984) and Rushton (1980; Rushton et al., 1986) have demonstrated that this is indeed so. (Broad heritability estimates for the five scales used were: Altruism (56%), empathy (68%), nurturance (70%), aggressiveness (72%), assertiveness (64%); correction for unreliability would increase these estimates.) These facts are important in ruling out of account the usual completely environmental theories of sociologists, and they are in line with the moderately high heritabilities of the relevant personality variables (Eaves, Eysenck &

Martin (1989), but they need complementation by hormonal and psychophysiological studies linking heredity and conduct. And of course they do not encourage absurd speculation about a "gene for crime".

One interesting consequence of the broad heritability for criminality is that **criminality is dysgenic** because criminals tend to have more children than non-criminals. Lynn (1995) found in British parents that criminals had an average of 3.91 children, non-criminals had 2.21. "The result suggests that heredity for criminal behaviour is dysgenic, involving an increase in the genes underlying criminal behaviour in the population" (p.405).

It seems useful here to mention the promises held out for the future of genetic research in the study of personality and criminal behaviour by **molecular genetics.** These new techniques are beginning to revolutionize behavioural genetics because they allow us to identify **specific genes** that contribute to genetic variance in behaviour. Consider a study by Brunner, Nelen, Breakefield, Rogers and van Dorst (1993). They carried out genetic and metabolic studies on a large kindred in which several males were affected by a syndrome of borderline mental retardation and abnormal behaviour that included impulsive aggression; attempted rage, and arson. Disturbed mono-amine oxidase A was discovered, showing a complete and selective deficiency of enzymnatic activity of MAOA. In each of the five affected males, a point mutation was discovered in the eighth exon of the MAOA structural gene, which changes a glutamine to a termination codon. They conclude by saying that "isolated completed MAOA deficiency in this family is associated with a recognizable behavioural phenotype that includes disturbed regulation of impulsive aggression" (p.578).

This study was complemented by experiments on mice in which the gene encoding MAOA was damaged, so that no MAOA could be made. The mice in question were found to be more likely to bite their human handlers, and fight more vigorously with their fellow mice, all signs of greater impulsive aggressiveness. MAOA, of course, is in part responsible for the control of neurotransmitters such as dopamine, serotonin and nor-epinephrene, all of which are likely to play a part in regulating aggressive behaviour. It is interesting that in the Brunner et al.. (1993) study, carrier females were not detectable by enzymatic activity in cultural fibroblasts. Whether this is due to high antivity of the normal allele, incomplete X-inactivation, or other factor is unknown.

Also of interest is a study by Ebstein et al. (1996) into a gene postulated to underlie the trait of **novelty-seeking**, which is the name given by Cloninger (Cloninger, Svrakic & Przybeck, 1993) to **extraversion**; Harm avoidance (neuroticism), reward dependence (psychoticism) and persistence are his other three variables. Arguing that novelty seeking behaviours are related to dopamine, as shown in animal studies. Ebstein and his colleagues looked at a particular exonic polymorphism, the 7 repeat alleles in the locus for the D4 dopamine receptor gene (D4DR). The predicted association was indeed found in 124 unrelated Israeli subjects; no such association was discovered for the other three personality traits. Benjamin, using the Costa & McCrae (1992) NBO inventory, found the same exonic polymorphism to be related to extraversion (positively) and to conscientiousness (negatively). The specific facets mediating these associations were **warmth, excitement-seeking, positive emotions, and deliberation** (negatively). These are traits often found associated with criminal behaviour, and it will be interesting to carry out similar studies contrasting criminals with high-altruism subjects. Another interesting single gene association is that with severe alcoholism (Blum et al. 1991). Alcoholics and normals were studied for their allelic association with the D_2 dopamine receptor (D2DR) gene, utilizing peripheral lymphocytes as the DNA source. "The combined alcoholic group compared to the non-alcoholic group shared a significantly greater association with the A1 'allele of the D2DR gene" (p.409). This is relevant because of the well-known association between alcoholism and crime, particularly violent crime; as Moir and Jessel point out, "all research points to the colossal influence of alcohol in virtually every brand of crime". (p.69). Finally, it is notable that psychoticism links significantly in the basal ganglia with dopamine 2 (Gray, Pickering & Gray, 1994). These associations deserve more detailed study.

The mediation between DNA and behaviour has been discussed in great detail by Zuckerman (1991) for personality, and for criminality by Raine (1993) in an academic and by Moir & Jessel (1995) in a popular book. Also of interest is a book by Masters and McGuire (1994) as "The Neurotransmitter Revolution", dealing more specifically with serotonin, social behaviour and the law. It cannot be the purpose of this chapter to summarize these extensive discussions, but a few important points may repay highlighting.

On the whole, MAO in low levels is associated with aggression, impulsivity, and novelty-seeking. **Dopamine** is related to aggressivity, **serotonin** to lack of aggressiveness. However, as Ginsburg (1994) has shown, differences in underlying genotypes are major modifying factors. "Where the effect on one genotype is to increase aggression, the effect on the other is to decrease it" (p.124). This research was done in mice, but it suggests complications as far as human research is concerned. In monkeys, there is a robust link between diminished serotonergic function and destructive aggression (Masters and McGuire, 1994). Clearly much remains to be discussed, particularly about the most relevant animal model for human conduct, and the conditions under which a given neurotransmitter functions in a particular manner.

Of particular interest is a recent study by Klinteberg (1995), concentrating on the development of antisocial behaviour in 82 male and 87 female subjects. Available were teacher ratings at age 13, self-ratings of normbreaking behaviours. Furthermore, in the male group, criminal offending was found to be related to adult psychopathy-related personality traits (high impulsiveness and monotony avoidance, and low socialization) and to an indicator of disturbances in the serotonergic transmitter system. The same personality pattern, even more accentuated, was characterizing the violent criminal offending male group. Among female subjects, criminal offenders were characterized by high cognitive-social anxiety (high psychasthenia), low guilt, and indications of low activity in serotonergic turnover. High normbreaking behaviours in adolescence were associated with adult impulsiveness, low socialization, and signs of low serotonergic activity in both the male and female subjects. Low MAO levels were also characteristic of offenders.

Testosterone is an obvious candidate for a proximal antecedent role in view of the large sex differences in criminality. A detailed summary of the literature leaves little doubt that one of the effects of exposing the hunan brain to high (male-typical) levels of androgens is to increase the probability of behaviour patterns that lead to criminality (Ellis & Coontz, 1990). Aggression in particular is associated with high plasma testosterone (e.g. Ehrenkranz, Bliss, Sheard, 1974; Dabbs et al., 1984, 1988); Olweus, Mattson, Schalling & Loew, 1980). Impulsiveness, P, E and N usually correlate positively with plasma testosterone, but usually not above .2; there is high daily variability in plasma testosterone so that the mean of a number of measures should be used to obtain an aggregate measure. It is of interest that at three different age levels, black men obtained higher serum testosterone levels than white men (all were male Vietnam era veterans). Whether this is relevant to the higher criminality levels of black men is not clear, nor why there should be such a difference (Ellis & Nyborg, 1992).

Finally, a brief discussion of autonomic measures may be appropriate. Arousal forms a central part of the theory of criminality, and arousal can be measured by autonomic indicators. Of particular interest are two studies by Raine, Venables & Williams

(1990a, b) in which they studied autonomic activity change at age 15 as a predictor of criminality at age 24. Arousal was specifically lower in 15-year-olds who were later identified as criminals at age 24; this was true of a number of different studies. The same authors found increased attentional processing characteristics in predicting inter criminality, using EEG and contingent negative variation measures; this may link up with psychoticism through the low latent inhibition activity of high P scorers (Lubow, 1989). Raine (1993) has given an excellent summary of the evidence, leaving little doubt about the tendency for criminality to be linked with low arousal.

The Eysenck theory centres on low arousal leading to poor Pavlovian conditioning and hence to lack of conscience and anti-social behaviour. I have noted that the evidence favours low arousal in criminals; how about poor conditioning? Hare (1978) reported on 14 such studies, and Raine (1993) summarized results from 5 further studies. With two dubious exceptions, all of these studies gave results in accordance with prediction. One study explicitly investigated Eysenck's (1977) prediction that children who were highly conditionable and who had antisocial parents would become "negatively socialized" into their parents' antisocial habits, whereas children who conditioned poorly would paradoxically avoid becoming antisocial. Raine and Venables (1981) successfully demonstrated that this was indeed so. The only study to relate the Gough So scale to conditioning found low scorers less responsive to verbal reinforcements (Sarbin, Allen & Rutherford, 1969).

Avoidance learning, too, is associated with (high) arousal, and may indeed be considered a form of conditioning; Raine (1993) has reviewed the literature and shown that criminals tend to show **poor** avoidance learning. They are also very sensitive to rewards, a behaviour linked with extraversion (Gray, 1982, 1987); thus failure to condition may be only in response to **negative** reinforcement. Other psychophysiological consequences of low arousal have been reviewed by Raine (1993); they tend to follow the lines of prediction in showing criminals to have **low arousal** patterns.

It may be noted that when we look at the four major interactive biosocial theories of delinquency and crime listed by Raine, Brennan & Farrington in their introductory chapter, those by Mednick (1977) and Buikhuisen (1988 clearly are included in the Eysenck (1977) formulation, as involving poor conditioning. They concentrate on one aspect of the conditioning process, namely **avoidance learning**, but this is only one aspect of a more complete picture which also includes the effects of positive reinforcement. Moffitt's (1993) theory adds perinatal complications, but is not otherwise incompatible with a low arousal-poor conditioning theory. There seems to be here the beginnings of a paradigm, so long missing in this field.

Moffitt lists poor nutrition as one of the social environment effects that might interact with biological causes. There is some evidence on this point that concerns the main subject of this book, namely violent behaviour. Schoenthaler (1991) has reviewed the studies which suggested very strongly that violence in prison could be reduced by some 40% by micro-nutrient supplementation. None of these studies was without flaws, but the wide-reaching agreement should not be disregarded. The most recent study puts the matter beyond reasonable doubt. A triple-blind, randomized, controlled study among 402 male prisoners aged 18 to 25 years in California showed a statistically significant change in serious rule violation (largely violence towards warders and fellow prisoners). Two formulae (100% and 300% RDA) were tried, and as with IQ studies, the 100% was the more successful, giving a rule violation decrease of 41%, as compared with 16% among the 300% RDA group. Among the placebo group, there was a slight (20%) rise during the same period. Looking at the difference between micro-nutrient and placebo, these were -61% and -36%, respectively. As expected, improvement was observed in those prisoners who had the most vitamin and mineral status as assessed by blood analysis. Clearly it is possible to modify violent conduct by means of micro-nutritional supplementation, and it would seem an urgent need to repeat the experiment with school children in deprived areas to assess the effect of such supplementation on school behaviour and antisocial conduct generally (Eysenck & Eysenck, 1991).

I would conclude that **personality is a concept that is an essential feature of any acceptable theory of criminality and antisocial behaviour.** Personality provides a taxonomy of human behaviour that includes antisocial, aggressive, violent and generally criminal behaviour, and relates to genetic and biological variables. Only by reference to the wider perspective offered by attention to the major dimensions of personality can we hope to understand the causes of criminality.

REFERENCES

Allsop, J., & Feldman, M. (1976) Item analyses of questionnaire measures of personality and antisocial behaviour in school. *British Journal of Criminology, 16,* 337–351.

Backorowski, J., & Newman, J. (1985) Impulsivity in adults: motor-inhibition and time- interval estimation. *Personality and Individual Differences, 6,* 133–136.

Benjamin, J., Li, L., Patterson, C., Greenberg, B., Murphy, D., & Hamer, D. (1996) Population familial association between the D4-dopamine receptor gene and measures of novelty-seeking. *Nature Genetics, 12,* 81–84.

Bernson, T., & Fairey, T. (1984) Personality in assaultive and non-assaultive juvenile male offences. *Psychological Reports, 54,* 527–530.

Birder, A. (1988) Juvenile delinquency. *American Review of Psychology, 39,* 253–282.

Blum, H., Noble, E., Sheridan, P., Finley, O., Montgomery, A., Richie, T., Oz Karajoz, T., Fitch, P., Sadlack, F., Sheffield, D., Dahlmann, T., Halbardian, S., & Nogami, H. (1991). Association of the A1 allele of the D_2 Dopamine receptor gene with severe alcoholism. *Alcohol, 8,* 409–416.

Bocker, W., & Haefner, H. (1973) *Gewalttatan Geistergestoerter.* Berlin: Springer Verlag.

Booraem, C., Flowers, J., Bordner, G., & Satterfield, D. (1977) Personal space variations as a function of criminal behavior. *Psychological Reports, 41,* 1115–1121.

Brunner, H., Nelen, M., Breakefield, D., Rogers, S., van Dorst, B. (1993) Abnormal behaviour associated with a point mutation in the structural gene for monoamine oxidase A. *Science, 262,* 578–580.

Buikhuisen, W. (1980) Chronic juvenile delinquency: A theory. In: W. Buikhuisen & R.A. Mednick (Eds.), *Explaining Criminal Behavior.* Pp. 27–51, 7. Leiden: E.J. Bride.

Chico, E., & Ferrando, P. (1995) A psychometric evaluation of the revised P-scale in delinquent and non-delinquent Spanish samples. *Personality and Individual Differences, 18,* 381–337.

Choynowski, M., (1995) Does expressiveness have a factorial structure? *Personality and Individual Differences, 18,* 167–187.

*Cloninger, C., Svrakic, D., & Przybeck, T., (1993) A psychological model of temperament and character. *Archway General Psychiatry, 50,* 975–990.

Costa, P., & McCrae, R. (1992) *Revised NEO-PI-R) and NEO Five Inventory (NEO-FFI) Professional Manual.* Odessa: Psychological Assessment Resources.

Dabbs, J., Frady, R., Carr, T., & Bosch, N. (1984) Saliva testosterone and criminal evidence in young adult prison inmates. *Psychosomatic Medicine, 49,* 174–182.

Dabbs, J., Ruback, R., Frady, R., Hopper, C., Spoutas, S. (1988) Saliva testosterone and criminal violence among women. Personality and Individual Differences, 2, 209–275.

Davey, G. (Ed.)(1983) *Animal Models of Human Behavior.* London, Wiley.

De Man, A., & Green, C. (1988) Selected personality correlates of assertiveness and aggressiveness. *Psychological Reports, 62,* 672–674.

Duckitt, J. (1988) The predictor of violence,. South African Journal of Psychology, 18, 10–18.

Eastwood, L. (1985) Personality, intelligence and personal space among violent and non-violent delinquents. *Personality and Individual Differences, 6,* 717–723.

Eaves, L., Eysenck, H.J., & Martin, N. (1989) *Genes, Culture and Personality: An Empirical Approach.* New York: Plenum Press.

Ebstein, R., Novick, O., Umansky, R., Priel, B., Oster, Y., Blaine, P., Bennett, E., Nemanor, L., Katz, M., & Belmaker, R. (1996) Dopamine D4 receptor (D4DR) exon III polymorphism associated with the human personality trait of novelty seeking. *Nature Genetics,* **12**, 79–81.

Ehrenkranz, J., Bliss, E., & Sheard, M. (1976) Plasma testosterone: Correlation with aggressive behaviour and social dominance in man. *Psychosomatic Medicine,* 36, 469–475.

Ellis, L., & Coontz, P. (1990) Androgens, brain functioning, and criminality: The foundations of anti-sociability. In: L. Ellis & H. Hoffman (Eds.), *Crime in Biological, Social, and Moral Contexts.* New York: Praeger.

Ellis, L., & Nyborg, H., (1992) Racial/ethnic variations in male testosterone levels: A probable contributor to group differences in health.

Steroids, 57, 72–75.

Eysenck, H.J. (1947) *Dimensions of Personality.* London: Routledge & Kegan Paul.

Eysenck, H.J. (1960) The development of moral values in children: 7. The contribution of learning theory. *British Journal of Educational Psychology,* 30, 11–21..

Eysenck, H.J. (1967) *The Biological Basis of Personality.* Springfield: C.C. Thomas.

Eysenck, H.J. (1976) *Sex and Personality.* London: Open Books.

Eysenck, H.J. (1977) *Crime and Personality.* London: Paladin.

Eysenck, H.J. (1981) *A Model for Personality.* New York: Springer Verlag.

Eysenck, H.J. (1983a) *A biometrical-genetical analysis of impulsive and sensation-seeking behavior.* In: M. Zuckerman (Ed.): *Biological Bases of Sensation-seeking, Impulsivity, and Anxiety.* London: Lawrence Erlbaum.

Eysenck, H.J. (1983b) Personality and ageing: An exploratory analysis, *Journal of Social Behavior and Personality,* 33, 11–21.

Eysenck, H.J. (1987) The definition of personality disorders and the criteria appropriate for their description. *Journal of Personality Disorders,* **1**, 211–219.

Eysenck, H.J. (1991) Dimensions of personality: 16, 5 or 37 - Criteria for a taxonomic paradigm. *Personality and Individual Differences,* **8**, 773–790.

Eysenck, H.J. (1992) The definition and measurement of psychoticism. *Personality and Individual Differences,* **13**, 757–785.

Eysenck, H.J. (1994) Personality: Biological foundations. In: P.A. Vernon (Ed.), *The Neuropsychology of Individual Differences.* Pp. 151–207. New York: Academic Press.

Eysenck, H.J. (1995a) *Genius: The Natural History of Creativity.* Cambridge: Cambridge University Press.

Eysenck, H.J. (1995b) Personality differences according to gender. *Psychological Reports,* **76**, 711–716.

Eysenck, H.J. (1995c) Some comments on the Gough socialization scale. *Psychological Reports,* **76**, 298.

Eysenck, H.J., & Eysenck, M.W. (1988) *Personality and Individual Differences: A natural science approach.* New York: Plenum Press.

Eysenck, H.J., & Eysenck, S.B.G. (1975) *Manual of the Eysenck Personality Questionnaire.* London: Hodder & Stoughton.

Eysenck, H.J., & Eysenck, S.B.G. (1991) Improvement of IQ and behaviour as a function of dietary supplementation: A symposium. *Personality and Individual Differences,* **12**, 329–365.

Eysenck, H.J., & Gudjonsson, G. (1989) *The Causes and Cures of Criminality.* New York: Plenum Press.

Eysenck, H.J., & Wakefield, J. (1981) Psychological factor as predictor of marital satisfaction. *Advances in Behaviour Research and Therapy.* 3, 151–192.

Eysenck, S.B.G., & Eysenck, H.J. (1971) A comparative study of criminals and matched controls in three dimensions of personality. *British Journal of Social and Clinical Psychology,* **10**, 302–366.

Eysenck, S.B.G., Rust, J., & Eysenck, H.J. (1977) Personality and the classification of adult offenders. *British Journal of Criminology,* **17**, 163–170.

Fahrenberg, J., Selg, H., & Hampel, R. (1978) *Der Freiburger Personalickbeitsinventer.* Goettingen: Hogrefe.

Farrington, D. (1986) Stepping stones to adult criminal careers. In: D. Olweus, J. Block, and M. Radke-Yarrow (Eds.), *Development of Antisocial and Prosocial Behavior.* New York: Academic Press.

Feldman, M. (1977) *Criminal Behaviour: A Psychological Analysis.* Chichester: John Wiley.

Garber, J., & Strassberg, Z. (1991) Construct validity: History and application to developmental psychopathology. In: W.A. Grove and D. Ciocati (Eds.), *Personality and Psychopathy.* Pp. 219- 258. Minneapolis: University of Minnesota Press.

Ginsburg, B. (1994) Ontogeny, social experience, and serotonergic functioning. In: R. Marberg, T. McGuire (Eds.), *The Neurotransmitter Revolution.* Pp. 113–128. Cartasdale: Southern Illinois University Press.

Gough, H. (1994) Theory, development, and interpretation of the CPI socialization scale. *Psychological Reports,* **75**, 651–700.

Gray, J. (1982) *The Neuropsychology of anxiety.* Oxford: Oxford University Press.

Gray, J. (1987) Perspective on anxiety and drug outcomes: A commentary. *Journal of Research in Personality,* **21**, 493–509.

Gray, N., Pickering, A., & Gray, J. (1994) Psychoticism and dopamine D2 binding in the basal ganglia, using single photen emission tomography. *Personality and Individual Differences*, 431–434.

Hare, R. (1978) Electrodermal and cardiovascular correlates of psychopathy. In: R. Hare and D. Schalling (Eds.), *Psychopathic Behaviour: Approaches to Research*. Pp. 107–144. New York: Wiley.

Heaven, P. (1993) Personality predictors of self-reported delinquency. *Personality and Individual Differences*, 14, 67–76.

Heaven, P. (1996) Personality and self-reported delinquency: Analysis of the "Big Five" personality dimension. *Personality and Individual Differences*, 20, 47–59.

Hormuth, S., Lamm, H., Michelitsch, I., Scheuermann, H., Trommsdorft, G., & Voegele, I. (1977) Impulskontrolle und einige Persoenlichkeits charakteristika bei delinquenten und nichtdelinquenten Jugendlichen. *Psychologische Beitraege*, 19, 340–354.

Jamison, R. (1980) Psychoticism, deviancy and perception of risk in normal children. *Personality and Individual Differences*, 1, 87–91.

Kinzel, A. (1970) Body buffer zone in violent prisoners. *American Journal of Psychiatry*, 127, 59–69.

Klinteberg, B. (1995) Biology, norms, and personality: A developmental perspective. *Reports from the Department of Psychology, Stockholm University*, No. 804..

Klinteberg, B., Humble, K., & Schalling, D. (1992) Personality and psychopathy of males with a history of early criminal behaviour. *European Journal of Personality*, 6, 245–266.

Lester, D. (1993) The effects of war on crime. *Psychological Reports*, 73, 381–382.

Loeber, R., & Dishion, T. (1983) Early prediction of male delinquency: A review. *Psychological Bulletin*, 94, 68–99.

Lossel, F., & Westendorfer, W. (1976) Personlickbeits korrelate delinquanten Verhaltens uder aftuzieller Delinquenz? *Zeitschrift fuer Sozial psychologie*, 7, 177–191.

Lubow, R., (1989) *Latent Inhibition and Conditioned Attention Theory*. Cambridge: Cambridge University Press.

Lynn, R. (1995) Dysgenic personality for criminal behaviour. *Journal of Biosocial Science*, 27, 405–408.

Mackintosh, N. (1974) *The Psychology of Animal Learning*. London: Academic Press.

Masters, R., & McGuire, M. (1994) *The Neurotransmitter Revolution*. Carbondale: Southern Illinois University Press.

McGurk, B., Davis, D., & Greham, J. (1981) Associative behaviour, personality, and personal space. *Behavior*, 7, 317–324.

Mednick, S.A. (1977) A biosocial theory of the learning of law-abiding behaviour. In: S.A. Mednick & K.O. Varintyansen (Eds.), *Biosocial Bases of Criminal Behaviour*. Pp. 1- 8, New York: Gardner.

Mitchell, J., Rogers, P., Cavanagh, J., & Wasyliw, M. (1980) The role of train anxiety in violent and non-violent delinquent behavior. *American Journal of Forensic Psychiatry*..

Moffitt, T.E. (1993) Adolescence-limited and life course-persistent anti- social behaviour: A developmental taxonomy. *Psychological Bulletin*, 100, 674–701.

Moir, A., & Jessel, D. (1995) *A Mind to Crime*. London: Michael Joseph.

Monahan, J. (1981) *Predicting violent behavior*. Beverly Hills: Sage.

Morrison, D. (1994) Assessing prediction of violence: Being accurate about accuracy. *Journal of Counselling and Clinical Psychology*. 68, 783–792.

Mossman, D. (1994) Assessing predictions of violence: Being accurate about accuracy. *Journal of Counselling and Clinical Psychology*, 62, 783–792.

Olweus, D. (1979) Stability of aggressive reaction patterns in males: A review. *Psychological Bulletin*, 81, 852–875.

Olweus, D. (1984) Stability in aggressive and withdrawn, inhibited behavior patterns. In: R.M. Kaplan, V.J. Konecni, & R. Novaco, (Eds.), *Aggression in Children and Youths*. The Hague: Eiphoft.

Olweus, D., Mattson, A., Schalling, D., & Loew, A. (1980) Testosterone, aggression, physical and personality dimensions in normal adolescent males. *Psychosomatic Medicine*, 42, 253–269.

Pena, M., & Luengo, M. (1993) Demara de la quatificasion y cunducta antisocial en los adolescentes. *Anolisis y Modificacion de Conducta*, 19, 643–663.

Platenov, K. (1959) *The Word as Physiological and Therapeutic Factor*. Mosser: Foreign Language Publishing House.

Powell, G.E. (1977) Psychosocial and social deficiency in children. *Advances in Behaviour Research and Therapy*. 1, 27–56.

Putsins, A. (1982) The Eysenck Personality Questionnaires and delinquency prediction. *Personality and Individual Differences*, 3, 335–340.

Raine, A. (1993) *The Psychopathology of Crime*. London: Academic Press.

Raine, A., & Venables, P. (1981) Classical conditioning and socialization in a biosocial interaction? *Personality and Individual Differences*, 2, 273–283.

Raine, A., Venables, P., & Williams, M. (1990a) Relationship between central and autonomic measures of arousal at age 15 years and criminality at age 24 years. *Archives of General Psychiatry,* **47,** 1003–1007.

Raine, A., Venables, P., & Williams, M. (1990b) Autonomic orienting responses in 15 year-old male subjects and criminal behaviour at age 24. *American Journal of Psychiatry,* **147,** 933–937.

Raleigh, M., & McGuire, M. (1994) Serotonin, aggression, and violence in marmot monkeys. In: D. Masters & M. McGuire (Eds.), *The Neurotransmitter Revolution.* Pp. 129–145.

Rice, F. (1992) *The Adolescent: Development, Relationships, and Culture.* Boston: Allyn & Bacon.

Rigby, H., Mak, A., & Slee, P. (1989) Compulsiveness, orientation, institutional authority, and gender as factors in self-reported delinquency among Australian adolescents. *Personality and Individual Differences,* **10,** 689–692.

Rotenberg, M., & Nachshon, I. (1979) Impulsiveness and aggression among Israeli delinquents. *British Journal of Social and Clinical Psychology,* **18,** 59–63.

Rushton, P. (1980) *Altruism, Socialization, and Society.* Englewood Cliffs, N.J.: Prentice Hall.

Rushton, J., Fulzer, D., Neale, M., Blizard, D., and Eysenck, H.J. (1994) Altruism and genetics. *Asta Genetica Media Eanellologica,* **33,** 205–271.

Rushton, P., Fulzer, D., Neale, M., Nias, D., & Eysenck, H. (1986) Altruism and aggression: The heritability of individual differences. *Journal of Personality and Social Psychology,* **50,** 1192- 1198.

Sakic, V., Knezovic, Z., & Zuzul, M. (1987) Popujera Eysenckove EPQ-C skale sklonosti Kriminalotetu na osobama orudenim za Krivicua djela "napuda na ljade". *Penoloske Teme,* **2,** 255–261.

Sarbin, T., Allen, V., & Rutherford, E. (1969) Social reinforcement, socialization, and chronic delinquency. *British Journal of Social and Clinical Psychology,* **4,** 73–78.

Schoenthaler, S. (1991) Abstracts of early papers on the effects of vitamin and mineral supplementation on IQ and behaviour. *Personality and Individual Differences,* **12,** 335–341.

Schoenthaler, S., Amos, S., Eysenck, H., Hughes, M., & Korda, D. (in press.) Nutritional Institutional rule violations. *Personality and Individual Differences,* in press.

Schwenkmetzger, P. (1983) Risikovenhalten Risikobereitschaft und Delinquenz: Theoretische Grundlagen und differential diagnostische Untersuchungen. *Zeitschrift fur Differentielle und Diagnostiche Psychologie,* **4,** 223–239.

Shaw, J., & Scott, W. (1991) Influences of parent discipline style on delinquent behaviour: The mediating role of control orientation. *Australian Journal of Psychology,* **43,** 61–67.

Staltin, H., & Magnusson, D. (1989) The role of early aggressive behaviour in the frequency, seriousness, and type of later crime. *Journal of Consulting and Clinical Psychology,* **57,** 710–718.

Steller, M., & Hunze, D. (1984) Zur Selbst boschreibung von Delinquenten in Freiburger Personenlichkeits inventor (FPI) - eine Sekundaeranalyse empirischer Unterruchnugen. *Zeischrift fue Differentielle und Diagnostische Psychologie,* **5,** 87–109.

Stewart, C., & Hemsley, D. (1984) Personality factors in the taking of criminal risks. *Personality and Individual Differences,* **5,** 119–122.

Taylor, I., & Watt, P. (1977) The relation of denial systems and behaviour in a normal population to subseuqnt delinquency and maladjustment. *Psychological Medicine,* **7,** 166–169.

Wardell, D., & Yendell, L. (1980) A multidimensional approach to criminal disorders: The assessment of impulsivity and its relation to crime. *Advances in Behaviour Research and Therapy,* **2,** 159–177.

Whitehill, M., Scott, I., & De Myer-Gapin, S. (1976) Stimulation-seeking in antisocial pre-adolescent children. *Journal of Abnormal Psychology,* **85,** 101–104.

Wolfgang, M., Thornberry, T., & Figlio, R. (1987) *From Boy to Man, from Delinquency to Crime.* Chicago: University of Chicago Press.

Zuckerman, M. (1991) *Psychobiology of Personality.* Cambridge: Cambridge University Press.

3

A MULTIDIMENSIONAL PSYCHOBIOLOGICAL MODEL OF VIOLENCE*

C. Robert Cloninger, Dragan M. Svrakic, and Nenad M. Svrakic

Center for Psychobiology of Personality and
Departments of Psychiatry and Genetics,
    Washington University Medical School
4940 Children's Place
St. Louis, Missouri 63110

Violence is an interpersonal activity influenced by a complex interplay of multiple psychosocial and neurobiological factors. In this chapter, we will describe a general psychobiological model of personality with multiple components, each of which has a distinct psychological description, neurobiological substrate, and genetic and environmental causes. This general model includes components of reactive aggression, like the impulsive-aggressive temperament traits described by Eysenck and Eysenck (1985) and Barratt and colleagues (in press), and components of predatary aggression, like the non-impulsive hostile attitudes described by Huesmann and others (Dodge & Newman, 1981; Heilbrun et al, 1978). First, we will describe the general model of personality briefly, and then I will relate the general model to violence and the causes of aggressive criminality in particular. The interaction of these components in the development of violence will be described as a complex adaptive system with nonlinear dynamics.

GENERAL PSYCHOBIOLOGICAL MODEL OF PERSONALITY

The way that people learn from experience and adapt their feelings, thoughts, and actions is what characterizes their personality. Specifically, personality can be broadly defined as the dynamic organization within an individual of the psychobiological systems that modulate adaptation to a changing environment (Cloninger et al, 1993). This includes systems regulating cognition, emotion and mood, personal impulse control, and social relations. Hence personality traits are enduring patterns of perceiving, relating to, and thinking about oneself, other people, and the world as a whole.

* Supported in part by NIH grants MH31302, MH46276, MH46280, MH54723, and AA08403.

Biosocial Bases of Violence, edited by Raine *et al.*
Plenum Press, New York, 1997

Personality is comprised of both temperament and character traits. Temperament refers to differences between individuals in their automatic responses to emotional stimuli, which follow the rules of associative conditioning or procedural learning of habits and skills. Temperament traits include basic emotional response patterns, such as fear, anger, and attachment. Temperament has been variously defined as those components of personality that are heritable, fully manifest in infancy, and stable throughout life (Goldsmith et al, 1987). Fortunately, each approach defines essentially the same traits, which are all present from infancy onwards and are moderately heritable and stable throughout life.

In contrast, character refers to individual differences in our voluntary goals and values, which are based on insight learning of intuitions and concepts about our self, other people, and other objects. Character traits describe individual differences in our self-object relationships, which begin with parental attachments in infancy, then self-object diffentiation in toddlers, and continue to mature in a step-wise manner throughout life. Whereas temperament refers to the way we are born (our emotional predispositions), character is what we make of ourselves intentionally. Some of the major differences between temperament and character in their rules of operation, brain substrate, and inheritance are summarized in Table 1. These differences support the suggestion that temperament involves procedural learning of habits and skills whereas character involves propositional learning of goals and values.

Factor analytic studies confirm that temperament and character are each multidimensional. For example, four dimensions of temperament and three dimensions of character have been distinguished in the most comprehensive available test, which is called the Temperament and Character Inventory (TCI). Each dimension is roughly normally distributed.

The four TCI temperament dimensions are called Harm Avoidance, Novelty Seeking, Reward Dependence, and Persistence. For example, individuals low in Harm Avoidance have little anticipatory anxiety or fear about danger; consequently, they are optimistic risk-takers or even reckless if they are not mature in character. Individuals who are high in Novelty Seeking are impulsive, quick-tempered whenever frustrated, and prone to break rules and regulations in order to pursue what they think will give them pleasure or thrills; those low in Novelty Seeking are reflective and law-abiding. Individuals who are low in Reward Dependence are aloof and insensitive to social cues; those who are high in Reward Dependence are more likely to form warm social attachments readily and to respond to sentimental appeals. Individuals low in Persistence are underachievers with labile moods, who give up easily when frustrated.

Table 1. Differences in learning between temperament and character

Learning variable	Temperament	Character
Level of awareness	Automatic	Intentional
Form of memory	Percepts	Concepts
	Procedures	Propositions
Type of activity	Habits, skills	Goals, values
Type of emotion	Reactive (unconscious)	Evaluative (conscious)
Learning principle	Associative	Conceptual
	Conditioning	Insight
Rate of acquisition	Gradual (quantitative)	Abrupt (qualitative)
Key brain system	Limbic system	Temporal cortex
	Striatum	Hippocampus

Each of these is inherited independently of one another, so all possible combinations of scores on these dimensions occur. Such different configurations predispose individuals to qualitatively distinct patterns of emotional response because of functional interactions among the dimensions. Traditional subtypes of temperament and personality disorder correspond to the possible configurations of the three temperaments Harm Avoidance, Novelty Seeking, and Reward Dependence, as depicted in Figure 1.

Different temperament configurations are associated with different risks of particular behaviors, such as violence or charity. However, the accuracy of predictions about behavior in individual cases is low unless the character configuration is also specified. For example, the adventurous temperament (i.e., low Harm Avoidance, high Novelty Seeking, and low Reward Dependence) may lead to either Antisocial Personality Disorder (when character is immature) or imaginative exploration and objective independence in scientific research (when character is mature). Persistence acts as a general moderator of self-control and influences character development, so it also has an important role in emotional regulation and personality development.

All possible combinations of the three character dimensions also occur, as depicted in Figure 2. Low Self-directedness is associated with irresponsible, aimless, and undisciplined behavior characteristic of poor impulse-control and personality disorder in general. Low Cooperativeness is associated with deficits in empathy, so that uncooperative individuals are described as hostile, aggressive, hateful, and revengeful opportunists. Low Self-Transcendence is associated with deficits in transpersonal identification or conscience; those low in Self-Transcendence show conventional and materialistic behavior with little or no concern for absolute ideals, such as goodness and universal harmony. Consequently, low character development, as in melancholic and disorganized character configurations in Figure 2, are associated with deficits in impulse-control, empathy, and

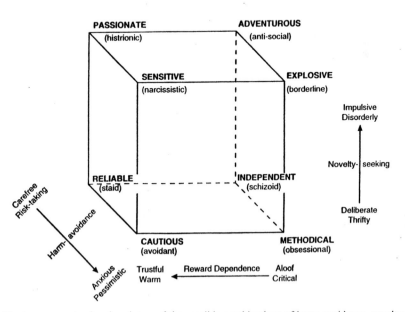

Figure 1. The temperament cube: descriptors of the possible combinations of harm avoidance, novelty seeking, and reward dependence (subtype in parenthesis if personality disorder is present).

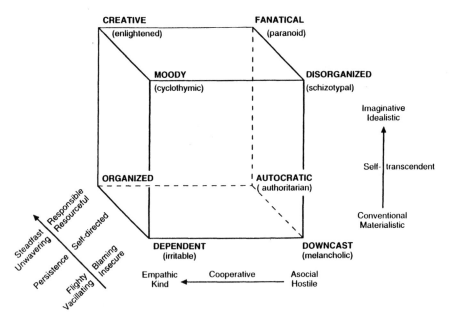

Figure 2. The character cube: descriptors of possible combinations of the three character dimensions and persistence.

conscience that may lead to violence and aggression depending on the situation and temperament of the individual.

Furthermore, temperament and character configurations are related to one another in complex but systematic ways. Temperament constrains character development but does not fully determine it because of the systematic effects of sociocultural learning and the

Table 2. Relative risk of immaturity (i.e., mild and severe personality disorder) as a function of temperament type in a sample from the general community (Cloninger et al, 1994)

Temperament type	Configuration	N	% Immature
High risk			
Borderline	NHr	39	72
Obsessional	nHr	44	59
Antisocial	Nhr	25	48
Passive-aggressive	NHR	30	40
Average	—	15	33
Low risk			
Avoidant	nHR	30	17
Schizoid	nhr	31	16
Histrionic	NhR	50	12
Reliable	nhR	36	6
Total		300	33

stochastic effects of experience. For example, the Explosive temperament configuration, which is also called Borderline Temperament (i.e., high Novelty Seeking and Harm Avoidance plus low Reward Dependence) predisposes to the underdevelopment of Self-directedness and Cooperativeness. Antisocial and Obsessional temperament are also likely to do the same, as shown in Table 2. However, temperament structure is not an inevitable or necessary determinant of character structure, which also depends substantially on social learning and luck (i.e., random environmental events that can be measured but not predicted in advance).

Fortunately, the complex patterns of relationship among temperament structure, character structure, and histories of behavioral conditioning and insight (social) learning can be quantified in a rigorous manner as a non-linear dynamic network of variables. Personality development, including the risk of violence in particular, cannot be well understood using categorical or unidimensional descriptive measures or linear statistical models.

THE MULTIPLE PERSONALITY COMPONENTS OF SOCIALIZATION VS. VIOLENCE

Research and discussion about the causes and prevention of violence have been handicapped by attempts to describe it by one or two categories (e.g., antisocial versus explosive personality disorders, or primary versus secondary psychopathy) or by one or two quantitative scales (e.g., reactive aggression or impulsivity versus predatory aggression or hostility or low cooperation). Such approaches capture important information, but are incomplete and even misleading from the perspective of the complex adaptive systems that are characteristic of personality development. Personality disorders usually involve many mixed and intermediate configurations, and only rarely pure prototypes like those described in DSM-IV. Most individuals with any personality disorder satisfy the criteria for multiple disorder subtypes (Cloninger et al, 1993). Likewise, personality development is characterized by abrupt transitions between different configurations of multiple dimensions of personality; such transitions give rise to the mixed and intermediate forms of personality disorder. It is useful to examine the contributions of each of the seven TCI dimensions to the degree of socialization and aggression or the risk of violence.

In college students, antisocial behavior is associated with high Novelty Seeking, low Reward Dependence, low Persistance, low Self-directedness, and low Cooperativeness (Table 3). Such relations have been considered in several other studies, as summarized below, which indicate that Harm Avoidance also plays a role in more severe antisocial behavior than is typical of college students.

Novelty Seeking: High Novelty Seeking is characterized by impulsive-aggressive traits, such as being quick-tempered and rule-breaking. It is what distinguishes impulsive personality disorders, such as borderline and antisocial personality disorders, from other personality disorders (Cloninger et al, 1994; Svrakic et al, 1993; Goldman et al, 1994). Most criminals are at least mildly impulsive, but most individuals who are high in Novelty Seeking are not criminal because most are mature in character. Furthermore, Novelty Seeking is not much higher in violent criminals than in criminals who commit only property crimes (Sigvardsson et al, 1987). Consequently, character traits have a stronger relationship to criminality in general than any temperament trait.

Harm Avoidance: High Harm Avoidance predisposes to poor character development in general, which in turn increases the risk of violence. High harm avoidance tends to inhibit dangerous acts, but when combined with high Novelty Seeking, explosive outbursts

Table 3. Correlations between the Temperament and Character Inventory (TCI) scales and the Zuckerman-Kuhlman Personality Questionnaire (ZKPQ), Eysenck Personality Questionnaire-Revised (EPQ-R), Jackson Personality Research Form (PRF), and Buss-Plomin EASI (Zuckerman & Cloninger, in press)

Other test scales	Correlations (x100) with TCI scales						
	NS	HA	RD	PS	SD	CO	ST
Zuckerman ZKPQ							
Impulsive Sen. Seeking	**68**	−39	−20	−15	−17	−11	28
Anxiety	−12	**66**	16	−11	**−49**	−8	4
Aggression-hostility	13	5	−27	−3	−32	**−60**	−18
Activity	0	−29	0	**46**	36	12	6
Sociability	37	−38	31	−1	9	10	11
Eysenck EPQ-R							
Neuroticism	−14	**59**	10	−6	**−45**	−17	2
Extraversion	**44**	**−53**	23	17	18	10	16
Psychoticism	**41**	−14	**−45**	−29	−31	**−42**	15
Lie	−21	−8	12	13	25	34	8
Jackson PRF							
Achievement	−19	−19	3	**67**	**46**	26	5
Aggression	8	2	−30	−9	−32	**−69**	−22
Autonomy	22	−15	**−55**	−9	−7	−20	4
Cognitive Structure	**−63**	30	8	27	10	1	−19
Succorance	−15	20	**53**	−4	−14	10	8
Buss-Plomin EASI							
Emotionality	− 6	**58**	18	−8	**−40**	−26	−5
Activity	5	−25	2	**45**	32	2	−1
Sociability	26	**−41**	31	14	25	21	6
Impulsivity	**67**	−22	−19	−24	−17	−12	22
Gough CPI							
Socialization	−26	−8	34	29	36	32	−6

(n=207 college students)
Correlations over .40 in bold;
all correlations over .2 are significant (P<.01)

are characteristic, as in borderline and passive-aggressive temperaments. Low Harm Avoidance is part of the antisocial personality configuration, and distinguishes violent from non-violent criminals in many, but not all cases (Sigvardsson et al, 1987).

Reward Dependence: Low Reward Dependence predisposes to personality disorder in general, especially low Cooperativeness characterized by lack of empathy, predatory aggression or revenge, and the secondary emotion of hate or scorn. It is a shared feature of both the antisocial and borderline personality disorders.

Persistence: Low Persistence predisposes to emotional lability and personality disorder. It makes a minor contribution to risk of substance abuse and criminality in most cases, but can make a difference depending on the configuration of other traits (Svrakic et al, 1996).

Self-directedness: Low self-directedness is the common feature of all personality disorders and has the largest effect size in the prediction of risk of substance abuse, which is charecteistic of most violent criminals. Low Self-directedness is associated with poor impulse-control and lack of discipline. When Self-directedness is high, an individual who is high in Novelty Seeking and low in Cooperativeness is characterized as a hostile auto-crat or authoritarian bully, but seldom violates conventions to commit crimes.

Cooperativeness: Low Cooperativeness involves intolerance, mistrust, and dislike of other people. It is associated with lack of empathy, which is related to the secondary emotions of hostility, contempt, and hatred, which often are necessary but insufficient antecedents of predatory aggression. This is a core feature of antisocial and explosive personality disorders, and of violent crime in particular. It quantifies concepts of predatory or hostile aggression in cognitive models.

Self-Transcendence: Low Self-transcendence is associated with lack of conscience and appreciation of moral ideals. However, high Self-transcendence increases concern for convention and material gain, so overall Self-transcendence makes only a minor contribution to risk of criminality and substance abuse.

In summary, the temperament configurations that are most likely to lead to childhood misconduct, substance abuse, and crime are the antisocial and explosive temperaments. These are both characterized by high Novelty Seeking and low Reward Dependence, and differ in Harm Avoidance, which is low in the antisocial and high in the borderline or explosive temperament. However, such temperaments are unlikely to lead to frequent violence unless there is a personality disorder, which is characterized by low Self-directedness and low Cooperativeness. The explosive and antisocial temperament configurations do usually lead to such personality disorder, but such developments are not inevitable.

TESTS OF PREDICTIONS ABOUT PROSPECTIVE RISK OF CRIMINALITY

This psychobiological model predicts that the antisocial temperament configuration in childhood increases the later risk of conduct disorders, substance abuse, and criminality. This has been confirmed in prospective (Sigvardsson et al, 1987; Tremblay et al, 1994) and cross sectional studies of child and adolescent temperament (Wills et al, 1994). These studies confirm that particular combinations of traits increase risk, more than expected from their average effects added together. That is, non-linear interactions among temperament are important in predicting risk of criminality. In a prospective study of Swedish adoptees from birth to 28 years of age, violent criminals were more likely to be high in Novelty Seeking and low in Harm Avoidance than non-violent criminals. In a large longitudinal study of boys from kindergarten to age 13 years, the delinquent boys were most likely to have the adventurous temperament type (high Novelty Seeking, low Harm Avoidance, and low Reward Dependence) than those without delinquency. In both these prospective studies, temperament traits were stably measurable from kindergarten through adolescence. Similarly, a cross-sectional study of adolescents showed that the interaction of these temperament dimensions was significantly non-linear in increasing risk of antisocial behavior and substance abuse (Wills et al, 1994). Also studies of adult criminals and substance abusers confirm that adults with these problems have impulsive personality disorders with high Novelty Seeking, and, in contrast to studies of antisocial conduct in children and adolescents, Harm Avoidance is also often high, not low (Nagoshi et al, 1992; Nixon et al, 1990). This suggests that it is essential to take stage of personality into account in evaluating the causes of antisocial behavior and violence.

INTERACTIONS OF TEMPERAMENT AND CHARACTER IN DEVELOPMENT

It is highly informative to consider problems in impulse control, working, and social adjustment from a developmental perspective. Fortunately, the subscale structure of the TCI takes steps of development into account, as outlined in Table 4. Each of the five subscales are numbered in the order of their development, which occur in the interlocking sequence shown in Table 4. This sequence of 15 steps provides a more detailed account of development than prior descriptions. As shown in Table 5, the TCI sequence corresponds to the stages of development distinguished by investigators such as Piaget from a cognitive perspective, Freud from a psychosexual perspective, and Erikson from a broad psychodynamic perspective. More generally, however, developmental sequences are expected to differ in individuals with different temperaments and life experiences, but the sequence described in Table 4 corresponds to the modal pathway that leads to full character development, that is, high scores on all three character dimensions.

Accordingly, the TCI provides a quantitative way of measuring the interrelations among each step in the developmental sequence with antecedent temperament traits as

Table 4. 15 Steps in full personality development, as measured by the temperament and character inventory

Steps	Self-directed	Cooperative	Self-transcendent
[1] co1		tolerant v suspicious (trust v mistrust)	
[2] sd1	responsible v blaming (confidence v doubt)		
[3] st1			obedient v intractable (respectful v judgmental)
[4] sd2	purposeful v aimless (moderate v indulgent)		
[5] co2		empathic v cruel (prudent v scornful)	
[6] st2			conscientious v unjust (worshipful v defiant)
[7] sd3	resourceful v inept (hopeful v helpless)		
[8] co3		generous v disagreeable (kind v hostile)	
[9] st3			spiritual v materialistic (contemplative v greedy)
[10] sd4	self-accepting v vain (humble v impatient)		
[11] co4		forgiving v revengeful (compassionate v callous)	
[12] st4			enlightened v possessive (joyfully free v controlling)
[13] sd5	integrity v conflict (peaceful v undisciplined)		
[14] co5		wise vs unprincipled (loving v harsh)	
[15] st5			creative v dualistic (idealistic vs practical)

Table 5. Comparison of different descriptions of personality development

TCI developmental step	Piaget	Freud	Erikson
[1] co1—trust	sensorimotor (reflexive)	oral (passive)	trust
[2] sd1—confidence	sensorimotor (enactive)	anal (negativistic)	autonomy
[3] st1—obedience	self-object differentiation	early phallic	
[4] sd2—purposefulness	intuitive	late phallic (exploratory)	initiative
[5] co2—empathy	operational (concrete)	latency (conforming)	
[6] st2—conscientiousness	operational (abstract)	early genital (conscientious work)	identity
[7] sd3—resourcefulness			
[8] co3—generosity		later genital (social maturity)	intimacy
[9] st3—spirituality			
[10] sd4—humility			
[11] co4—compassion			generativity
[12] st4—enlightenment/joy			
[13] sd5—integrity/peace			integrity
[14] co5—wisdom/pure love			
[15] st5—creativity/goodness			

Table 6. Correlations of each step in character development (designated as step N) with temperament dimensions and the prior step in character development (designated as step N-1)

Developmental step N	Correlation (x 100) with TCI scales				
[N] Subscale—Label	NS	HA	RD	PS	N-1
Level I = "Walking Together"					
[1] co1—trust	-2	-25	**35**	13	--
[2] sd1—confidence	1	**-45**	15	18	**30**
Level II = "Working Together"					
[3] st1—obedience	17	-7	11	18	-13
[4] sd2—purposefulness	-2	**-47**	22	**40**	7
[5] co2— empathy	-6	-22	**55**	20	**36**
Level III = "Hearts Beating as One"					
[6] st2—conscientiousness	-5	-19	25	**30**	11
[7] sd3—resourcefulness	7	**-58**	12	**55**	10
[8] co3—generosity	-9	-19	**41**	19	**32**
Level IV = "Spirits Feeling as One"					
[9] st3—spirituality	2	-8	**33**	15	14
[10] sd4—humility	-24	-12	12	-7	7
[11] co4—compassion	-21	-12	**40**	9	**35**
[12] st4—enlightenment/joy	-11	-1	**33**	8	**31**
Level V = "Minds thinking as One"					
[13] sd5—integrity/peace	-19	**-43**	23	**43**	16
[14] co5—wisdom/pure love	-16	-4	**30**	1	21
[15] st5—creativity/goodness	-16	5	**34**	8	25

Data based on 593 individuals representative of general population of metropolitan St. Louis (Cloninger et al, in press)
Correlations of .30 or higher in bold

well as prior character development. The correlations of each step in character development with the temperament dimensions and the prior step is shown in Table 6. These steps can be organized into 5 groups, making divisions at the 5 points where a step is moderately correlated with the prior step but not the subsequent step. In other words, such points of division correspond to frequent fixation points in development. Self-directedness and Cooperativeness are positively correlated, so that development of one facilitates the development of the other. However, the subscales of Self-Transcendence progress from a negative correlation between ST-1 to increasingly positive correlations with Self-directedness and Cooperativeness, particularly for ST-4 and ST-5. As a result, the transitions from one level of Self-transcendence to the next, which reflect increasingly more inclusive concepts of self, are the most difficult steps in personality development.

Problems in the first group, involving problems in basic trust and confidence, are characteristic of prepsychotic individuals. Problems in the second group, involving disobedience, lack of purposefulness and empathy, are characteristic of individuals with severe personality disorders who have problems in working, socialization, and impulse-control. Problems in the third level, involving little conscience, resourcefulness, and generosity, are typical of individuals with mild personality disorders and problems in social intimacy or group identification. The fourth and fifth tiers involve progressive steps in cognitive and spiritual development among socially mature individuals.

Problems with violence are most characteristic of individuals with severe personality disorders who are rebellious, selfish, and lacking in goals and values to organize and direct their impulses with self-control. Novelty Seeking has little impact on whether a person becomes fixated at this immature level of development. Such problems are more determined by disturbances that occur in basic trust and confidence, as a result of parental abuse and neglect and temperament factors, such as high Harm Avoidance and low Reward Dependence. The combination of high Harm Avoidance and low Reward Dependence predisposes to poor impulse-control and psychopathic behaviors like lying, cheating, and stealing because such people are insecure, inept, have little sensitivity to social cues (Cloninger, 1987). Also problems in development in the second tier, such as problems in developing a sense of meaning and purpose in life or empathy for others, are associated with high Harm Avoidance, low Reward Dependence, and low Persistence. This means that individuals with psychopathic temperament traits (i.e., high Harm Avoidance and low Reward Dependence) are at high risk for conduct and personality disorders associated with violence.

THE BIOSOCIAL BASIS OF VIOLENCE AS A COMPLEX ADAPTIVE SYSTEM

Personality involves a dynamic organization of multiple personal and social needs as an adaptive response to a changing environment, as defined earlier. The adaptive nature of personality suggests the operation of some optimization principle in the course of personality development. Viewed as a dynamic optimization process, personality development involves a process of self-organizing change that must satisfy multiple constraints. These constraints may include opposed needs, such as a desire to approach novelty competing with a desire to avoid novelty for sake of security in someone who is high in both Novelty Seeking and Harm Avoidance. The optimization of such multiple constraints results in non-linear dynamics, which is consistently characteristic of all systems involving growth and development of multidimensional adaptive systems in biology, neuroscience, psychology, and sociology (Kauffman, 1993). Such multidimensional dynamic systems are usually called complex adaptive systems.

The dynamic evolution of complex adaptive systems can be represented as a walk through a landscape with hills and valleys. The height of hills and depth of values corresponds to the fitness value of a point in a multidimensional landscape, such that the highest points are those that are most fit, most satisfying, or optimal. Progress from one hill (i.e., local optimum) to a still higher hill corresponds to the process of personality development toward a global optimum. However, such progress usually requires passage through an intermediate valley, which involves an intial decrease in fitness. Consequently, individuals may become fixed at points that are not optimal overall, but are better than anything achievable without considerable personal effort, social support, and possibly guidance or a leap of faith.

A precise quantitative formulation of personality development as a complex adaptive system based on the seven dimensional TCI model has recently been presented in detail elsewhere (Svrakic et al, 1996). The basic specification of temporal development states that the rate of change of each personality trait is proportional to the gradient of its fitness function and stochastic variation describing random effects on each personality component. The fitness function was specified as a fourth-degree polynomial with a U-shaped curvature and was estimated from the observed correlations among the TCI personality dimensions, which is summarized in Table 7.

Table 7 shows that different personality components may act in synergy (i.e., are positively correlated) or in opposition (i.e., be negatively correlated. This simply means that certain temperament traits may promote the increase of some charcter traits while simultaneously promoting the decrease of others. For example, Novelty Seeking is negatively correlated with Cooperativeness and Self-directedness, but positively correlated with Self-transcendence. Consequently, individuals who are high in Novelty Seeking will tend to develop low Cooperativeness, low Self-directedness, and high Self-transcendence. On the other hand, high Reward Dependence favors high values of all three character dimensions, and high Harm Avoidance favors low values of all three charcter dimensions. In addition, Self-directedness favors increases in Cooperativeness and decreases in Self-transcendence. In order to predict the simultaneous evolution of all the dimensions, this joint network of relationships must be taken into account. Furthermore, particular configurations may enhance the stability of the system more than expected from their average effects in the population as a whole. In other words, it is an empirical question whether the patterns of change are linear or non-linear, even though such complex adaptive systems are usually expected to have a non-linear dynamics.

Table 7. Correlations among 7 dimensions of the Temperament and Character Inventory (x 100) in 300 individuals from the general population (Cloninger et al, 1994)

		NS	HA	RD	PS	SD	CO	ST
Novelty seeking	(NS)	100						
Harm avoidance	(HA)	−8	100					
Reward dependence	(RD)	8	−16	100				
Persistence	(PS)	−14	−27	3	100			
Self-directedness	(SD)	−26	**−47**	21	28	100		
Cooperativeness	(CO)	−10	−28	**54**	18	**57**	100	
Self-transcendence	(ST)	20	−8	28	11	−10	15	100

Correlations of .30 or higher in bold

EMPIRICAL TESTS OF NON-LINEAR DYNAMICS OF PERSONALITY DEVELOPMENT

Empirical observations about personality development confirm that it behaves as a nonlinear adaptive system. Several of these observations will be summarized here because they show the need for psychobiologists to think of personality development as a multidimensional adaptive process rather than to continue to use linear models of quantitative traits or categories.

Specifically, there is no one-to-one correspondence between temperament types and character types, as would be expected with a linear system. Each temperament type may develop into different character configurations with varying probabilities. The observed patterns of association between temperament type and character type are summarized in Table 8 for a sample of 593 individuals in the general population (Cloninger et al, in press).

Furthermore, when change in character configuration was observed in a prospective follow-up study, the change was not random at all. Rather, character configurations change consistently toward an alternative configuration that is predicted to be most probable according to the expected dynamics of a complex adaptive system. The mean values of the population remain nearly stationary, but the individuals in the population explore available alternative configurations in order to optimize their individual adjustment. Many remain stable, but those who change move systematically in directions constrained by the correlations among all the dimensions of personality. In a one year follow-up of a community sample of 593 individuals (Cloninger et al, in press), 121 or about 20% of individuals made substantial changes in their character configuration. Of these 121 individuals, 116 or 96% changed in the specific direction predicted by the nonlinear model dynamics; that is, their change was to the most probable alternative configuration.

For example, let us consider the complex dynamics of the explosive (borderline) temperament and the adventurous (antisocial) temperaments, who are most strongly at risk to become low in Cooperativeness and to be violent. Figure 3 depicts the probability of each of the 8 possible character types actually observed in individuals with explosive temperaments

Table 8. Observed proportion of different character types according to the temperament type of individuals in the general population (n=593; Cloninger et al, in press)

| | % of character types in each temperament type | | | | | | | |
| | Temperaments with high NS | | | | Temperaments with low NS | | | |
Character types	Expl	Adve	Sens	Pass	Meth	Avoi	Inde	Reli
Uncooperative								
Melancholic	44	20	17	4	37	9	12	6
Schizotypal	35	24	19	10	13	9	5	2
Fanatical	0	6	0	4	6	5	10	9
Autocratic	0	2	0	4	4	1	17	1
Cooperative								
Cyclothymic	4	7	21	12	10	12	0	11
Dependent	4	2	19	4	7	28	5	8
Organized	4	13	10	22	16	19	24	21
Creative	7	15	10	32	5	11	20	38
Column totals	100	100	100	100	100	100	100	100

Types are classified according to median splits as high or low, and labelled according to the three dimensional configurations depicted in Figures; Persistence is ignored in these classifications.

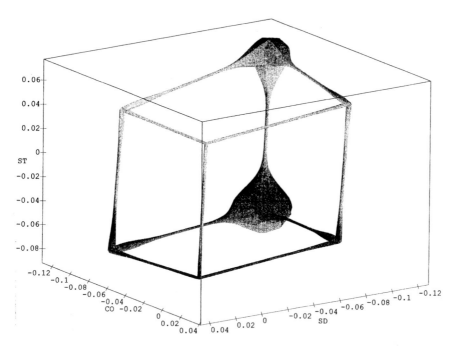

Figure 3. Probability of all possible character types observed in individuals with explosive (borderline) temperaments.

(High Novelty Seeking and Harm Avoidance with low Reward Dependence), as summarized in Table 8. The larger the protuberance at each corner of the character cube, the higher the probability of that configuration. Among the 8 possible three-dimensional character configurations, 44% have the melancholic configuration (which is low in Cooperativeness, Self-directedness, and Self-transcendence) and 35% have the schizotypal configuration (which is low in Cooperativeness and Self-directedness, but high in Self-transcendence). Thus in neutral social environments, explosive temperaments nearly always are expected to become uncooperative (i.e. low in Cooperativeness), just as was observed (see Table 8).

In contrast, Figure 4 depicts the observed probabilities of each of the 8 possible character types in individuals with adventurous temperaments. The character development of adventurous individuals is much more variable and less extreme than that of people with explosive temperaments. The greater variability and lower deviance of character in adventurous temperaments is expected because of the opposing effects of low Harm Avoidance (which favors maturity) and low Reward Dependence (which favors immaturity) on all three characters. Consequently, small variation in the relative influence of Harm Avoidance and Reward Dependence can lead to substantial differences in optimal character configuration. The adventurous temperament (high Novelty Seeking with low Harm Avoidance and Reward Dependence) may develop develop any of the possible character configurations with an appreciable probability. the four most stable configurations are melancholic (20% probability), autocratic (11%), schizotypal (24%), and fanatical (6%), which all have low Cooperativeness. The remaining four configurations are less attractive and stable. Consequently, the probability is about 61% that an a person with an adventurous temperament will become low in Cooperativeness. The full set of predicted probabilities for character as a non-linear function of the underlying temperament type, as

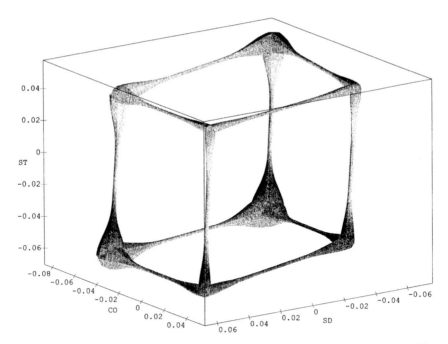

Figure 4. Probability of all possible character types observed in individuals with adventurous (antisocial) temperaments.

described elsewhere (Svrakic et al, 1996), is summarized in Table 9, and agrees quite well with the observed pattern summarized in Table 8.

In summary, personality development can be well described as a complex adaptive system with non-linear dynamics. This allows a quantitative prediction of transitions in configuration over time that is sensitive to quantitative differences in multiple personality dimensions simultaneous (the internal milieu) as well as to external influences, such as social learning and random events. Linear models do not account for the frequency and systematic direction of abrupt transitions observed empirically.

IMPLICATIONS FOR FUTURE BIOSOCIAL RESEARCH ON VIOLENCE

The observations about personality structure, inheritance, and development here have several important implications for future work on violence. First, the decomposition of personality into quantitative measures of dissociable aspects of learning and memory, namely temperament and character, is crucial to future biosocial research into criminality because these domains operate according to distinct principles. Temperament and character differ in fundamental ways in their mechanisms of learning and their patterns of inheritance and development. Currently, the TCI is the only personality inventory that distinguishes these domains. The TCI has also been validated for differential diagnosis of personality disorder and its subtypes in systematic studies of clinical and general population samples. Other tests like the EPQ with only a few factors may be partially informative in group contrasts but do not have

Table 9. Predicted proportion of different character types according to the temperament type of individuals in the general population (Cloninger et al, in press)

	% of character types in each temperament type							
	Temperaments with high NS				Temperaments with low NS			
Character types	Expl	Adve	Sens	Pass	Meth	Avoi	Inde	Reli
Uncooperative								
Melancholic	46	13	20	0	42	4	2	2
Schizotypal	36	16	10	4	2	8	2	0
Fanatical	2	17	8	8	2	7	5	3
Autocratic	5	14	8	1	30	1	10	1
Cooperative								
Cyclothymic	3	7	17	24	1	21	0	5
Dependent	4	6	14	1	10	18	2	2
Organized	2	12	16	5	10	15	57	24
Creative	2	15	7	57	3	26	22	63
Column totals	100	100	100	100	100	100	100	100

specificity in individual cases. For example, high Neuroticism scores occur in both anxiety disorders without personality disorder and in violent personality disorders.

Second, the decomposition of temperament into independently inherited dimensions, rather than factor-analytically derived phenotypic factors, may be particularly useful now that specific genes are being mapped for heritable personality traits. Such mapping should help to specify risk groups objectively, unconfounded by environmental factors.

Third, the decomposition of character into dissociable steps in development, rather than factor-analytically derived phenotypic factors, may be particularly helpful in unravelling the complex interactions among multiple influences on personality development. Social learning influences are important in character development regardless of the temperament configuration. Social learning can be formulated as a shift in the relative attractiveness of alternative character outcomes in the context of quantitative models of complex adaptive systems (Svrakic et al, 1996).

Fourth, genetic and social determinants of violence are often correlated because parental personality disorder and substance dependence interferes with parenting of children. Adoption results suggest that the intergenerational cycle of violence may be broken by improved child rearing of vulnerable children. This might be implemented by extensions of programs like Head Start, which would provide support to families in terms of teaching coping skills and effective child rearing rather than simply financial support, which is often inappropriately used.

Multidimensional (complex) adaptive systems with non-linear dynamics result in extensive nonlinear gene-environment interaction. Consequently, the inconsistencies observed in neuroendocrine and psychosocial studies are an expected characteristic of complex adaptive systems. Consistent results are only expected when analyses are directed to more homogeneous components of violence, such as heritable quantitative personality traits like Novelty Seeking, Harm Avoidance, and Reward Dependence.

REFERENCES

Bayon, C., Hill, K., Svrakic, D.M., Przybeck, T.R., Cloninger, C.R. (1996). Dimensional assessment of personality in an outpatient sample: relations of the systems of Millon and Cloninger. J Psychiatric Research, in press.

Buss, A.H., & Plomin, R. (1975). *A Temperament Theory of Personality Development.* New York: Wiley.

Cloninger, C.R. (1987). A systematic method for clinical description and classification of personality variants. *Archives of General Psychiatry, 44,* 573–588.

Cloninger, C.R. (1994). Temperament and personality. *Current Opinion in Neurobiology, 4,* 266–273.

Cloninger, C.R., & Gottesman, I.I. (1987). Genetic and environmental factors in antisocial behavior disorders. In S.A. Mednick, T.E. Moffitt, S.A. Stack (Eds.). *The Causes of Crime: New Biological Approaches.* Cambridge: Cambridge University Press, pp. 92–109.

Cloninger, C.R., Sigvardsson, S., Bohman, M., and von Knorring, A.-L. (1982). Predisposition to petty criminality in Swedish adoptees: II. Cross-fostering analysis of gene-environment interaction. *Archives of General Psychiatry, 39,* 1242–1247.

Cloninger, C.R., Svrakic, D.M., Przybeck, T.R. (1993). A psychobiological model of temperament and character. *Archives of General Psychiatry, 50,* 975–990.

Cloninger, C.R., Bayon, C., Przybeck, T.R. (in press). Epidemiology and Axis I comorbidity of antisocial personality. In D.M. Stoff, J. Breiling, J.D. Maser (Eds.). *Handbook of Antisocial Behavior.* New York: John Wiley & Sons.

Cloninger, C.R., Przybeck, T.R., Svrakic, D.M., Wetzel, R.D. (1994). *The Temperament and Character Inventory (TCI): A Guide to Its Development and Use.* St. Louis: Washington University, Center for Psychobiology of Personality.

Costa, P.T. Jr., & McCrae, R.R. (1992). *NEO PI-R: Revised NEO Personality Inventory (NEO PI-R).* Odessa, FL: Psychological Assessment Resources Inc.

Dodge, K.A., Newman, J.P. (1981). Biased decision-making processes in aggressive boys. *Journal of Abnormal Psychology, 90,* 375–379.

Eysenck, H.J. (1964). *Crime and Personality.* Boston: Houghton Mifflin.

Eysenck, S.B.G., & Eysenck, H.J. (1977). Personality and the classification of adult offenders. *British Journal of Criminology, 17,* 169–179.

Eysenck, H.J., Eysenck, M. (1985). *Personality and Individual Differences.* New York: Plenum Press.

Goldman, R.G., Skodal, A.E., McGrath, P.J., Oldham, J.M. (1994). Relationship between the Tridimensional Personality Questionnaire and DSM-III-R personality traits. *American Journal of Psychiatry, 151,* 274–276.

Goldsmith, H.H., Buss, A.H., Plomin, R., Rothbart, M.K., Thomas, A., Chess, S., Hinde, R.A., McCall, R.B. (1987). What is temperament? Four Approaches. *Child Development, 58,* 505–527.

Heilbrun, A.B., Heilbrun, L.C., Heilbrun, K.L. (1978). Impulsive and premeditated homicide: An analysis of subsequent parole risk of the murderer. *Journal of Criminal Law and Criminology, 69,* 108–114.

Jackson, D.N. (1974). *Personality Research Form Manual.* Goshen, NY: Consulting Psychologists.

Kauffman, S.A. (1993). *The origins of order: Self-organization and selection in evolution.* New York: Oxford University Press.

Knowlton, B.J., Mangels, J.A., Squire, L.R. (1996). A neostriatal habit learning system in humans. *Science, 273,* 1399–1402.

Nagoshi, C.T., Walter, D., Muntaner, C., Haertzen, C.A. (1992). Validation of the Tridimensional Personality Questionnaire in a sample of male drug users. *Personality and Individual Differences, 13,* 401–409.

Nixon, S.J., Parsons, O.A. (1990). Application of the tridimensional personality theory to a population of alcoholics and other substance abusers. *Alcoholism: Clinical and Experimental Research, 14,* 513–517.

Robbins, T.W. (1996). Refining the taxonomy of memory. *Science, 273,* 1353–1354.

Sigvardsson, S., Bohman, M., Cloninger, C.R. (1987). Structure and stability of childhood personality: Prediction of later social adjustment. *Journal of Child Psychology and Psychiatry, 28,* 929–946.

Svrakic, D.M., Whitehead, C., Przybeck, T.R., Cloninger, C.R. (1993). Differential diagnosis of personality disorders by the seven factor model of temperament and character. *Archives of General Psychiatry, 50,* 991–999.

Svrakic, N.M., Svrakic, D.M., Cloninger, C.R. (1996). A general quantitative theory of personality development: Fundamentals of a self-organizing psychobiological complex. *Development and Psychopathology, 8,* 247–272.

Tremblay, R.E., Pihl, R.O., Vitaro, F., Dobkin, P.L. (1994). Predicting early onset of male antisocial behavior from preschool behavior. *Archives of General Psychiatry, 51,* 732–739.

Wills, T.A., Vaccaro, D., McNamara, G. (1994). Novelty seeking, risk taking, and related constructs as predictors of adolescent substance use: An application of Cloninger's theory. *Journal of Substance Abuse, 6,* 1–20.

Zuckerman, M., Cloninger, C.R. (in press). Relationships between Cloninger's Zuckerman's, and Eysenck's dimensions of personality. *Personality and Individuals Differences.*

4

INDIVIDUAL DIFFERENCES AND LEVELS OF ANTISOCIAL BEHAVIOR

Michael Rutter

MRC Child Psychiatry Unit and
Social, Genetic and Developmental Psychiatry Research Centre
Institute of Psychiatry, De Crespigny Park, Denmark Hill
London SE5 8AF, United Kingdom

In considering the interplay between biological and psychosocial processes in the causation of antisocial behavior, the usual approach has been to focus on the causal mechanisms involved in the determination of individual differences. It is important to recognize, however, that that is by no means the only causal question of importance. Policy makers tend to be more concerned with the prevention of the disorders causing social disability than with whether the disorder is shown by this individual or that one. Scientists, too, have needed to study the causal factors involved in either changes in level over time or differences between groups such as those based on gender, age, or country of residence or of origin (Rutter & Smith, 1995). Criminologists sometimes assume that differences in level apply only to aggregated data but, of course, that is not so. Thus, the reduction in level of crime during early adult life applies to individuals as well as to groups. But the differences among individuals in liability to antisocial behaviour provides quite different information. The individual differences concern population variance whereas levels concern overall rates or frequencies.

Thus, with respect to differences in level, it has been very important that over the course of this century the infantile mortality in all industrialized nations has fallen greatly and life expectancy has increased markedly. It is evident that improved living conditions constitute the single most important explanation for this change. That that is so serves as a reminder that the factors involved in differences in level are not necessarily synonymous with those involved in individual differences. Although, in the United Kingdom today, social factors play some role in individual differences in infantile mortality, they are not the predominant causal factor. Similarly, the major increase in the height of London school children over the course of this century is almost certainly mainly due to improved nutrition, but nutritional factors play a relatively minor role in individual differences in London children's height today (Tizard, 1975). In this chapter, the very substantial rise in crime rates over the last 50 years are used to examine the implications for the interplay between nature and nurture, for that between persons and their environments, and for the causation of antisocial behavior.

Biosocial Bases of Violence, edited by Raine *et al.*
Plenum Press, New York, 1997

1. CRIME RATES SINCE WORLD WAR II

Official crime statistics show a massive rise in crime since 1950 (Smith, 1995). In England and Wales, for example, the total crime rate increased by a factor of 5.5 over this time period. The figures for other industrialized countries are mostly of the same general order although there are some nations (such as Sweden and Norway) exhibiting a much steeper rise (a factor of 13 to 14 in these two countries). Care has to be taken in the interpretation of these figures because they are open to influences from both changes in the law and the reporting of crimes by the general public. Nevertheless, a careful examination of the evidence makes clear that measurement errors could not possibly explain changes over time of this magnitude. Moreover, the rise is as evident for crimes such as serious assault (which are much less subject to measurement error) as for more minor crimes. Homicide stands out as the one crime that exhibits a rather different pattern. Although there has been a marked rise in homicide in some countries, in many (including the UK) there has been little change in rate over the last 50 years. There has also been no appreciable rise in killings by children and adolescents in the UK, although of course this is a much rarer phenomenon (Justice, 1996). The US data serve as an important reminder that the causal influences in one country may not apply in others. The rise in killings by young people in the US is largely confined to those involving the use of guns (Snyder, Sickmund & Poe-Yamagata, 1996).

Victim surveys provide an alternative way of examining crime trends over time (Mirrlees-Black, Mayhew & Percy, 1996). These data are based on asking members of the general population whether a number of specified criminal events have happened to them over a particular reference period, typically one year. These data are available for England only since 1981 but they confirm the rise in crime, although the degree of change overall is not as great as that shown by official crime statistics.

A third approach to the assessment of crime trends and crime is provided by self reports. The Epidemiological Catchment Area Study in the United States asked people about various aspects of their behavior over a recent period and, with respect to antisocial behavior, for their lifetime from early childhood onwards (Robins, Tipp & Przybeck, 1991). The difference between cohorts in the lifetime prevalence of antisocial personality disorder (a severe and persistent disorder involving antisocial behavior) strongly suggested a substantial increase over time. Thus the rate was 3.8% for those under 30, 3.7% for those aged 30 to 44, 1.4% in the 45 to 64 year age group, and 0.3% among those aged over 65. Because of the way that antisocial personality disorder is defined, the ultimate rate in the youngest cohort is likely to be much higher, probably around 6.4%. Of course, there are substantial problems with retrospective recall and it may well be that the changes over time are exaggerated by these figures. Nevertheless, a detailed examination of the evidence suggest that it is unlikely that the cohort differences are entirely an artifact of variations in recall, and it is striking that the evidence points in the same direction as that from victim surveys and from official crime statistics.

Several features of the time trends warrant mention (Smith, 1995). First, the sex ratio in England and Wales dropped from 11 to 1 in 1957 to around 5 to 1 in 1977; since then it has remained fairly stable but has perhaps dropped a little further. A similar pattern of change was evident in the United States, as shown both by official statistics and also by the Epidemiological Catchment Area Study findings. Second, although detailed analyses have yet to be published, it appears that the rise in crime in the United Kingdom has been at least as great in rural areas as in the inner cities. As in other countries, of course, the crime rate in inner cities is higher in absolute terms but the rise over time has been one that applies to the country as a whole. Third, the increase in crime rates over the last 50 years cannot be attributed to a changing population structure. Although many European countries had an increase in immigration in the period following the Second World War, and although the current level of crime is higher among some (but by no means all) ethnic minorities than in the white population, it is completely implausible that crime by ethnic minorities accounts for all of the rise over the last half century. To begin with, ethnic minorities are responsible for a fairly small proportion of crimes; the rate of crime is lower than average among certain ethnic minorities (notably those of Asian origin, so far as the UK is concerned), and the rise in crime has been at least as marked in rural areas where the proportion of ethnic minorities is low as in the inner cities where it is much higher. The situation in the United States is different in that the proportion of the population who are ethnic minorities is much higher and the proportion of crimes committed by black people is also very much greater. On the other hand, the black population in the United States has been present there for almost as long as the white population and the rise in crime could not possibly be due to the minor changes in the ethnic composition of the general population in that country.

It has sometimes been assumed that the rise in crime is in some way a product of urbanization and industrialization. This explanation, however, does not hold up in that the changes over time in crime and in these features do not run in parallel.

Finally, it is important to note that there are exceptions to the rise in crime since World War II. Japan stands out most obviously as the country that is different—both with respect to an overall low rate of crime and also for a complete lack of an increase over this half century -at least up to the end of the 1980s. Clearly, it would be extremely helpful to determine why Japan stands out as so different. Unfortunately, the pattern of life in Japan is different from that in Europe and North America in so many respects that it is difficult to know which are relevant with respect to crime. Among the factors that have been thought to be possibly significant are: the much greater likelihood in Japan that offenders will be caught and punished; a much more effective control of firearms compared with the United States, a more even distribution of income, the relative lack of poverty ghettos, and apparently much stronger informal social controls.

With the exception of Japan, cross-national comparisons in crime rates are not particularly informative both because of the difficulties in ensuring that the measurement of crime is comparable across countries and because the variations are less striking than those over time. There is one very notable exception, however, and that is the very much higher rate of homicide in the United States compared with all European countries. Aggregating the data for 1987 to 1990, the annual deaths by homicide of people aged 15 to 24 years were 15.3 per 100,000 in the United States, compared with 0.9 in the UK and the Netherlands, 0.7 in France, and 0.4 in Japan (UNICEF, 1993). Obviously, there must be a somewhat different explanation for this striking difference in the homicide rate compared with the much smaller differences for other crimes. The lack of effective gun control in the United States stands out as the most likely explanatory factor.

1.1. Possible Reasons for the Rise in Crime

Before turning to possible explanations for the rise in crime since World War II, several points require emphasis. First, it is not the case that crime rates have always been going up. Statistics over a long time period are inevitably unsatisfactory but it is probable that rates of crime were particularly high in the early part of the nineteenth century, that the rates fell in the latter half of the nineteenth century and the early part of this century, and have only risen again during the last 50 years. Second, it is not just crime that has increased; there have been broadly comparable increases in suicide, attempted suicide, alcohol and drug abuse, and depressive disorders (Rutter & Smith, 1995). It should not necessarily be assumed that all these rises have the same explanation but it would appear parsimonious to assume that at least part of the explanation extends to disorders including, but extending beyond, crime. Third, at least so far as the other psychosocial disorders are concerned, the rise applies particularly to younger age groups, and not older ones. Thus, suicide rates have been coming down in old people over the same period of time that they have been going up in young males. It is not possible to determine whether the same applies to crime because that is more predominantly a feature of younger age groups. Finally, whatever the importance of genetic factors with respect to individual differences in the liability to antisocial behavior, changes in the gene pool could not possibly explain the marked and rapid rise in crime over such a short time period. Whether the factors involved in the rise in crime over time apply to groups that are more vulnerable as a result of individual predisposition is, of course, another matter.

In searching for explanations for the rise in crime, it is rather easier to rule out factors than to rule in strong contenders. Thus, at an individual level, poverty/social disadvantage and unemployment constitute important risk factors for antisocial behavior. But they cannot possibly account for the rise in crime because, in Europe, in the 1950s to the 1970s (a period when crime rates were soaring) there was a marked increase in the standard of living and persisting low unemployment in most countries. It is possible that increasing affluence produced increased opportunities for theft but, for a variety of reasons, that doesn't seem likely to be a prime explanatory factor. On the other hand, it is possible that opportunities for crime have risen as a result of increasing proliferation of high rise apartment blocks where surveillance is more difficult and the growth of large shopping centres where theft is possibly somewhat easier than in the old style local shop where most customers were likely to be known to the shopkeeper. Nevertheless, although there is evidence that situational factors influence crime, there have been no systematic analyses to address the hypotheses that these have played a part in the rise in crime over time.

At an individual level, a lower IQ and poorer educational attainment than average constitute risk factors for antisocial behavior (Moffitt, 1993a). Clearly, if their effects are direct (and some evidence suggests that they may be) they cannot be implicated in the rise of crime because both IQ levels and educational attainments have risen over the last 50 years. On the other hand, in so far as a failure to meet educational goals may constitute a risk, it is conceivable that rising educational expectations could have created relevant stress experiences.

Obstetric factors play no substantial role in the risk for antisocial behavior although it has been suggested that they may be of some relevance in relation to the risk for violent crime, when combined with genetic risks (Raine, Brennan & S. A. Mednick, 1994, in press-b; Raine, Brennan, B. Mednick & S. A. Mednick, in press-a). Until that finding is replicated in independent samples, it is difficult to know the extent to which it has validity. However, it seems highly unlikely that obstetric factors have played any role in the rise of crime because, in the population as a whole, obstetric risks have diminished over this time period. To some extent, this has been accompanied by an increased survival of babies who would have died in a previous era, so it is uncertain how far the benefits outweigh the losses when the situation is considered in total population terms. The main remaining risks, however, apply to very low birthweight babies and these constitute such a tiny proportion of the population that they could not conceivably play any significant role in the change over time in levels of crime.

There has been much hype in the media about the possible adverse effects of food additives on children's behavior and it is highly likely that the use of such additives has increased substantially since World War II. On the other hand, although allergic-type responses can have behavioral effects in individual children, the evidence certainly does not support the view that these effects are widespread in the population (Taylor, 1991).

The last half century has been one in which there has been a very major rise in the rate of divorce in North America and in most European countries (Hess, 1995). Probably, too, this has been associated with a rise in population levels of family discord and disruption, although that is more difficult to assess. Because of the close parallels in timing and because family discord and disruption constitute important risk factors for antisocial behavior at the individual level, it is certainly plausible that this has played a role in the rise in crime. Causal factors can be put to the test most satisfactorily when there is a reversal in time trends or swings in both directions. Those are largely lacking in the relevant time period and accordingly no firm causal inference can be made.

There is a complex two-way set of interconnections between alcohol and drug use on the one hand and antisocial behavior on the other (Rutter, 1996-a; Kaplan, 1995). Longitudinal studies suggest that each constitutes a risk factor for the other. To some extent this may simply be a function of the fact that both represent manifestations of the same underlying liability to a broader range of socially disapproved behavior. On the other hand, it is likely that a heavy use of alcohol and drugs do play some contributory role in the risk for antisocial behavior. In part, this probably comes about through the deleterious effects on employment and in part through the effects of alcohol and drugs in reducing inhibitions to violent behavior. Thus, it may well be that the very substantial rise in levels of alcohol and drug use may have contributed to the rise in crime although it does not seem likely that this constitutes the main explanation.

The final main contender for a possible risk factor involved in the rise of crime is provided by a major change in the meaning of adolescence that has taken place over this century, but especially in the last 50 years. There has been a great increase in the commercial focus on the adolescent age period in terms of music, clothes and advertising generally. As the age of puberty has come down, this has been accompanied by a major increase in sexual behavior during the teenage years, with the accompanying pressures and need for decisions that were much less evident in the earlier parts of this century. The great prolongation of education has meant an increasing period of economic dependence on families at a time when the young people are independent in so many other ways. Also, the age of marriage has risen considerably over the same period of time. Thus, biological maturity is occurring earlier than it used to; some adult social roles (such as sexual relations and participation in the commercial market) have also been taken up earlier than in the past; but others (such as marriage, employment and economic independence) have been occurring much later. Moffitt (1993b) has termed this a "maturity gap". It is quite difficult to know to what extent these changes in pattern have constituted risk factors but it is possible that they may be of some relevance.

Because the rapid and major rise in crime could not plausibly be accounted for by any change in the gene pool, it is obvious that environmental factors must be invoked to account for it. In view of the evidence that the rise has been very substantial, it is clear that environmental factors must be able to play a major role in determining levels of antisocial behavior. What is less apparent, is just which environmental factors have been mainly responsible.

It should not be assumed, however, that environmental factors are operating in isolation. Caspi (1996 - personal communication) has suggested that over the last 50 years, at least in the United States, there has been an increasing tendency for children who are personally most vulnerable to be especially likely to be reared in high risk environments. He argues that this may have come about through the combination of selective outmigration from the inner cities and the concentration of biological and social risks (such as poor obstetric care, family disruption, and economic privation) in subgroups in populations, especially ethnic minorities in inner cities. It is difficult to know how far this postulated increased correlation between risks has actually occurred but clearly it is an important possibility to consider.

2. HETEROGENEITY OF ANTISOCIAL BEHAVIOR

In the discussion so far, the rise in crime has been discussed as if antisocial behavior constitutes a homogenous entity, but clearly it does not (Rutter, 1996-b, in press; Rutter et al., in press-a). At one extreme, antisocial behavior constitutes something engaged in to a minor degree, for transient periods, by almost all young males and a high proportion of females. At the other extreme, antisocial behavior constitutes a key part of serious personality disorders extending from childhood into adult life and accompanied by a pervasive and severe social impairment. In addition, there are varieties of crime committed as a consequence of serious mental disorders such as schizophrenia; rarely, antisocial behavior has its origins in some medical condition; and occasionally chromosome anomalies may play a subsidiary contributory role in the liability to antisocial behavior. Potentially, genetic findings should be very helpful in sorting out the origins of heterogeneity. The evidence to date is limited (Bock & Goode, 1996; Carey, 1994) but three main distinctions seem likely to prove important. First, the evidence is consistent in showing that hyperactivity associated with antisocial behavior involves a very strong genetic component (Silberg et al., 1996 & in press). Because of variations in the ways in which hyperactivity has been defined and diagnosed, it is not possible to derive a precise quantitative estimate of the influence of genetic factors but it seems likely that they account for some two thirds of the population variance. A caveat, however, is necessary as a result of the evidence that the very low correlation within dizygotic pairs is likely to be much influenced by a sibling contrast effect, due either to rating bias or to the interaction between the twins (Eaves et al., 1996; Silberg et al., in press). By contrast, antisocial behavior that is unaccompanied by hyperactivity, poor peer relationships, and other manifestations of social impairment seems to be largely environmentally determined (Silberg et al., 1996).

Second, twin studies suggest that antisocial behavior that persists into adult life involves a stronger genetic component than that which is confined to childhood and adolescence. Again, a caveat is necessary in view of the potential influence of a lower base rate for antisocial behavior in adult life but the finding seems to be valid. Simonoff (personal communication), in her longitudinal study from childhood to adult life, found that antisocial behavior in childhood was a very powerful predictor of antisocial behavior in adult life. Nevertheless, despite this strong continuity over time, individual differences in antisocial behavior in childhood were largely a function of environmental factors whereas genetic factors were strong in adult life (Rutter et al., in press-a). The age difference arises because, although most crime in adult life is preceded by antisocial behavior in childhood, quite a lot of antisocial behavior in childhood does *not* persist into adult life. The Simonoff findings are based on a small sample but they are in keeping with the results of other studies with larger samples, albeit with weaker data (DiLalla & Gottesman, 1989; Lyons et al., 1995). At first sight, it might be thought that this age effect is in contradiction to the strong genetic component for early onset antisocial behavior accompanied by hyperactivity. The contradiction, however, is apparent rather than real because it is just such early onset antisocial behavior, that is associated with hyperactivity, that is most likely to persist into adult life. It also appears that the male predominance is much stronger for adult crime than for juvenile delinquency (Graham & Bowling, 1995).

Third, adoptee studies suggest that the genetic component in violent crime is weaker than that involved in theft and other forms of nonviolent crime (Bohman, 1996). The caveat here is that adoptee studies (because they rely on intergenerational connections) will be influenced by the trends over time in levels of antisocial behavior. Nevertheless, although much has still to be learned about the nature of heterogeneity in antisocial behavior, it is already clear that there is meaningful heterogeneity that will have to be taken into account in understanding its origins.

3. MULTIPLICITY OF CAUSAL QUESTIONS

All too often, it is assumed that the only causal question is that concerned with the origins of individual differences, but that is a seriously mistaken view. As already noted, it is as necessary to examine the causes of differences in level between groups or in individuals over time, as it is to ask why there are individual differences within each of those groups during any single time period. The causal mechanisms involved may be similar or overlapping, or they may be quite different. Thus, the political, economic, and social forces that play a predominant role in the variations in unemployment levels over time have little to do with the reasons why this man is without a job and that one is not (Rutter, 1994a). But, the contrast between the causes of differences in levels and the causes of individual differences within populations by no means encompasses the range of important causal questions.

So far as the individual differences question is concerned, two points need to be made. First, because risk factors for antisocial behavior can be identified, this does not necessarily mean that they operate directly on crime. Thus, for example, the mediation might lie in personality variables or cognitive features that are involved only indirectly (albeit possibly strongly) in the liability to antisocial behavior. Second, most studies of risk factors depend on comparisons between individuals, rather than the much more powerful strategy of the study of causal processes of examining changes within individuals over time. Longitudinal studies, combined with appropriate forms of statistical analysis, carry very important advantages (Rutter, 1994b).

The causal question with respect to differences in level of antisocial behavior might be thought to constitute one causal question but it involves at least two. On the one hand, there are causes that may influence the number of individuals who engage in antisocial behavior— the "how many?" question - and there are the causes involved with differences in the amount of antisocial activity engaged in by those individuals involved in crime at all—the "how much?" question. Thus, Farrington (1986) showed that the marked fall in level of crime in early adult life was primarily a function of a reduction in the number of individuals committing antisocial acts rather than with the amount of crime committed. That is to say, those who continued in crime tended to do so at much the same level as they had shown when younger.

A fourth causal question concerns the severity of antisocial behavior. Some individuals never progress beyond petty theft whereas others go on to much more serious crime. Similarly, developmental studies show that there is a pathway leading from oppositional behavior in early childhood to more serious conduct disorder in adolescence, a pathway followed by only some individuals (Loeber & Hay, 1994). Does the increase over time and severity simply reflect an initial difference in the overall liability to antisocial behavior or are different factors involved in the progression across all the stepping stones in a criminal career, as they are sometimes termed?

A further, possibly related, distinction is between transient antisocial behavior and persistent or recurrent recidivist crime. Part of that question, of course, also involves the persistence from the years of childhood into adult life. The evidence from longitudinal studies suggests that, in addition to initial differences in liability, later experiences may play a role in whether individuals persist or desist in criminal behavior in adult life. For example, a harmonious marriage (perhaps particularly to a nondeviant partner) predisposes to a discontinuation of criminality whereas incarceration, alcoholism, and unemployment all predispose to continuation (Quinton, Pickles, Maughan & Rutter, 1993; Sampson & Laub, 1993).

These findings underline the very important point that the origins of a risk factor and its mode of risk mediation are not necessarily synonymous (Rutter, Silberg & Simonoff, 1993). Thus, people's own behavior plays a major role in whether or not they are incarcerated, and also in the choice of marital partner. Nevertheless, rigorous statistical analyses, using a variety of approaches, made it quite clear that even after taking full account of initial differences in behavior, these experiences do indeed make an impact on the course of antisocial behavior (Sampson & Laub, 1993).

Finally, there is the question of the factors involved in the actualization of a behavioral propensity. Whether or not an individual with a liability to engage in antisocial behavior actually commits some crime will be influenced by situational constraints and opportunities as well as by individual factors (Clarke, 1992; Tonry & Farrington, 1995). This is, of course, a consideration that applies to a much broader range of behaviors than crime. Thus, whether or not someone who is depressed seeks to kill herself will be influenced in part by the availability at the time of some appropriate means of bringing about her own death. The dramatic fall among older people in the rate of suicide following the detoxification of domestic gas in the United Kingdom provides an illustration of this point. Similarly, the major rise in drug taking during the last 50 years is almost certainly due as much to the greater availability of drugs as to any changes in the individual predisposition to use drugs.

4. PSYCHE-SOMA INTERPLAY

There is an unfortunate tendency in some quarters to assume that if a biological factor related to some behavior can be identified then the soma must provide an explanation for the psyche, but any such assumption is a mistake. That is not how biology works and there is abundant evidence of a two-way interplay. Thus, male sex hormones have a causal impact on aggression, assertiveness, and dominance but the causal process also operates in the opposite direction. In animals, manipulation of dominance hierarchies produces consequent changes in hormonal status (Rose, Holaday & Bernstein, 1971). Similarly, in humans, winners of closely fought chess (Mazur, Booth & Dabbs, 1992) or tennis matches (Mazur & Lamb, 1980) show a rise in sex hormones whereas the losers show a fall. Decades ago, animal studies showed that stress experiences led to structural and functional changes in the neuroendocrine system (Hennesey & Levine, 1979). Animal studies, too, have shown that learning, in the form of imprinting, is accompanied by regular neural changes (Horn, 1990). In addition, numerous studies have shown that visual input plays a determinative role in the development of visual systems in the brain (Blakemore, 1991). Research to examine the functional connections between the brain and behavior are immensely important but we have got to be careful to avoid any simplistic assumption that the causal processes operate in only one direction. They do not. Perhaps particular care is going to be needed now in the interpretation of brain imaging studies, where already there are signs that premature inferences are being drawn.

5. REPRESENTATIVENESS OF GROUPS STUDIED

For obvious practical reasons, many biological studies have been undertaken with incarcerated or hospitalized samples, sometimes chosen because of their extreme or unusual characteristics. As a consequence, there must be great caution in generalizing findings to the general population of antisocial or violent individuals.

6. STRENGTH OF CAUSAL EFFECT

There is a widespread assumption, perhaps particularly marked among some behavior geneticists, that statistics showing 'the proportion of population variance accounted for' provides a direct measure of the strength of causal effect being exerted. For several different reasons, this is a seriously fallacious assumption. To begin with, as already discussed, it addresses only the question of the causes of individual differences and this may have little or nothing to do with the causes of variations in level, and various other important causal issues (Rutter & Smith, 1995). But, even with respect to individual differences, the proportion of population variance explained has only an extremely weak connection with the strength of causal effects at the individual level (Rutter, 1987). That is because the variance explained is hugely dependent upon the percentage of the population who could be influenced by the risk factor. For example, the American National Collaborative Study showed that Down syndrome resulted in a 60 point IQ deficit on average at the individual level but, nevertheless, it accounted for only 0.03% of population variance in IQ. Of course, that constitutes an extreme example because of the relative rarity of Down syndrome but the point applies much more generally. Risk factors that are hugely important at the individual level (i.e., in those individuals who have the risk factor) will always account for only a small proportion of population variance if the number of individuals affected by the risk factor is low. A further consideration, but this time one that is widely appreciated, is that the quantitative estimates are population-specific (Rutter, 1991). If circumstances change, so will the quantitative estimates alter. If the samples studied include very few individuals who have been reared in high risk environments, genetic effects will be greater than when the sample spans a wide range of environments including a high proportion with a high risk. This is an important consideration because many twin and adoptee studies fall into the former category, so overestimating genetic effects - at least as applied to clinic populations from high risk environments (Baumrind, 1993). The implication is that demonstration of a very strong genetic component in the population variance for antisocial behavior is of very little use in predicting the effects of environmental change, if the change of environment is pervasive and marked. It has to be added, too, that if an environmental change affects the whole population, it may bring about a major change in the level of a behavior without having any appreciable impact on the figures for genetic effects on population variance. Thus, genetic factors have always predominated in individual differences in height at any one point in time but, nevertheless, improvements in nutrition brought about a 12cm increase in the height of London schoolboys over the first half of this century (Tizard, 1975).

A somewhat different consideration is that the findings tend to be rather different when dealing with a measured environmental risk factor rather than inferring environmental risk from that which is not genetic. Thus, with respect to both depression and alcoholism, Kendler and his colleagues (Kendler, Heath, Neale, Kessler & Eaves., 1993, Kendler et al., 1996) failed to find any significant shared environmental effect when the population variance was partitioned in the usual way. By contrast, when parental loss was introduced into the models as a measured variable, it could be shown to have a shared environmental effect (Kendler, Neale, Kessler, Heath & Eaves, 1992; Kendler et al., 1996). Unfortunately, most genetic studies continue to treat the environment as an unmeasured "black box" variable. A further consideration is that, ordinarily, most genetic analyses do not take into account either gene-environment correlations or gene-environment interactions. In view of their probable importance, at least in some circumstances, that may be a rather serious limitation (Rutter et al., in press-b).

Finally, the usual approach to statistical modelling introduces a further source of distortion. The principle of parsimony is used to derive as simple a model as possible that includes the smallest number of factors that still provide a good fit to the data. The principle sounds inherently reasonable and the process of eliminating variables only if their removal does not result in any significant reduction in the fit of the model seems logical. Nevertheless, the procedure may well mean that there is no significant difference in the impact of a variable eliminated and one retained in the model. It comes back to the very basic statistical point that the fact that the effect of one variable is statistically significant and another is not, is not at all the same as saying that there is a significant difference between the effects of the two variables. The inclusion or exclusion of variables is strongly dependent on where thresholds happen to fall and it is important to bear in mind the distortions that inevitably follow in some cases.

7. PERSON-ENVIRONMENT INTERPLAY

Up to now, there has been very little research that has focused explicitly on the different forms of person-environment interplay. Nevertheless, such little research as there has been emphasizes its likely importance with respect to antisocial behavior, as well as other forms of psychopathology (Rutter et al., in press-b). First, attention needs to be paid to the effects of passive gene-environment correlations. That is, the parental qualities that mediate genetic risks to the offspring are often ones that also bring about adverse environments that carry environmentally mediated risks. Thus, as already noted, the association between antisocial behavior in parents and antisocial behavior in their children reflects genetic mediation in part. However, there is also much research indicating that families in which one or both parents are antisocial have a very much increased likelihood of being characterized by discord, disruption, scapegoating, and the provision of deviant models of behavior. There is also evidence that these constitute risks for antisocial behavior in the children, even when neither parent shows antisocial behavior themselves. Joanne Meyer et al.'s (1996) analysis of the Virginia Twin Study of Adolescent Behavioral Development data, using an extended twin-family design, has provided evidence that there is both genetic and environmental mediation of the risks associated with antisocial personality disorder in parents. Kendler et al. (1996) used a similar approach to indicate that parental alcoholism created a genetic risk for alcoholism in the offspring, that it also increased the likelihood of parental loss, and that such loss created a shared environmental risk effect, even after account had been taken of genetic influences.

Evocative and active person-environment correlations are also likely to be important. That is to say, through their behavior, people elicit responses from other people (which may be either risky or protective in their effects) and also shape and select their environments. Thus, Robins' (1966) long-term follow-up study of antisocial boys, together with a general population comparison group, showed the very strong tendency for antisocial behavior in childhood to be followed by a much increased risk of adverse environments in adult life. These included features such as multiple divorces, very frequent changes of job, unemployment, and a lack of close friendships. Similarly, Champion and her colleagues (Champion, Goodall & Rutter, 1995) showed that antisocial behavior at age 10 was associated with a doubling of the risk of severely negative life events and experiences some two decades later. Quinton et al. (1993) also demonstrated that antisocial youngsters (especially girls) showed a much increased tendency to have a deviant partner in adult life, partnerships that were much more likely to be characterized by discord and breakdown.

Moreover, as already noted, longitudinal analyses have indicated that these risk experiences, that are at least partially self induced, play a significant role in the perpetuation of criminal behavior (Sampson & Laub, 1993).

Finally, there are the effects brought about as a result of person characteristics, whether genetically influenced or not, that influence people's vulnerability to environmental hazards of one kind or another. It is usually argued that it has been very rare to be able to demonstrate gene-environment interactions. Although true as applied to normally distributed dimensions of behavior or cognition, the implication is nevertheless misleading. The point is that there are numerous well documented examples of gene-environment interactions in biology and medicine but they apply to specific genetic and environmental risks, often operating only in subsegments of populations (Rutter & Pickles, 1991). By contrast, behavior genetics has tended to look for overall interactions between unmeasured genetic and environmental factors in relation to characteristics where there are no particular grounds for expecting interactions to be operative. In addition, the usual multivariate statistical approaches are rather insensitive for the detection of interaction effects (Wahlsten, 1990). In the field of antisocial behavior, however, several adoptee studies (see Cadoret, Yates, Troughton, Woodworth & Stewart, 1995) have provided evidence that adoptees who are genetically at risk by virtue of having antisocial biological parents are more susceptible to adversities in the environments in which they were reared. Effects have sometimes been quite strong. Thus, for example, in the Bohman (1996) adoptee study, the rate of antisocial behavior was about 3% when both a genetic risk and a rearing risk were absent; it doubled in the presence of an adverse rearing environment (but in the absence of a genetic risk); it rose to 12% in the presence of a genetic risk (but in the absence of a rearing risk); but it rose to 40% when there was both a genetic and an environmental risk. It should be added that, with respect to the proportion of population variance explained, the adoptee design will always greatly underestimate the effect of interaction processes just because the design is based on removing the usual overlap between nature and nurture.

8. CONCLUSIONS

It is all too apparent that we are a very long way indeed from having achieved any integration of biological and psychosocial perspectives in the processes leading to antisocial behavior. On the face of it, evidence on the major changes in levels of crime over time suggest more powerful environmental effects than seem to be implied by some twin study data. As has been emphasized, however, factors involved in differences in level are not necessarily synonymous with those involved in individual differences, although they may coincide. It seems quite likely that the relative importance of genetic and environmental influences on antisocial behavior may vary according to the type of antisocial behavior being considered. But does the heterogeneity reflect different classes of behavior or, rather, does it represent different mixes of dimensional risk factors? To what extent do the genetic and environmental risks operate in parallel, in additive fashion, and to what extent do they involve a more complex interplay resulting from gene-environment correlations and interactions? The prevailing theories are more eclectic than used to be the case, in so far as they recognize the need to invoke a multiplicity of causal mechanisms, but they are not much more successful than their predecessors in their delineation of the specific processes involved in the several different types of causation that have to be considered. The research agenda for the future is as overfull as ever but the outlook now is much better than it was some years ago. Probably, three features have been most critical in strengthening the research potential. First, there has been a considerable improvement in the conceptualization of the different mechanisms involved in the interplay between nature and nurture. Second, there have been advances in the conceptualization and measurement of specific environmental risk factors, with an appreciation of the need to distinguish between indirect risk indicators and basic risk mechanisms, and between environmental risks that affect all children in a family and those that impinge differentially on each child. Third, the advances in molecular genetics have made it possible to identify particular genes, a possibility that greatly enhances the power to examine gene-environment correlations and interactions directly.

REFERENCES

Baumrind, D. (1993). The average expectable environment is not good enough: A response to Scarr. *Child Development, 64,* 1299–1317.

Blakemore, C. (1991). Sensitive and vulnerable periods in the development of the visual system. In G. R. Bock & J. Whelan (Eds.), *The childhood environment and adult disease* (pp. 129–146). Chichester, England: Wiley.

Bock, G. R. & Goode, J. A. (Ed.). (1996). *Genetics of Criminal and Antisocial Behaviour. Ciba Foundation Vol. 194.* Chichester, England & New York: Wiley.

Bohman, M. (1996). Predisposition to criminality: Swedish adoption studies in retrospect. In G. R. Bock & J. A. Goode (Eds.), *Genetics of criminal and antisocial behaviour. Ciba Foundation Vol. 194* (pp. 99–114). Chichester, England & New York: Wiley.

Cadoret, R. J., Yates, W. R., Troughton, E., Woodworth, G., & Stewart, M. A. (1995). Genetic-environmental interaction in the genesis of aggressivity and conduct disorders. *Archives of General Psychiatry, 52,* 916–924.

Carey, G. (1994). Genetics and violence. In A. J. Reiss Jr, K. A. Miczek, & J. A. Roth (Eds.), *Understanding and Preventing Violence: Vol. 2: Biobehavioral influences,* (pp. 21–58). Washington, DC: National Academy Press.

Champion, L. A., Goodall, G. M., & Rutter, M. (1995). Behavioural problems in childhood and stressors in early adult life: A 20-year follow-up of London school children. *Psychological Medicine, 25,* 231–246.

Clarke, R. V. (Ed.). (1992). *Situational crime prevention: Successful case studies.* New York: Harrow and Heston.

DiLalla, L. F. & Gottesman, I. I. (1989). Heterogeneity of causes for delinquency and criminality: Lifespan perspectives. *Development and Psychopathology, 1,* 339–349.

Eaves, L. J., Silberg, J. L., Meyer, J. M., Maes, H. H., Simonoff, E., Neale, M. C., Pickles, A., Reynolds, C. A., Erickson, M. T., Heath, A. C., Loeber, R., Rutter, M., Truett, K. R., & Hewitt, J. K. (1996). *Genetics and developmental psychopathology: 2. The main effects of genes and environment on behavioral problems in the Virginia Twin Study of Adolescent Behavioral Development.* Manuscript submitted for publication.

Farrington, D. P. (1986). Age and crime. In M. Tonry & N. Morris (Eds.), *Crime and justice*. Chicago: Chicago University Press.

Graham, J., & Bowling, B. (1995). *Young people and crime*. Home Office Research Study 145. London: Home Office.

Hennesey, J. W., & Levine, S. (1979). Stress, arousal, and the pituitary-adrenal system: A psychoendocrine hypothesis. In J. M. Sprague & A. N. Epstein (Eds.), *Progress in psychobiology and physiological psychology* (pp. 133–178). New York: Academic Press.

Hess, L. E. (1995). Changing family patterns in Western Europe: Opportunity and risk factors for adolescent development. In M. Rutter & D. J. Smith (Eds.), *Psychosocial disorders in young people: Time trends and their causes* (pp. 104–193). Chichester, England: Wiley.

Horn, G. (1990). Neural bases of recognition memory investigated through an analysis of imprinting. *Philosophical Transactions of the Royal Society, 329*, 133–142.

Justice (1996). *Children and homicide: Appropriate procedures for juveniles in murder and manslaughter cases*. London: Author.

Kaplan, H. B. (ed) (1995). *Drugs, crime, and other deviant adaptations: Longitudinal studies*. New York: Plenum Publications Corporation.

Kendler, K. S., Heath, A. C., Neale, M. C., Kessler, R. C., & Eaves, L. J. (1993). Alcoholism and major depression in women: A twin study of the causes of comorbidity. *Archives of General Psychiatry, 50*, 690–698.

Kendler, K. S., Neale, M. C., Kessler, R. C., Heath, A. C., & Eaves, L. J. (1992). Childhood parental loss and adult psychopathology in women: A twin study perspective. *Archives of General Psychiatry, 49*, 109–116.

Kendler, K. S., Neale, M. C., Prescott, C. A., Kessler, R. C., Heath, A. C., Corey, L. A., & Eaves, L. J. (1996). Childhood parental loss and alcoholism in women: A causal analysis using a twin-family design. *Psychological Medicine, 26*, 79–95.

Loeber, R., & Hay, D. F. (1994). Developmental approaches to aggression and conduct problems. In M. Rutter & D. F. Hay (Eds.), *Development through life: A handbook for clinicians* (pp. 488–516). Oxford, England: Blackwell Scientific.

Lyons, M. J., True, W. R., Eisen, S. A., Goldberg, J., Meyer, J. M., Faraone, S. V., Eaves, L. J. & Tsuang, M. T. (1995). Differential heritability of adult and juvenile antisocial traits. *Archives of General Psychiatry, 52*, 906–915.

Mazur, A., Booth, A., & Dabbs, J. M. (1992). Testosterone and chess competition. *Social Psychology Quarterly, 55*, 70–77.

Mazur, A., & Lamb, T. A. (1980). Testosterone, status, and mood in human males. *Hormones and Behavior, 14*, 236–246.

Meyer, J. M., Rutter, M., Simonoff, E., Shillady, C. L., Silberg, J. L., Pickles, A., Hewitt, J. K., Maes, H. H. & Eaves, L. J. (1996). *Familial aggression for conduct disorder symptomatology: The role of genes, marital discord, and family adaptability*. Manuscript in preparation.

Mirrlees-Black, C., Mayhew, P. & Percy, A. (1996). *The 1996 British Crime Survey : England and Wales*. Home Office Statistical Bulletin, Issue 19/96. London : Home Office Research and Statistics Directorate.

Moffitt, T. E. (1993a). The neuropsychology of conduct disorder. *Development and Psychopathology, 5*, 135–152.

Moffitt, T. E. (1993b). Adolescence-limited and life-course-persistent antisocial behavior: A developmental taxonomy. *Psychological Review, 100*, 674–701.

Quinton, D., Pickles, A., Maughan, B., & Rutter, M. (1993). Partners, peers, and pathways: Assortative pairing and continuities in conduct disorder. *Development and Psychopathology, 5*, 763–783.

Raine, A., Brennan, P., Mednick, B., & Mednick, S. A. (in press-a). High rates of crime, violence, academic problems, and behavioral problems in those with both early neuromotor deficits and negative family environments. *Archives of General Psychiatry*.

Raine, A., Brennan, P., & Mednick, S. A. (1994). Birth complications combined with early maternal rejection at age 1 year predispose to violent crime at age 18 years. *Archives of General Psychiatry, 51*, 984–988.

Raine, A., Brennan, P., & Mednick, S. A. (in press-b). The biosocial interaction between birth complications and early maternal rejection in predisposing to adult violence: Specificity to serious, early onset violence. *Criminology*.

Robins, L. (1966). *Deviant children grown up*. Baltimore: Williams and Wilkins.

Robins, L. N., Tipp, J., & Przybeck, T. (1991). Antisocial personality. In L. Robins & D. A. Regier (Eds.), *Psychiatric disorders in America: The epidemiologic catchment area study* (pp. 258–290). New York: Free Press.

Rose, R. M., Holaday, J. W., & Bernstein, I. S. (1971). Plasma testosterone, dominance risk and aggressive behavior in male rhesus monkeys. *Nature, 231*, 366–368.

Rutter, M. (1987). Continuities and discontinuities from infancy. In J. Osofsky (Ed.), *Handbook of infant development* (2nd ed., pp. 1256–1296). New York: Wiley.

Rutter, M. (1991). Nature, nurture, and psychopathology: A new look at an old topic. *Development and Psychopathology, 3,* 125–136.

Rutter, M. (1994b). Beyond longitudinal data: Causes, consequences, changes and continuity. *Journal of Consulting and Clinical Psychology, 62,* 928–940.

Rutter, M. (1994a). Concepts of causation, tests of causal mechanisms, and implications for intervention. In A. L. Petersen & J. T. Mortimer (Eds.), *Youth unemployment and society* (pp. 147–171). New York: Cambridge University Press.

Rutter, M. (1996-a). Commentary: Testing causal hypotheses about mechanisms in comorbidity. *Addiction, 91,* 495.

Rutter, M. (1996-b). Introduction: Concepts of antisocial behaviour, of cause, and of genetic influences. In G. Bock & J. Goode (Eds.), *Genetics of Criminal and Antisocial Behaviour: Ciba Foundation Vol. 194* (pp. 1–15). Chichester, England & New York: Wiley.

Rutter, M. (in press). Antisocial behavior: Developmental psychopathology perspectives. In D. Stoff, J. Breiling, & J. D. Maser (Eds.), *Handbook of Antisocial Behavior.* New York: Wiley.

Rutter, M., Dunn, J., Plomin, R., Simonoff, E., Pickles, A., Maughan, B., Ormel, J., Meyer, J., & Eaves, L. (in press-b). Integrating nature and nurture: Implications of person-environment correlations and interactions for developmental psychopathology. *Development and Psychopathology.*

Rutter, M., Maughan, B., Meyer, J., Pickles, A., Silberg, J., Simonoff, E. & Taylor, E. (in press-a). Heterogeneity of antisocial behavior: Causes, continuities, and consequences. In D. W. Osgood (Ed.), *Motivation and delinquency.* Lincoln, NE: University of Nebraska Press.

Rutter, M., & Pickles, A. (1991). Person-environment interactions: Concepts, mechanisms, and implications for data analysis. In T. D. Wachs & R. Plomin (Eds.), *Conceptualization and measurement of organism-environment interaction* (pp. 105–141). Washington, DC: American Psychological Association.

Rutter, M., Silberg, J., & Simonoff, E. (1993). Whither behavior genetics? A developmental psychopathology perspective. In R. Plomin & G. E. McClearn (Eds.), *Nature, Nurture, and Psychology* (pp. 433–456). Washington, DC: APA Books.

Rutter, M., & Smith, D. J. (Ed.). (1995). *Psychosocial Disorders in Young People: Time trends and their causes.* Chichester, England: Wiley.

Sampson, R. J., & Laub, J. H. (1993). *Crime in the making: Pathways and turning points through life.* Cambridge, MA: Harvard University Press.

Silberg, J., Meyer, J., Pickles, A., Simonoff, E., Eaves, L., Hewitt, J., Maes, H., & Rutter, M. (1996). Heterogeneity among juvenile antisocial behaviors: Findings from the VTSABD. In G. Bock & J. Goode (Eds.), *Genetics of Criminal and Antisocial Behaviour: Ciba Foundation Vol. 194* (pp. 76–86). Chichester, England & New York: Wiley.

Silberg, J. L., Rutter, M. L., Meyer, J., Maes, H., Simonoff, E., Pickles, A., Hewitt, J., & Eaves, L. (in press). Genetic and environmental influences on the covariation among symptoms of hyperactivity and conduct disturbance in juvenile twins. *Journal of Child Psychology and Psychiatry.*

Smith, D. J. (1995). Youth crime and conduct disorders: Trends, patterns and causal explanations. In M. Rutter & D. J. Smith (Eds.), *Psychosocial disorders in young people: Time trends and their causes* (pp. 389–489). Chichester, England: Wiley.

Snyder, H. N., Sickmund, M., & Poe-Yamagata, E. (1996). *Juvenile offenders and victims: 1996 update on violence.* Washington, DC: Office of Juvenile Justice and Delinquency Prevention.

Taylor, E. (1991). Toxins and allergens. In M. Rutter & P. Casaer (Eds.), *Biological risk factors for psychosocial disorders* (pp. 199–232). Cambridge, England: Cambridge University Press.

Tizard, J. (1975). Race and IQ: The limits of probability. *New Behaviour, 1,* 6–9.

Tonry, M., & Farrington, D. P. (Ed.). (1995). *Building a Safer Society: Strategic approaches to crime prevention.* Chicago & London: University of Chicago Press.

UNICEF (1993). *The progress of nations 1993.* New York: Author.

Wahlsten, D. (1990). Insensitivity of the analysis of variance to heredity-environment interaction. *Behavioral and Brain Sciences, 13,* 109–161.

5

OBSERVATIONAL LEARNING OF VIOLENT BEHAVIOR

Social and Biosocial Processes

L. Rowell Huesmann

Research Center for Group Dynamics
Institute for Social Research
University of Michigan
Ann Arbor, Michigan 48106-1248

1. INTRODUCTION

Although habitual aggressive and violent behaviors seldom develop in children unless there is a convergence of multiple predisposing and precipitating biosocial and contextual factors, there is compelling evidence that early observation of aggression and violence in the child's environment or in the mass media contributes substantially to the development of aggressive habits that may persist throughout the life course (Bandura, 1986; Berkowitz, 1993; Paik & Comstock, 1994; Eron, Huesmann, Lefkowitz & Walder, 1972; Huesmann, 1986; Huesmann & Eron, 1986; Huesmann & Miller, 1994). The empirical evidence concerning the importance of observational learning has been accumulating for decades but has been given added relevance by the emergence of social/cognitive process models to explain individual differences in aggression. In this chapter I provide an overview of an unified cognitive/information-processing model of social behavior within which aggression can be understood, I elaborate on the key role that observational learning plays in the development of the cognitive/information-processing structures that control social behavior in general and aggressive behavior in particular; and I discuss the biosocial processes that seem to be involved in observational learning of these cognitive/information-processing structures.

In the context of this chapter, aggressive behavior is any behavior intended to injure or irritate another person (Berkowitz, 1993; Eron et al., 1972). Excluded from this definition is the "assertive" behavior of dynamic sales people and executives that is often called "aggressive" by the public. Psychologists have usually distinguished between the kind of aggressive behavior that is directed at the goal of obtaining a tangible reward for the aggressor (*instrumental or proactive aggression*) and the kind of aggressive behavior that is simply intended to hurt someone else (at different times denoted *hostile, angry, emotional,*

Biosocial Bases of Violence, edited by Raine *et al.*
Plenum Press, New York, 1997

or reactive aggression) (Berkowitz, 1993; Feshbach, 1964). While some scholars have argued that all aggressive behavior is instrumental in some way, rageful assaults out of anger are often so different in character from violent acts committed for tangible gain that the distinction seems valuable. Furthermore, one can often detect individual differences in arousal predispositions associated with people who habitually engage in the different types of aggression (Baker, Hastings, & Hart, 1984; Craven & Lochman, 1997; Raine, Venables, & Williams, 1990). Nevertheless, an examination of the underlying cognitive processes involved (e.g., Dodge & Coie, 1987) has led to a realization that many of the same mechanisms are involved in both types of aggression. Clearly, anger plays a more important role in hostile aggression, but that does not mean that anger does not play any role in instrumental aggression. Clearly, lack of self-control plays a role in instrumental aggression, but that does not mean that self-control does not play any role in hostile aggression.

Before proceeding to the major themes of this chapter, I need to review three well-established facts about aggression and violence. *First*, habitual aggressive behavior usually emerges early in life, and early aggressive behavior is very predictive of later aggressive behavior and even of aggressive behavior of offspring (Farrington, 1982; 1995; Huesmann, Eron, Lefkowitz & Walder, 1984; Loeber & Dishion, 1983; Magnusson, Duner, & Zetterblom, 1975; Olweus, 1979). Process models for aggressive behavior need to explain this continuity over time and across generations. *Second*, as the title of this book suggests, severe aggression is most often a product of multiple interacting social and biological factors (Coie & Dodge, in press) including genetic predispositions (Bouchard, 1984; Cloninger & Gottesman, 1987; Mednick, Gabrielli, & Hutchins, 1984; Rushton et al., 1986), environment/genetic interactions (Lagerspetz & Lagerspetz, 1971; Lagerspetz & Sandnabba, 1982), CNS trauma and neurophysiological abnormalities (Moyer, 1976; Nachson & Denno, 1987; Pontius, 1984), early temperament or attention difficulties (Kagan, 1988; Moffitt, 1990), arousal levels (Raine & Jones, 1987; Raine, Venables, & Williams, 1990; in press), hormonal levels (Olweus, Mattsson, Schalling, & Low, 1988), family violence (Widom, 1989), cultural perspectives (Staub, 1996), poor parenting (Patterson, 1995), inappropriate punishment (Eron, Walder, & Lefkowitz, 1971), environmental poverty and stress (Guerra, Huesmann, Tolan, Eron, & VanAcker, 1995), peer-group identification (Patterson, Capaldi, & Bank, 1991), and other factors. No one causal factor by itself explains more than a small portion of individual differences in aggressiveness. *Third*, early learning and socialization play a key role in the development of habitual aggression. From a social cognitive perspective the variety of predisposing factors discussed above may make the emergence of certain specific cognitive routines, scripts, and schemas more likely, but these cognitions are learned through interactions of the child with the environment (Bandura, 1973; Berkowitz, 1974; Eron, Walder, & Lefkowitz, 1971). Aggression is most likely to develop in children who grow up in environments that reinforce aggression, provide aggressive models, frustrate and victimize them, and teach them that aggression is acceptable.

To best understand the role that environmental variations play in this process, one must distinguish between *situational instigators* that may precipitate, motivate, or cue aggressive cognitions/responses and those more lasting components of the child's *socializing environment* that mold the child's cognitions (schemas, scripts, normative beliefs) and therefore their responses to these stimuli over time, i.e., that socialize the child. An environment rich with environmental deprivations, frustrations and provocations is one in which aggressive behavior is socialized in children over time and then regularly stimulated in children across situations.

2. COGNITIVE PROCESSES, INFORMATION PROCESSING, AND AGGRESSIVE BEHAVIOR

Over the past 15 years two general cognitive/information processing models have emerged of how humans acquire and maintain aggressive habits. One, developed by Huesmann and his colleagues (Huesmann 1982; 1986; 1988; Huesmann and Eron, 1984) initially focused particularly on scripts, beliefs, and observational learning, and the other developed by Dodge and his colleagues (1980; 1986; Crick & Dodge, 1994; Dodge & Frame, 1982) focused particularly on perceptions and attributions. However, both hypothesize a similar core of information processing, both draw heavily on the work of cognitive psychologists and information processing theory, and both draw from Bandura's (1977; 1986) earlier formulations of cognitive processing in social learning as well as Berkowitz's (1990) neoassociationist thinking.

In Figure 1 I present Huesmann's (in press) recent integration of the key elements of these social/cognitive models into an unified information processing model that explains the role of cognition in aggressive behavior. This model begins with the premise that social behavior is controlled to a great extent by cognitive *scripts* (Abelson,1981) that are stored in a person's memory and are used as guides for behavior and social problem solving. A script incorporates both procedural and declarative knowledge and suggests what events are to happen in the environment, how the person should behave in response to these events, and what the likely outcome of those behaviors would be. It is presumed that while scripts are first being established they influence the child's behavior through *"controlled"* mental processes (Schneider & Shriffrin, 1977; Shriffrin & Schneider, 1977), but these processes become *"automatic"* as the child matures. Correspondingly, scripts that persist in a child's repertoire, as they are rehearsed, enacted, and generate consequences, become increasingly more resistant to modification and change. *Causal schemas* are a second kind of cognition assumed to influence behaviors. Causal schemas are the data base that the individual employs to evaluate environmental cues and make attributions about others' intentions. These attributions in turn will influence the search for a script for behaving. *Normative beliefs* are a third kind of cognitive schema hypothesized to play a central role in regulating aggressive behavior. Normative beliefs are cognitions about the appropriateness of aggressive behavior. They are related to perceived social norms but are different in that they concern what's "right for you." Normative beliefs are used to interpret other's behaviors, to guide the search for social scripts, and to filter out inappropriate scripts and behaviors.

Within this model that there are four possible loci at which individual differences and situational variations can influence aggressive behavior. First, the objective situation primes both cognitive and emotional reactions. However, which environmental cues are given most attention and how they are interpreted may vary from person to person and may depend on a person's neurophysiological predispositions, current mood state, and previous learning history as reflected in activated schemas including normative beliefs. Negative affect, e.g. a bad mood, primes the interpretation of environmental events to be more negative. Very aversive situations and frustration will produce negative affect in almost everyone, but the intensity of the affect depends on the interpretation given to the situation. Environmental stimuli may directly trigger conditioned emotional reactions and may cue the retrieval from memory of cognitions that define the current emotional state. For example, the "sight of an enemy" or the "smell of a battlefield" may provoke both instantaneous physiological arousal and the recall of thoughts about the "enemy" that give meaning to the aroused state as anger. That emotional state may influence both which cues

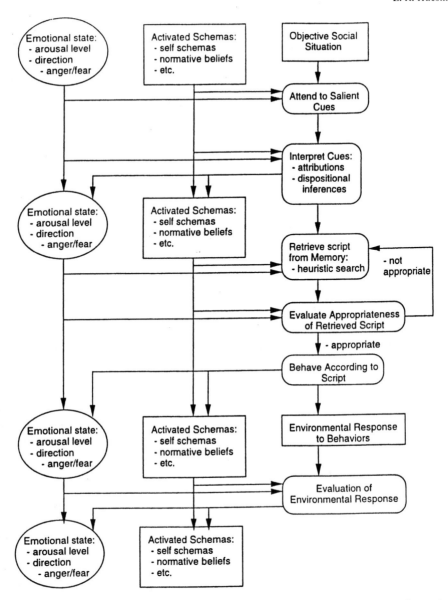

Figure 1. A unified social information processing model for aggressive behavior (Huesmann, in press).

the person attends to and how the person evaluates the cues to which he or she does attend. A highly aroused, angry person may focus on just a few highly salient cues and ig-
nore others that convey equally important information about the social situation. Then the
angry person's evaluation of these cues may be biased toward perceiving hostility when
none is present. A person who finds hostile cues the most salient or who interprets ambiguous cues as hostile will be more likely to experience anger and activate schemas and
scripts related to aggression.

A second locus for individual differences is the script generation process. It is presumed that the more aggressive individual has encoded in memory a more extensive, well-
connected network of social scripts emphasizing aggressive problem solving. Therefore,

such a script is more likely to be retrieved during any search. However, the search for a script is also strongly affected by one's interpretation of the social cues, one's activated schemas including normative beliefs, and one's mood state and arousal. For example, bad moods, even in the absence of supporting cues, will make the retrieval of scripts previously associated with bad moods more likely; the presence of a weapon, even in the absence of anger, will make the retrieval of scripts associated with weapons more likely; and the perception that another person has hostile intentions will activate scripts related to hostility. Additionally, the schemas that have been activated, particularly the self-schema and normative beliefs, will influence the direction of the search for a script. The man who believes in "an eye for an eye" and perceives himself as "an avenger" is more likely to retrieve a script emphasizing aggressive retaliation. Finally, less direct, more subtle, prosocial approaches to solving social problems may require greater search. It is hypothesized that the angry, aroused person is less likely to engage in broad search and is more likely to retrieve scripts including aggressive, retaliatory actions.

A third locus for the expression of individual differences and situational variation occurs after a script is activated. Before acting out the script, it is proposed that one evaluates the script in light of internalized activated schemas and normative beliefs to determine if the suggested behaviors are socially appropriate and likely to achieve the desired goal. Different people may evaluate the same script quite differently. The habitually aggressive person is expected to hold normative beliefs condoning more aggression and thus will employ more aggressive scripts. For example, if a man suddenly discovers that his wife has been unfaithful, he may experience rage and access a script for physical retribution. However, whether or not the man executes that script will depend on his normative beliefs about the appropriateness of "hitting a female." Even within the same person, different normative beliefs may be activated in different situations and different mood states. The person who has just been to church may have activated quite different normative beliefs than the person who has just watched a fight in a hockey game on TV. Although evaluation of the script on the basis of one's normative beliefs is the most important filtering process according to Huesmann (1988; Guerra, Huesmann, & Hanish, 1994), he suggests that two other evaluations also play a role. First, one needs to be able to predict the desirability of the consequences of utilizing such a script. Scripts include predictions about likely outcomes, but people differ in their capacities to think about the future, in their concern with the future, and in their evaluation of the desirability of consequences. The more a person focuses on immediate consequences and the less the person is concerned with the future, the more palatable an aggressive solution to a social problem may seem (Huesmann & Levinger, 1976). In addition, people may consider different dimensions of the outcome in evaluating its desirability. Some may focus on tangible rewards; others may focus on interpersonal relations, for example. Second, even if people predict the consequences of an action accurately and agree on its desirability, they may differ a lot in their evaluation of the possibility of performing according to the script that produces the outcome. A person with low-perceived self-efficacy for non-aggressive behavior may reject most prosocial scripts.

The fourth locus for individual differences in this model is a person's interpretation of society's responses to their behaviors and how that interpretation affects the person's schemas and mood. With the "right" interpretation of society's responses, one may maintain aggressive scripts even in the face of strong negative responses from society. For example, a child who is severely beaten for behaving aggressively may attribute the beating to being disliked by the punisher rather than to anything he did. An aggressive teen-age male, rather than change his aggressive behaviors, which perhaps provide immediate gratification on some dimensions, may alter his normative beliefs to make the feedback he is receiving seem less negative. He might integrate some of the readily available aphorisms about aggression into his regulatory schemata. The boy who is told he is bad because he pushed others out of the way may shrug his shoulders and think, "Nice guys finish last." The boy who shoves a child who bumped into him may think, "An eye for an eye." Alternatively, he may mitigate society's negative reinforcements for his aggressive behavior by choosing environments in which aggression is more accepted. Thus, the more aggressive adolescent male may spend more time interacting with other aggressive peers who accept his behaviors as a way of life. Not only do such social networks provide adolescents with environments in which aggression is not discouraged, such social networks promote the internalization of normative beliefs favoring aggression.

The model described above explains how social scripts that have been acquired by an individual may be accessed and used to guide behavior, and how certain individual and environmental factors could promote greater or lesser use of aggressive scripts. According to this model an habitually aggressive person is one who regularly retrieves and employs scripts for social behavior that emphasize aggressive responding. What might promote the retrieval and utilization of aggressive scripts? Clearly, the regular retrieval and use of aggressive scripts would suggest above all that a large number of aggressive scripts have been stored in memory. Similarly, the regular execution of such scripts would suggest that normative beliefs and other schemas supporting aggression have been acquired and encoded. Thus, we must examine how schemas and scripts are acquired and maintained.

3. THE ROLE OF OBSERVATIONAL LEARNING IN THE ACQUISITION OF AGGRESSIVE SCRIPTS, SCHEMAS, AND BELIEFS

While a variety of constitutional factors, ranging from body size to brain structure, may predispose individuals toward acquiring particular social beliefs, schemas, and scripts, there is every reason to expect that they must be "acquired." Learning plays the key role in the acquisition of scripts and schemas for social behavior just as learning plays the key role in the acquisition of procedural and declarative knowledge relevant to intellectual life.

Thirty-five years of child development research beginning with Bandura's seminal studies in the early 1960s have suggested that *observational learning and conditioning interact* as the child develops to enable the child to acquire scripts and schemas (Bandura, 1973, 1977, 1986; Bandura, Ross, & Ross, 1961; Coie & Dodge, in press; Eron, Lefkowitz, & Walder, 1971; Huesmann & Eron, 1986; Lefkowitz, Eron, Walder, & Huesmann, 1977). We will review the empirical evidence in more detail below, but the conclusions are clear. A belief, schema, or script is most likely to first be suggested by observing others and then more firmly established by having its use reinforced. Observation of parents, siblings, peers all are important, but so is observation of characters in the mass media.

An important encoding principle applying to observational learning is what is known as *encoding specificity* (Tulving & Thompson, 1973). This refers to the empirical fact that the specific context in which information appears when it is encoded becomes associated with the encoded information and can trigger its activation in memory better than other semantically related information. Thus, for example, the color of a room in which a violent act is observed may later trigger memories of that act. A variety of other characteristics of the observed scene enhance or diminish the likelihood of a child encoding the observed scripts, or adopting the inferred schemas or beliefs. Observed scripts that are not very salient and have observed consequences that are not very desirable are not very likely to be encoded as possible scripts for future use. The less children identify with the people being observed and the more unrealistic their actions seem to the children, the less likely children are to encode what they saw. The observer may also experience *vicariously* the reinforcements and consequences that the observed model experiences (Bandura, Ross, & Ross, 1963), and encode these outcomes as part of the social script derived from the observation. The more social approbation the observed people receive, the less likely is observational learning.

During the observational learning process the schemas that have been primed and are activated influence how well the observed scripts can be encoded and integrated into memory as well as the kinds of inferences that will be made. Both emotional states and situational stimuli may prime schemas. If the activated schemas are discrepant with the observed script, encoding is difficult; if they are consistent it is easier. When highly aroused and angry, for example, persons may view a physically aggressive sequence of behaviors as more appropriate than they would otherwise. A young boy who can only recall seeing aggressive behaviors is more likely to encode a newly observed aggressive behavior then is a boy whose mind is filled with memories of prosocial solutions. A child with normative beliefs accepting of aggression is much more likely to encode new aggressive scripts for behavior.

Once encoded through observational learning, the maintenance of a script or belief in memory will be influenced by instrumental learning. Observed scripts that are imitated but never produce a desirable outcome are likely to extinguish. Observed beliefs that are encoded but never confirmed are less likely to persist. Unfortunately, many beliefs and schemas are likely to be self-fulfilling. The individual who attributes hostility to everyone else, is sooner or later likely to be surrounded by people who really do feel hostile. Similarly, one might think that, because aggressive behavior very often produces negative consequences for the aggressor, the retrieval of aggressive scripts might extinguish. However, such instrumental learning depends on how the individual interprets society's response to the behavior. Often, because of the schemas that the aggressor has activated, the aggressor does not attribute the negative reaction of society to the specific script that the aggressor employed, and no learning takes place. The boy who is harshly punished by a teacher for taking another child's toy without asking will not unlearn the behavior if he interprets the cause of his punishment as dislike by the teacher.

Even if aggressive scripts are not used and reinforced they may become more accessible to a child if the child rehearses them. The rehearsal may take several different forms from simple recall of the original scene, to fantasizing about it, to play acting. The more elaborative, ruminative type of rehearsal characteristic of children's fantasizing is likely to generate greater connectedness for the script, thereby increasing its accessibility in memory. Also, through such elaborative rehearsal the child may abstract higher-order scripts representing more general strategies for behavior than the ones initially stored. Of course, rehearsal also provides another opportunity for reevaluation of any script. It may be that some scripts initially accepted as appropriate (under specific emotional and memory states) may be judged as inappropriate during rehearsal.

4. EMPIRICAL DATA ON INFORMATION PROCESSING, OBSERVATIONAL LEARNING, AND AGGRESSION

The information processing model for the development of aggression described in this chapter is based on the presumption that predisposing personal factors and environmental context interact through observational and enactive learning to lead to the emergence of cognitive processes (including emotional processes) and cognitive schemas that promote aggression. Cue attention and evaluation, script retrieval, script evaluation, and evaluation of the environment's responses to one's actions are the four key parts of the social/cognitive performance model as outlined in Figure 1. In this section I will review the subset of key studies that provide evidence about the variety of cognitive processes implicated within this model and the biological processes that interact with them.

4.1. Hostile Attributional Bias and Observational Learning

It is now well established that aggressive individuals tend to perceive hostility in others where there is no hostility, i.e. display a hostile attributional bias (Dodge, 1980; Dodge & Coie, 1987; Dodge & Frame, 1982; Dodge,Price, Bachorowski, & Newman, 1990; Graham & Hudley, 1994; Nasby, Hayden & DePaulo, 1979; Slaby & Guerra, 1988; Steinberg & Dodge, 1983). A long history of research on social perception (e.g. see Fiske, 1982; Fiske & Taylor, 1991; Schneider, 1991), as well as recent research on aggression, suggest that this hostile attributional bias is a product of the schemas that an individual has encoded and activating cues. Dodge and Tomlin (1987) reported evidence that aggressive children are relying on their own encoded aggressive self-schemas and stereotypes in making intent attributions. Zelli and Huesmann (1993) have found that college students with greater ingrained persecution beliefs are more likely to perceive hostility when none is there. There is also strong evidence that these hostile cue interpretations become an automatic cognitive process. Bargh (1989) and Winter and Uleman (1984) have shown that inferences about the dispositions of others occur automatically without conscious awareness. More recently Zelli, Huesmann, & Cervone (1995) have shown that more aggressive individuals automatically encode ambiguous sentences with an aggressive interpretation and then are more likely to recall them when prompted with an aggressive cue.

How do schemas promoting hostile attributional bias develop? There is good reason to think that observational learning plays a major role. A variety of evidence exists to support the conclusion that those who see more violent behavior in real life or in the mass media begin to falsely perceive more violence around them than do others. Gerbner and his colleagues (Gerbner & Gross, 1980) have reported that high exposure to media violence in adults makes them see the world as a more hostile place. Similarly both Bryant, Carveth, and Brown (1981) and Tyler & Cook (1984) showed specifically that exposure to media violence increased viewers estimates of the frequency of aggression in society. These schemas representing the frequency of aggressive behavior in one's surroundings can then be expected to influence the attributions one makes about those around them.

4.2. Observational Learning of Deviant Scripts

It is methodologically difficult to assess the kinds of scripts that individuals have encoded; however, one can assess the kinds of scripts they are most likely to retrieve and make inferences from those data. The available evidence suggests that, in fact, the most accessible social scripts for aggressive children are aggressive scripts. For example, the scripts retrieved by more aggressive children to solve hypothetical problems tend to incorporate more physical aggression and manipulation actions (Rubin, Bream & Rose-Krasnor, 1991; Rubin, Moller, & Emptage, 1987; Waas, 1988). Priming by negative intent cues is more likely to activate an aggressive script in aggressive children (Graham & Hudley, 1994). Aggressive children are less likely to generate more subtle prosocial scripts to solve social problems (Deluty, 1981; Taylor & Gabriel, 1989), and there is some evidence that, as hypothesized, a narrower search process for a script is associated with more aggressive behavior (Shure & Spivack, 1980).

Researchers have also shown that the observation of aggressive scripts in real life or in the mass visual media leads to the encoding of such scripts. Children growing up observing violence around them behave more violently (e.g. Guerra, Huesmann, Tolan, VanAcker & Eron, 1995), and children whose parents physically aggress against them are more likely to physically aggress against their own children later in life (Widom, 1989). However, it is hard to show that such effects are due to the acquisition by the children of specific scripts through observation. The research on media violence and aggression provides more compelling evidence of that process.

Both the well-known contagion of suicide and copycat crimes (Berkowitz, 1993) provide some of the clearest examples of specific aggressive scripts being acquired by adults through observation from the media. More importantly from a scientific standpoint perhaps, numerous laboratory and field experiments (see Paik & Comstock, 1994; Huesmann, 1982; Huesmann, Moise, & Podolski, in press) have demonstrated the encoding of specific scripts from such observations. The typical paradigm is that randomly selected children who are shown either a violent or non-violent short film are observed as they play afterwards (Bandura, Ross, & Ross, 1961, 1963a, 1963b). The consistent finding is that children who see the violent film clip behave more aggressively immediately afterwards. Such results have been obtained both for aggression directed at inanimate objects (e.g., "Bobo" dolls) and for aggression directed at peers (Bjorkqvist, 1985; Josephson, 1987). In one very typical study Bjorkqvist (1985) in Finland exposed 5 to 6 year old children to either violent or non-violent films. These children were then observed playing together in a room by two observers who did not know which type of film each child had seen. Children who had just seen the violent film ended up being rated higher on physical aggression (hitting other children, wrestling, etc), verbal aggression (screaming at others, threatening others), and aggression at objects (intentional destruction of toys, etc.). The empirical data are also compelling that new aggressive scripts are abstracted out of the elements of specific scripts being observed. Thus, the aggressive scripts that children display after being exposed to violent scenes are not exactly the same as the scripts observed (Bjorkqvist, 1985).

The generality of these experimental conclusions has been confirmed by a substantial body of field research showing that early childhood exposure to violence is correlated with childhood aggression and predictive of adult aggression. For example, in a study initiated in 1960 on 870 youth in Columbia County, New York, Eron and his colleagues found that boys who watched more violence in the mass media in elementary school were statistically more likely to be aggressive ten years later (after graduating from high school), even controlling for initial aggressiveness, social class, education, and other relevant variables (Eron, Huesmann, Lefkowitz, & Walder, 1972; Lefkowitz, Eron, Walder, & Huesmann, 1977). A 22-year follow-up of these same subjects revealed that their early violence viewing also related to their adult criminality at age 30 (Huesmann, 1986).

These observational learning studies have also confirmed the validity of the encoding specificity principle with regard to aggressive scripts. Even a neutral cue that is present in an observed aggressive script may trigger the retrieval of that script. For example, Josephson (1987) showed that a walkie-talkie present in an aggressive video could trigger aggressive behavior in boys who had watched that video when they later saw a walkie-talkie.

Information processing theory suggests that even quite different specific aggressive scripts and schemas are linked together in one's memory network by a common "hostility" node and thus should be primed by other aggressive ideas or cues, even if they have no substantive connection. There is significant evidence supporting this view. The classic example of such an effect is the Berkowitz and LePage (1967) "gun" experiment. In this study subjects gave larger shocks to punish a partner who was not learning if there was a gun in the experimental room. The gun cues the hostility node which leads to the utilization of a more aggressive script for behaving in the experiment. In fact, the same effect can be obtained when the cue activating the "hostility" node is anger or another negative emotion (Berkowitz, Cochran, & Embree, 1981). In addition, the observation of violence in the mass media or environment has been shown to activate a wide variety of hostile thoughts (Bushman & Geen, 1990).

A substantial body of research in cognitive psychology has shown that rehearsal of information enhances its connectedness in the memory network and makes it more accessible (Klatzky, 1980). Thus, rehearsal of an aggressive script should make its retrieval more likely in the future. One common type of rehearsal of social behavior is fantasizing, and the empirical evidence shows that fantasizing about aggressive behaviors is positively related to behaving aggressively. For example, in a longitudinal study of early elementary school children conducted in five different countries, Huesmann and Eron (1984; Rosenfeld, Huesmann, Eron, & Torney-Purta, 1982) reported that self-reported fantasizing about aggression was correlated with peer-nominated aggressive behavior in all five countries. Both Viemero (Viemero & Paajanen, 1992) and Huesmann (1986) have also found in field studies that TV violence viewing predicts fantasizing about aggression which in turn predicts later aggressive behavior.

4.3. Observational Learning of Deviant Normative Beliefs

A script may not be employed, even if it has been retrieved, if the script is evaluated as inappropriate when filtered through an individual's normative beliefs about aggression (See Figure 1). Huesmann and Guerra (1997; Guerra, Huesmann, & Hanish, 1994; Huesmann, Guerra, Zelli, & Miller, 1992; Huesmann, Zelli, Fraczek, & Upmeyer, 1993) have developed a reliable measure of normative beliefs about general aggression (e.g. "Is it usually O.K. to push and shove other people around?") and retaliation (e.g. "If a girl screams at you, is it O.K. to hit her?") and have shown that children and adults who are more aggressive have normative beliefs that are more approving of aggression. More important, recent longitudinal studies have shown that normative beliefs about aggression seem to crystalize during early childhood (Huesmann & Guerra, 1997). For children age 6 or 7 such beliefs are very unstable and do not predict much about subsequent aggressive behavior. However, such beliefs are predicted by the child's own previous behavior. For children age 10 and 11 the picture changes. Normative beliefs are now stable and predict subsequent aggressive behavior. Thus, age 6 to 9 seems to be a period during which normative beliefs and other schemas relating to aggressive social behavior are being developed through interactions with the environment.

How do children acquire normative beliefs? Again substantial evidence supports a key role for observational learning. Children hear the beliefs expressed by their parents and peers, but they also observe their parents and peers behaviors and draw inferences about the acceptability of aggression and violence from what they see. Thus, their beliefs tend to be correlated with those of parents (Huesmann et al., 1984; Miller, 1991) and peers (Henry, Guerra, Huesmann, & VanAcker, 1996). In addition, observation of violence in the mass media influences beliefs about the acceptability of violence. For example, it has been shown that the more televised violence a child watches, the more accepting is the child's attitude toward aggressive behavior (Dominick & Greenberg,

1972). Longitudinal studies also show that adult normative beliefs about violence are related to observation of violence in childhood (Huesmann, Moise, Podolski, & Eron, 1996). The causal direction of this effect has been established by experimental studies which have demonstrated that children and young adults become more tolerant of aggression immediately after even very brief exposures to violence (Drabman & Thomas, 1974; Thomas & Drabman, 1975; Linz, Donnerstein, & Penrod, 1988; Malamuth & Check, 1981).

Normative beliefs are not the only schemas relevant to script selection and evaluation. Self-schemas provide an internal context within which scripts must be evaluated as well. Heightened activation of self-schemas decreases the likelihood of aggression when the self-schema is non-aggressive (Carver, 1974), probably by filtering out potential aggressive scripts. On the other hand, as Baumeister (1996) has shown, a self-schema that includes an extremely positive evaluation of oneself can promote the selection of aggressive scripts when a person threatens that self-evaluation. Perceptions of self-efficacy for executing the script in question would also be expected to be important in the evaluation of a script (Bandura, 1986; McFall, 1982), and the implication for aggressive behavior would seem to be that those with high self-efficacy for prosocial behavior would be less likely to behave aggressively. Finally, schemas about others may be as important as schemas about oneself in affecting aggressive behavior. Schemas about others which promote disindividuation allow the utilization of aggressive scripts which might otherwise be unacceptable (Diener, 1976; Prentice-Dunn & Rogers, 1983).

What role does observational learning play in the acquisition of schemas about others and the self? Obviously, schemas about others must be mainly acquired through observation. However, there is also a strong argument to be made that observation of the self plays a major role in the acquisition of self-schemas (Bem, 1967). According to self-perception theories, schemas and beliefs about oneself are based to some extent on inferences drawn from observations of one's own behavior. We have already noted that in young children, aggressive behavior is *predictive* of the adoption of normative beliefs accepting of aggression (Huesmann & Guerra, 1997). There is significant evidence that other kinds of self-schemas are also influenced by self-perceptions. For example, subjects who observe themselves describing themselves in flattering terms are more likely to score higher on self-esteem afterwards (Jones, Rhodewalt, Berglas, & Skelton, 1981).

5. BIOSOCIAL PROCESSES IN OBSERVATIONAL LEARNING

These empirical data all point to the importance of observational learning in acquiring the schemas that underlie hostile attributional bias, in acquiring social scripts that control behavior, and in acquiring normative beliefs that filter out inappropriate behaviors. Yet relatively little is known about the biological processes that underlie observational learning. Let us review what is known.

5.1. Neurotransmitter Effects

Two kinds of animal studies have shown that the observation of violence or the perception of the threat of violence produce detectable neuronal activation and neurotransmitter changes. Welch and Welch (1971) showed that mice placed near where other mice were fighting showed increased MAO production. Such production would inhibit norepinephrine and dopamine thereby readying the animal more for aggression. Ogawa and associates (Ishikawa, Hara, Ohdo, & Ogawa, 1992) more recently have shown that observation of fighting produces stress reactions in rats and increased plasma corticosterone. Equally interesting are the recent experiments of Miczek (1995) on threats of fighting. Animals who are faced with confrontational situations, even if no fighting or overt aggressive behaviors occur, show very specific neuronal activation patterns in the brain stem, specifically c-fos expression in the periaqueductal grey area. What is particularly relevant to this discussion is that these patterns of activation can subsequently be triggered simply by letting the animal view the physical context in which the confrontation took place. For example, Miczek (1995) has demonstrated this phenomenon in rats by first placing a rat for a brief period in a confrontational situation in which the rat is not touched. On the next day he brings the rat back to the same locale. Despite the fact that no other rat is present, and there is no confrontation, the rat displays the brain stem activation pattern characteristic of confrontation. The physical context apparently has "reminded" the rat of the confrontation and "activates" the rat for confrontation. Such studies suggest a biological basis for encoding specificity with anger.

Taken together these animal studies suggest a neurophysiological basis for observational learning. Observation of violence or confrontation produces innate neuronal activation patterns which become associated with cues observed in the environment even in the absence of any reinforcement to the animal. These cues then can trigger the same activation patterns which ready the animal for aggression. Although there are some potential alternative explanations for the effects in all of these studies that need to be explored more, most notably the role of pheromones, they clearly suggest that exposure to violence around the organism produces neurotransmitter changes that ready the organism for violence. It remains to be seen if similar relations can be demonstrated in humans.

5.2. Hormones and Observational Learning

Given the level of evidence for a correlation in humans between testosterone level and dominating others or winning a competition, it would not be surprising to find that the observation of violence under some circumstances would also stimulate increased testosterone production in humans. Unfortunately, only one relevant study seems to have been conducted, and its results were ambiguous. Hellhammer (1985) exposed young, adult males to different kinds of film scenes and then measured their testosterone levels. They found increases in testosterone after exposure to erotic films and decreases after exposure to films generating anxiety but no changes after exposure to aggressive films. Part of the problem may be that, theoretically, the testosterone response of the viewer should depend on with whom the viewer identifies, the victim or the aggressor. This is an important issue because it is well established that hormones can alter the perception of social signals between cospecifics (Brain, 1983). In particular, heightened testosterone can make a organism more sensitive to threatening stimuli. If the observation of violence or the threat of violence produces learned associations between various cues and testosterone responses that distort social signalling, then subsequent aggressive interactions become more likely in similar situations.

5.3. Arousal, Hostile Attributional Bias, and Script Retrieval

These hormonal and neurotransmitter processes related to the observation of violence can be seen as providing a biological contribution to both hostile attributional biases and to the retrieval of aggressive scripts. Cues associated with observing violence trigger physiological responses that prepare one for violence and testosterone sensitizes one to threatening stimuli. Similarly the arousal associated with observation of violence would be expected to have an effect on the information processing operations governing social behavior. The organism that has been exposed more to violence could be expected to be more generally aroused in the presence of cues previously associated with observation of violence. Too high a general level of arousal makes the difficult task of interpreting ambiguous cues, even more difficult. Thus hostile attributional bias, as predicted, is more likely under conditions of high emotional arousal (Dodge & Somberg, 1987). It also promotes a shallow, quick search for scripts with the best learned scripts dominating retrieval. Finding an appropriate script to respond to a threat without escalating it to violence is not an easy cognitive task and successful performance on difficult cognitive tasks diminishes as arousal becomes very high (Anderson, 1980). If testosterone is also stimulated and distorts the perception of social stimuli, the risk of hostile bias increases and retrieval of aggressive scripts increases. The activation of specific neurotransmitters or neuronal patterns associated with aggression can serve as an additional cue for an aggressive interpretation and retrieval.

Individual differences in arousal can be expected to play another role as well in the acquisition of scripts through observational learning. Those who have a low baseline level of arousal, e.g. extroverts (Eysenck, 1977), can be expected to seek stimulation to raise their level of arousal to an optimal level. In our modern society, observation of violence, in person or through violent movies and films, provides an obvious opportunity to increase arousal. Unfortunately, contrary to this prediction, introverts seem to spend more time watching television in general (Huesmann, 1986); however, this phenomenon could simply be a consequence of introverts' more restricted social lives. It may well be that extroverts when they are exposed to violence are more aroused by it and that arousal is reinforcing. Therefore, they attend more to violence and are more at risk to acquire violent scripts through observational learning. To complicate the picture more, one must consider the valence of the arousal produced by many scenes of violence. Neiss (1988) has argued against the construct of general arousal on the basis that different types of positive and negative arousal have little in common. Arousal produced by horrible scenes of carnage may be experienced as quite unpleasant while arousal produced by erotic scenes of domination may be experienced as quite pleasant by the same person. In addition, different individuals are known to respond to the same scenes with quite different levels of arousal (Malamuth & Donnerstein, 1984). It is therefore plausible to expect that repeated exposures to scenes of violence resulting in habituation could have different consequences for different individuals. For the individual who experiences violence as unpleasantly arousing, habituation could reduce the aversive consequences of behaving violently and make the learning of aggressive scripts more likely. For the individual who experiences violence as pleasantly arousing, habituation could reduce the rewarding consequences of behaving violently and make the learning of aggressive scripts less likely. Clearly, more research is needed in this area.

5.4. Enactive Learning and Biosocial Processes

While the focus of this chapter is on observational learning, observational learning clearly interacts with conditioning. Equally important as observational learning to the acquisition of appropriate scripts should be a child's responsiveness to the consequences of employing the scripts, e.g., rewards and punishments. One would expect that those children who are less easily conditioned by social approbation would be more at risk for acquiring inappropriate aggressive scripts in a normal environment. The child who experiences less anxiety in response to social disapproval and less gratification in response to social rewards would be expected to be less conditionable. Empirical studies on arousability, as mentioned previously, suggest that difficult to arouse individuals may indeed be less easily conditioned and more at risk for violent and aggressive behavior (Raine & Venables, 1981; Raine, Venables, & Williams, in press). However, information processing theory again suggests that it is important to distinguish between instrumental aggression and hostile aggression in this regard. While more easily arousable and thus conditionable children may be less at risk for instrumental aggression, they should be more at risk for hostile aggression, and, indeed, there is some evidence that they are more at risk (Baker et al., 1984) and commit more violent crimes (Hare & McPherson, 1984).

6. HOSTILE AGGRESSION AND INSTRUMENTAL AGGRESSION REVISITED

Throughout the review and analysis in this chapter, we have seen a number of ways in which the information processing, the schemas, and the biosocial processes underlying hostile and instrumental aggression differ. On the biosocial level, low arousability should be a risk factor for instrumental aggression and high arousability a risk factor for hostile aggression, and we reported some evidence of that. At the information processing level instrumental aggression should be more a function of having encoded a large repertoire of aggressive scripts for solving social problems and of having acquired normative beliefs approving of aggression. Hostile aggression should be more a function of high emotional responsivity and hostile attributional bias. Crick and Dodge (in press) have recently reported empirical data that seem to be consistent with these predictions.

At the same time the basic social/cognitive processing for all social behavior is the same, and the basic information processing model presented in this paper applies to both kinds of aggression. Similarly, the process of observational learning operates similarly for both types of aggression. One particular biosocial interaction that could be expected to influence both kinds of aggression, therefore, is the interaction been mood and learning or behavior. From the social cognitive perspective, one would expect intense bad moods to enhance the likelihood that previously observed scripts and schemas would be activated. The arousal makes complex cognitive processing less likely. The dysphoric valence of the emotion activates schemas related to hostility, and the arousal narrows memory activation. The result is that observed antisocial, aggressive scripts are more likely to be encoded during such emotional states, and they are more likely to be retrieved and utilized during such emotional states. Thus, it is not surprising that high temperatures (Anderson & Anderson, 1984), crowding (Matthews, Paulus, & Baron, 1979), and other irritators and stressors (Guerra, Huesmann, Tolan, VanAcker, & Eron, 1995) increase all kinds of aggression, while cognitive reflection on the cause of irritation reduces aggression at unrelated targets (Berkowitz & Troccoli, 1990).

7. CONCLUSIONS

Extensive empirical research on social information processing coupled with theoretical elaborations from cognitive science constructs has led to the emergence of a unified model of social information processing in aggressive behavior (Huesmann, in press). The model identifies four processes in social problem solving at which point emotional arousal, activated schemas, and situational cues interact to affect aggression: 1) cue attention and interpretation, 2) script retrieval, 3) script evaluation and selection, and 4) evaluation of society's response to one's behavior. Although these processes may first require cognitive control in the developing child, they eventually seem to operate as relative automatic cognitive processes. It is argued that the cognitive programs, scripts, schemas, and beliefs that comprise this information processing system are acquired through a process of observational learning followed by conditioning.

The evidence suggests that humans attend to environmental cues differentially and interpret the cues differently as a function of predisposing neurophysiological factors, their emotional arousal, the kinds of cognitive schemas they have acquired, and which schemas are activated. More aggressive individuals tend to focus on fewer cues and cues that are more frequently symptomatic of hostility, tend to interpret ambiguous cues more readily as symptomatic of hostility, and tend to believe that the world is more hostile. This is particularly true when the individual is angry, either because of situational factors or a predisposition toward more general hostility. More aggressive individuals also have a greater proportion of aggressive scripts encoded in memory with more accessible links to everyday cues. They have been found to rehearse their aggressive scripts more through aggressive fantasizing and to recall more aggressive scripts from ambiguous cues. It has been shown that, while young children do not have well defined or stable normative beliefs about the appropriateness of aggression, older children do have well formed beliefs, and those beliefs influence how they evaluate retrieved scripts.

Each of these processes depends on cognitive scripts, schemas, and beliefs that must be acquired by the child through interactions with the environment. While evolutionary forces operating through genetic influences on neurophysiology may predispose individuals to process information in one way or another, the existing empirical evidence suggests that learning from observing others is a key process in acquiring scripts and schemas for social behavior. The acquisition through observational learning of scripts, schemas, and beliefs — and the use of these cognitions — is influenced by and influences at least three biological systems implicated in aggression: neurotransmitter processes and specific neuronal activation patterns, arousal patterns and individual differences in arousal, and hormonal responses and individual differences in hormonal responses. Animal data reveal that observation of violence does in fact activate specific neurotransmitter and activation systems. While the evidence on hormonal responses to the observation of violence in humans is inconclusive, it is clear that what human males observe influences their testosterone level, and it is clear that testosterone level can influence information processing. Substantial evidence also suggests that arousal level and individual differences in arousal influence both observational learning and the expression of scripts acquired through observational learning. Some individuals may be predisposed to acquire observed scripts and schemas more easily and to be conditioned more easily. Depending on the social environment in which that individual is raised and that individual's predisposing biological factors (particularly arousal), the individual may learn to be more or less aggressive.

In summary, from the social/cognitive, information-processing perspective, it is easy to see that once a child begins to perceive the world as hostile, to acquire scripts and schemas emphasizing aggression, and to believe that aggression is acceptable, the child enters a vicious cycle that will be difficult to stop. Biological predispositions place some children much more at risk for entering this cycle, and interact with observational learning to promote aggression. The biological responses of the organism in turn may promote the continuation of the cycle. If not interrupted, the cycle can be expected to continue into adulthood, maintaining aggressive behavior throughout the life span.

8. REFERENCES

Abelson, R. P. (1981). The psychological status of the script concept. *American Psychologist, 36,* 715–729.

Anderson, C. & Anderson, D. (1984). Ambient temperature and violent crime: Tests of the linear and curvilinear hypotheses. *Journal of Personality and Social Psychology, 46,* 91–97.

Anderson, J. R. (1980). *Cognitive Psychology.* San Francisco: Freeman.

Baker, L., Hastings, J., & Hart, J. (1984). Enhanced psychophysiological responses of type A coronary patients during type A-relevant imagery. *Journal of Behavioral Medicine, 7,* 287–306.

Bandura, A. (1973). *Aggression: A social learning analysis.* Englewood Cliffs, NJ: Prentice-Hall.

Bandura, A. (1977). *Social Learning Theory.* Englewood Cliffs: Prentice Hall.

Bandura, A. (1986). *Social foundations of thought and action: A social-cognitive theory.* Englewood Cliffs, NJ: Prentice-Hall.

Bandura, A., Ross, D., & Ross, S.A. (1961). Transmission of aggression through imitation of aggressive models. *Journal of Abnormal Social Psychology, 63,* 575–582.

Bandura, A., Ross, D., & Ross, S.A. (1963a). Imitation of aggression through imitation of film-mediated aggressive models. *Journal of Abnormal and Social Psychology, 66,* 3–11.

Bandura, A., Ross, D., & Ross, S.A. (1963b). Vicarious reinforcement and initiative learning. *Journal of Abnormal and Social Psychology, 67,* 601–607.

Bargh, J.A., (1989). Conditional automacity: Varieties of automatic influence in social perception and cognition. In J.S. Uleman & J.A. Bargh (Eds.), *Unintended Thought.* New York: Guilford Press.

Baumeister, R. F., Smart, L., & Boden, J. M. (1996). Relation of threatened egotism to violence and aggression: The dark side of high self-esteem. *Psychological Review, 103,* 5–33.

Bem, D. J. (1967). Self-perception. An alternative interpretation of cognitive dissonance phenomena. *Psychological Review, 74,* 183–200.

Berkowitz, L. (1974). Some determinants of impulsive aggression: The role of mediated associations with reinforcements for aggression. *Psychological Review, 81,* 165–176.

Berkowitz, L. (1990). On the formation and regulation of anger and aggression: A cognitive-neoassociationistic analysis. *American Psychologist, 45,* 494–503.

Berkowitz, L. (1993). Pain and aggression: Some findings and implications. Special Issue: The pain system: A multilevel model for the study of motivation and emotion. *Motivation and Emotion, 17,* 277–293.

Berkowitz, L. & LePage, A. (1967). Weapons as aggression-eliciting stimuli. *Journal of Personality and Social Psychology, 7,* 202–207.

Berkowitz, L. & Troccoli, B. T. (1990). Feelings, direction of attention and expressed evaluations of others. *Cognition and Emotion, 4,* 305–325.

Berkowitz, L., Cochran, S., & Embree, M. (1981). Physical pain and the goal of aversively stimulated aggression. *Journal of Personality and Social Psychology, 40,* 687–700.

Bjorkqvist, K. (1985). *Violent films, anxiety and aggression.* Helsinki: Finnish Society of Sciences and Letters.

Bouchard, T. J. (1984). Twins reared together and apart: What they tell us about human diversity. In S.W. Fox (Ed.), *Individuality and determinism: Chemical and biological basis.* New York: Plenum.

Brain, P. F. (1983). Pituitary-gonadal influences and intermale aggressive behavior. In B. B. Svare, (Ed.), *Hormones and Aggressive Behavior* (pp. 3–15). New York: Plenum Press.

Bushman, B. J., & Geen, R. (1990). Role of cognitive-emotional mediators and individual differences in the effects of media violence on aggression. *Journal of Personality and Social Psychology, 58*(1), 156–163.

Bryant, J., Carveth, R. A., & Brown, D. (1981). Television viewing and anxiety: An experimental examination. *Journal of Communication, 31,* 106–119.

Carver, C. S. (1974). Facilitation of physical aggression through objective self-awareness. *Journal of Experimental Social Psychology, 10,* 365–370.

Coie, J. D. & Dodge, K. A. (in press). Aggression and antisocial behavior. In N. Eisenberg (Ed.) *Handbook of Child Psychology: Volume 3: Social, Emotional, and Personality Development,* New York: John Wiley & Sons.

Cloninger, C. R., & Gottesman, A. (1987). Genetic and environmental factors in antisocial behavior disorders. In Mednick, S. A., Moffitt, T. E., Stack, S. A. (Eds.), *The causes of crime: New biological approaches.* New York: Cambridge University Press.

Comstock, G. A., & Paik, H. (1991). The effects of television violence on aggressive behavior: A meta-analysis. In *A preliminary report to the National Research Council on the understanding and control of violent behavior.* Washington, D.C.: National Research Council.

Crick, N. R., & Dodge, K. A. (1994). A review and reformulation of social information processing mechanisms in children's adjustment. *Psychological Bulletin, 115,* 74–101.

Crick, N.R. & Dodge, K.A. (in press). Social information-processing mechanisms in reactive and proactive aggression. *Child Development.*

Diener, E. (1976). Effects of prior destructive behavior, anonymity, and group presence on deindividuation and aggression. *Journal of Personality and Social Psychology, 33,* 497–507.

Deluty, R. H. (1981). Alternative thinking ability of aggressive, assertive, and submissive children. *Cognitive Therapy and Research, 5,* 309–312.

Dodge, K.A. (1980). Social cognition and children's aggressive behavior. *Child Development, 53,* 620–635.

Dodge, K.A. (1986). A social information processing model of social competence in children. In M. Perlmutter (Ed.), *The Minnesota symposium on child psychology* (pp. 77–125). Hillsdale, NJ: Erlbaum.

Dodge, K. A., & Coie, J. D. (1987). Social information processing factors in reactive and proactive aggression in children's peer groups. *Journal of Personality and Social Psychology, 53,* 1146–1158.

Dodge, K.A., & Frame, C.L. (1982). Social cognitive biases and deficits in aggressive boys. *Child Development,* 53, 620–635.

Dodge, K.A., & Somberg, D.A. (1987). Hostile attributional biases among aggressive boys are exacerbated under conditions of threats to the self. *Child Development, 58,* 213–224.

Dodge, K.A., Price, J.M., Bachorowski, J.A. & Newman, J.P. (1990). Hostile attributional biases in severely aggressive adolescents. *Journal of Abnormal Psychology, 99,* 385–392.

Dodge, K.A., & Tomlin, A. (1987). Utilization of self-schemas as a mechanism of attributional bias in aggressive children. *Social Cognition,* 5(3),

Dominick, J. R. & Greenberg, B. S. (1972). Attitudes toward violence: The interaction of television exposure, family attitudes, and social class. In G. A. Comstock & E. A. Rubinstein (Eds.), *Television and social behavior (Vol. 3). Television and adolescent aggressiveness.* Washington: U.S. Government Printing Office.

Drabman, R. S., & Thomas, M. H. (1974). Does media violence increase children's toleration of real-life aggression? *Developmental Psychology, 10,* 418–421.

Eron, L. D., Walder, L. O., & Lefkowitz, M. M. (1971). *The Learning of Aggression in Children.* Boston: Little Brown.

Eron, L. D., Huesmann, L. R., Lefkowitz, M. M., & Walder, L. O. (1972). Does television violence cause aggression? *American Psychologist, 27,* 253–263.

Eysenck, H. J. (1964). *Crime and personality.* St. Albans, UK: Paladin.

Farrington, D. P. (1982). Longitudinal analyses of criminal violence. In M.E. Wolfgang & N.A. Weiner (Eds.), *Criminal Violence,* Beverly Hills, CA: Sage.

Farrington, D.P. (1985). The development of offending and antisocial behavior from childhood: Key findings from the Cambridge study in delinquent development. *Journal of Child Psychology and Psychiatry, 36,* 1–36.

Farrington, D. P. (1991). Childhood aggression and adult violence: Early precursors and later life outcomes. In D. J. Pepler & K. H. Rubin (Eds.), *The development and treatment of childhood aggression* (pp. 5–29). Hillsdale, NJ: Lawrence Erlbaum Associates.

Feshbach, S. (1964). The function of aggression and the regulation of aggressive drive. *Psychological Review, 71,* 257–272.

Fiske, S.T. (1982). Schema-triggered affect: Applications to social perception. In M.S. Clark & S.T. Fiske (Eds.), *Affect and cognition: The 17th Annual Carnegie Symposium on Cognition.* Hillsdale, NJ: Erlbaum.

Fiske, S. T., & Taylor, S. E. (1991). *Social Cognition.* New York, N.Y.: McGraw Hill, Inc.

Graham, S. & Hudley, C. (1994). Attributions of aggressive and nonaggressive African-American male early adolescents: A study of construct accessibility. *Developmental Psychology, 30,* 365–373.

Guerra, N. G., Huesmann, L. R., & Hanish, L. (1994). The role of normative beliefs in children's social behavior. N. Eisenberg (Ed.), *Review of Personality and Social Psychology, Development and Social Psychology: The Interface.* London: Sage.

Guerra, N.G., Huesmann, L.R., Tolan, P.H., VanAcker, R. & Eron, L.D. (1995). Stressful events and individual beliefs as correlates of economic disadvantage and aggression among urban children. *Journal of Consulting and Clinical Psychology 63(4),* 518–528.

Hare, R. D. & McPherson, L. M. (1984). Violent and aggressive behavior by criminal psychopaths. *International Journal of Law and Psychiatry, 7,* 35–50.

Hellhammer, D. H. (1985). Changes in saliva testosterone after psychological stimulation. *Psychoneuroendocrinology, 10,* 1, 77–81.

Henry, D., Guerra, N., Huesmann, R., VanAcker, R. (1996). Normative influences on aggression in urban elementary school classrooms. Manuscript submitted for publication.

Huesmann, L.R. (1982). Television violence and aggressive behavior (pp.126–137). In D. Pearl, L. Bouthilet & J. Lazar (Eds.), *Television and Behavior: Ten Years of Scientific Programs and Implications for the 80's: Vol. 2.* Washington, D.C.: U.S. Government Printing Office.

Huesmann, L. R. (1986). Psychological processes promoting the relation between exposure to media violence and aggressive behavior by the viewer. *Journal of Social Issues, 42*, 3, 125–139.

Huesmann, L. R. (1988). An information processing model for the development of aggression. *Aggressive Behavior, 14*, 13–24.

Huesmann, L. R. (in press). The role of social information processing and cognitive schema in the acquisition and maintenance of habitual aggressive behavior. In R. G. Geen & E. Donnerstein (Eds.), *Huesmann Aggression: Theories, Research and Implications for Policy*, New York: Academic Press.

Huesmann, L. R., & Eron, L. D. (1984). Cognitive processes and the persistence of aggressive behavior. *Aggressive Behavior, 10*, 243–251.

Huesmann, L. R., & Eron, L. D. (Eds.). (1986). *Television and the Aggressive Child: A Cross-National Comparison*. Hillsdale: N.J.: Erlbaum.

Huesmann, L. R., Eron, L. D., Lefkowitz, M. M., & Walder, L. O. (1984). The stability of aggression over time and generations. *Developmental Psychology, 20*, 1120–1134.

Huesmann, L.R., & Guerra, N.G. (1997). Normative beliefs about aggression and aggressive behavior. *Journal of Personality and Social Psychology*, 72(2), 408–419.

Huesmann, L. R., Guerra, N. G., Zelli, A., & Miller, L. (1992). Differing normative beliefs about aggression for boys and girls. In K. Bjorkqvist & P. Niemela (Eds.). *Of mice and women: Aspects of female aggression*. Orlando, Florida: Academic Press.

Huesmann, L.R. & Levinger, G. (1976). Incremental exchange theory: a formal model for progression in dyadic social interaction. In L. Berkowitz (Ed.), *Advances in Experimental Social Psychology, 9*, 1071–1078.

Huesmann, L. R., & Miller, L. S. (1994). Long-term effects of repeated exposure to media violence in childhood. In L. R. Huesmann (Ed.), *Aggressive Behavior: Current Perspectives*, New York: Plenum.

Huesmann, L. R., Moise, J. & Podolski, C. P. (in press). The effects of media violence on the development of antisocial behavior. In D. Stoff, J. Breiling, & J. Masser (Eds.) *Handbook of Antisocial Behavior*.

Huesmann, L. R., Moise, J., Podolski, C., & Eron, L. D. (1996). The roles of normative beliefs and fantasy rehearsal in mediating the observational learning of aggression. Paper presented at the meetings of the *American Psychological Society*, San Francisco.

Huesmann, L. R., Zelli, A., Fraczek, A., & Upmeyer, A. (1993). Normative attitudes about aggression in American, German, and Polish college students. *Third European Congress of Psychology*, Tampere, Finland.

Ishikawa, M., Hara, C., Ohdo, S., Ogawa, N. (1992). Plasma corticosterone response of rats with sociopsychological stress in the communication box. *Physiology & Behavior*, 52, 475–480.

Jones, E. E., Rhodewalt, F., Berglas, S., & Skelton, J. A. (1981). Effects of strategic self-presentation on subsequent self-esteem. *Journal of Personality and Social Psychology, 41*, 407–421.

Josephson, W. L. (1987). Television violence and children's aggression: Testing the priming, social script, and disinhibition predictions. *Journal of Personality and Social Psychology, 53*, 882–890.

Kagan, J. (1988). Temperamental contributions to social behavior. *American Psychologist, 44*, 668–674.

Lagerspetz, K., & Lagerspetz, K. M. J. (1971). Changes in aggressiveness of mice resulting from selective breeding, learning and social isolation. *Scandinavian Journal of Psychology*, 12, 241–278.

Lagerspetz, K., & Sandnabba, K. (1982). The decline of aggression in mice during group caging as determined by punishment delivered by cagemates. *Aggressive Behavior, 8*, 319–334.

Lefkowitz, M. M., Eron, L. D., Walder, L. O., & Huesmann, L. R. (1977). *Growing up to be violent: A longitudinal study of the development of aggression*. New York: Pergamon.

Linz, D. G., Donnerstein, E. & Penrod, S. (1988). Effects of long-term exposure to violent and sexually degrading depictions of women. *Journal of Personality and Social Psychology, 55*, 758–768.

Loeber, R., & Dishion, T. J. (1983). Early predictors of male delinquency: A review. *Psychological Bulletin, 94*, 68–94.

McFall, R. M. (1982). A review and reformulation of the concept of social skills. *Behavioral Assessment, 4*, 1–35.

Magnusson, D., Duner, A. & Zetterblom, G. (1975). *Adjustment: A longitudinal study*. Stockholm: Almqvist & Wiksell.

Malamuth, N. M. & Check, J. V. P. (1981). The effects of mass media exposure on acceptance of violence against women: A field experiment. *Journal of Research in Personality*, 15, 436–446.

Malamuth, N. M., & Donnerstein, E. (Eds.), (1984). *Pornography and sexual aggression*. Orlando, FL: Academic Press.

Matthews, R., Paulus, P., & Baron, R. A. (1979). Physical aggression after being crowded. *Journal of Nonverbal Behavior, 4*, 5–17.

Mednick, S. A., Gabrielli, W. F., & Hutchings, B. (1984). Genetic influences in criminal convictions: Evidence from an adoption cohort. *Science, 224*, 891–894.

Miczek, K. (1996). Technical Report, Tufts University.

Miller, L. S. (1991). Mothers' and children's attitudes about aggression. Dissertation Abstracts.

Moffitt, T. E. (1990). Juvenile delinquency and attention-deficit disorder: Developmental trajectories from age 3 to 15. *Child Development, 61*, 893–910.

Moyer, K. E. (1976). *The psychobiology of aggression.* New York: Harper & Row.

Nachson, I., & Denno, D. (1987). Violent behavior and cerebral hemisphere dysfunctions. In Mednick, S. A., Moffitt, T. E. & Stack, S. A. (Eds.), *The causes of crime: New biological approaches,* New York: Cambridge University Press.

Nasby, H., Hayden, B., & DePaulo, B.M. (1979). Attributional bias among aggressive boys to interpret unambiguous social stimuli as displays of hostility. *Journal of Abnormal Psychology,* 89, 459–468.

Neiss, R. (1988). Reconceptualizing arousal: Psychobiological states in motor performance. *Psychological Bulletin, 103*, 345–366.

Olweus, D. (1979). The stability of aggressive reaction patterns in males: A review. *Psychological Bulletin,86*, 852–875.

Olweus, D., Mattsson, A., Schalling,D., & Low, H. (1988). Circulating testosterone levels and aggression in adolescent males: A causal analysis. *Psychosomatic Medicine, 50*, 261–273.

Paik, H. and Comstock, G.A. (1994). The effects of television violence on antisocial behavior: A meta-analysis. *Communication Research, 21*, 516–546.

Patterson, G.R. (1995). Coercion - a basis for early age of onset for arrest. In J.McCord (Ed.), *Coercion and punishment in long-term perspective.* New York: Cambridge University Press.

Patterson, G.R., Capaldi, D.M. & Bank, L. (1991). An early starter model for predicting delinquency. In D.J. Pepler & K.H. Rubin (Eds.), *Systems and Development: Symposia on Child Psychology.* Hillsdale, NJ: Lawrence Erlbaum.

Pontius, A. A. (1984). Specific stimulus-evoked violent action in psychotic trigger reaction: A seizure-like imbalance between frontal lobe and limbic system? *Perceptual and Motor Skills 59*, 299–333.

Prentice-Dunn, S., & Rogers, R. (1983). Deindividuation in aggression. In R.G. Geen & E. Donnerstein (Eds.), *Aggression: Theoretical and empirical reviews* (Vol 2, pp. 155–171).

Raine, A. & Jones, F. (1987). Attention, autonomic arousal, and personality in behaviorally disordered children. *Journal of Abnormal Child Psychology, 15*, 583–599.

Raine, A. & Venables, P. H. (1981). Classical conditioning and socialization—A biosocial interaction? *Personality and Individual Differences, 2*, 273–283.

Raine, A., Venables, P. H., & Williams, M. (1990). Relationships between central and autonomic measures of arousal at age 15 and criminality at age 24 years. *Archives of General Psychiatry, 47*, 1003–1007.

Raine, A., Venables, P. H., & Williams, M. (in press). Better autonomic conditioning and faster electrodermal half-recovery times at age 15 as possible protective factors against crime at age 29 years. *Developmental Psychology.*

Rosenfeld, E., Huesmann, L.R., Eron, L.D., & Torney-Purta, J.V. (1982). Measuring patterns of fantasy behavior in children. *Journal of Personality and Social Psychology, 42,* 347–366.

Rubin, K.H., Bream, L.A. & Rose-Krasnor, L. (1991). Social problem solving and aggression in childhood. In D.J. Pepler & K.H. Rubin (Eds.), *The Development and Treatment of Childhood Aggression.* Hillsdale, NJ: Lawrence Erlbaum.

Rubin, K.H., Moller, L. & Emptage, A. (1987). The preschool behavior questionnaire: A useful index of behavior problems in elementary school-age children. *Canadian Journal of Behavioral Science, 19,* 86–100.

Rushton, J. P., Fulker, D. W., Neale, M. C., Nias, D. K. B. & Eysenck, H. J. (1986). Altruism and aggression: The heritability of individual differences. *Journal of Personality and Social Psychology, 50,* 1192–1198.

Schneider, D.J. (1991). Social cognition. *Annual Review of Psychology, 42,* 527–561.

Schneider, W., & Shiffrin, R. M. (1977). Controlled and automatic human information processing: I. Detection, search, and attention. *Psychological Review, 84,* 1–66.

Shiffrin, R.M. & Schneider, W. (1977). Controlled and automatic human information processing: II. Perceptual learning, automatic attending, and general theory. *Psychological Review, 84,* 127–190.

Shure, M. B., & Spivack, G. (1980). Interpersonal problem-solving as a mediator of behavioral adjustment in preschool and kindergarten children. *Journal of Applied Developmental Psychology, 1,* 45–57.

Slaby, R. G., & Guerra, N. G. (1988). Cognitive mediators of aggression in adolescent offenders: I. Assessment. *Developmental Psychology, 24,* 580–588.

Staub, E. (1996). Cultural-societal roots of violence: The examples of genocidal violence and of contemporary youth violence in the United States. *American Psychologist, 51,* 117–132.

Steinberg, M.D. & Dodge, K.A. (1983). Attributional bias in aggressive boys and girls. *Journal of Social and Clinical Psychology, 1,* 312–321.

Taylor, A.R. & Gabriel, S.W. (1989). Cooperative versus competitive game-playing strategies of peer accepted and peer rejected children in a goal conflict situation. Paper presented at the biennial meeting of the *Society for Research in Child Development,* Kansas City, MO.

Thomas, M. H. & Drabman, R.S. (1975). Toleration of real life aggression as a function of exposure to televised violence and age of subject. *Merrill-Palmer Quarterly*, *21*, 227–232.

Tulving, E., & Thompson, D.M. (1971). Retrieval processes in recognition memory. *Journal of Experimental Psychology*, *87*, 359–380.

Tyler, T. & Cook, F. L. (1984). The mass media and judgments of risk: Distinguishing impact on personal and societal level judgments. *Journal of Personality and Social Psychology*, *47*, 693–709.

Viemero, V. & Paajanen, S. (1992). The role of fantasies and dreams in the TV viewing-aggression relationship. *Aggressive Behavior*, *18(2)*, 109–116.

Waas, G.A. (1988). Social attributional biases of peer-rejected and aggressive children. *Child Development*, 59,969–992.

Welch, A. S., & Welch, B. L. (1971). Isolation, reactivity and aggression: Evidence for an involvement of brain catecholamines and serotonin (pp. 91–142). In B. E. Eleftheriou & J. P. Scott (Eds.), *The physiology of aggression and defeat*. New York: Plenum.

Widom, C. S. (1989). Does violence beget violence? A critical examination of the literature. *Psychological Bulletin*, *106*(1), 3–28.

Winter, L. & Uleman, J. S. (1984). When are social judgements made? Evidence for the spontaneousness of trait inference. *Journal of Personality and Social Psychology*, *4*, 904–917.

Zelli, A. & Huesmann, L. R. (1993). Accuracy of social information processing by those who are aggressive: The role of beliefs about a hostile world. Prevention Research Center, University of Illinois at Chicago.

Zelli, A., Huesmann, L. R. & Cervone, D. P. (1995). Social inferences in aggressive individuals: evidence for automatic processing in the expression of hostile biases. *Aggressive Behavior*, *21*, 405–418.

6

THE RELATIONSHIP BETWEEN LOW RESTING HEART RATE AND VIOLENCE

David P. Farrington

Institute of Criminology
Cambridge University
Cambridge, United Kingdom

1. HEART RATE AND VIOLENCE

According to Raine (1993, pp. 166–172), one of the most replicable findings in the literature is that antisocial and violent youth tend to have low resting heart rates. A possible explanation of this is that a low heart rate indicates fearlessness. Conversely, high heart rates, especially in infants and young children, are associated with anxiety, behavioral inhibition, and a fearful temperament (Kagan, 1994). Fearful people are unlikely to commit violent acts. Another possibility is that a low heart rate reflects autonomic under-arousal. Low autonomic arousal, like boredom, leads to sensation-seeking and risk-taking in an attempt to increase stimulation and arousal levels. People who take risks are more likely to be violent than others.

In the British National Survey of Health and Development, which is a prospective longitudinal survey of over 5,300 children born in England, Scotland, or Wales in March 1946, heart rate was measured at age 11. A low resting heart rate predicted convictions for violence and sexual offenses up to age 21; 81% of violent offenders and 67% of sexual offenders had below-average heart rates (Wadsworth, 1976, p.249). A low heart rate was especially characteristic of boys who had experienced a broken home before age 5, and among these boys it was not related to violence or sexual offenses. A low heart rate was significantly related to violence and sexual offenses among boys who came from unbroken homes.

In the Cambridge Study in Delinquent Development, which will be described below, resting heart rate was measured at age 18. The boys who were convicted of violence before age 25, and those who were chronic offenders (with six or more convictions), had significantly low heart rates (Farrington, 1987, p.55). Similarly, Raine *et al.* (1990, p.1005) found that boys with low resting heart rates at age 15 were more likely to be convicted up to age 24. Another result which is worth mentioning was obtained in the Montreal longitudinal-experimental study, which is a follow-up of over 100 children originally selected at age 5. Low heart rate at age 11 was significantly associated with teacher ratings of fighting and bullying at the same age (Kindlon *et al.*, 1995).

Biosocial Bases of Violence, edited by Raine *et al.*
Plenum Press, New York, 1997

Using more recent data collected in the Cambridge Study, this chapter addresses three main questions:

a. How is resting heart rate related to measures of violence?
b. Is heart rate related to violence independently of other (individual, family, and social) variables?
c. Are there interactions between heart rate and other variables in relationships with violence?

2. THE CAMBRIDGE STUDY IN DELINQUENT DEVELOPMENT

The Cambridge Study in Delinquent Development is a prospective longitudinal survey of the development of offending and antisocial behavior in 411 London males. At the time they were first contacted in 1961–62, these males were all living in a working-class inner-city area of South London. The sample was chosen by taking all the boys who were then aged 8–9 and on the registers of 6 state primary schools within a one-mile radius of a research office that had been established. Hence, the most common year of birth of these males was 1953. In nearly all cases (94%), their family breadwinner at that time (usually the father) had a working-class occupation (skilled, semi-skilled or unskilled manual worker). Most of the males were white (97%) and of British origin. The study was originally directed by Donald J. West, and it has been directed since 1982 by David P. Farrington, who has worked on it since 1969. It has been mainly funded by the Home Office. The major results can be found in four books (West, 1969, 1982; West & Farrington, 1973, 1977), in more than 60 papers listed by Farrington and West (1990), and in a summary paper by Farrington (1995). These publications should be consulted for more details about the variables included in this chapter.

The original aim of the Study was to describe the development of delinquent and criminal behavior in inner-city males, to investigate how far it could be predicted in advance, and to explain why juvenile delinquency began, why it did or did not continue into adult crime, and why adult crime usually ended as men reached their twenties. The main focus was on continuity or discontinuity in behavioral development, on the effects of life events on development, and on predicting future behavior. The Study was not designed to test any one particular theory about delinquency but to test many different hypotheses about the causes and correlates of offending.

One reason for casting the net wide at the start and measuring many different variables was the belief that theoretical fashions changed over time and that it was important to try to measure as many variables as possible in which future researchers might be interested. It is essential to measure as many theoretical constructs as possible in order to investigate the independent, additive, interactive, and sequential effects of different risk factors on key outcomes such as offending. Another reason for measuring a wide range of variables was the fact that long-term longitudinal surveys were very uncommon, and that the value of this particular one would be enhanced if it yielded information of use not only to delinquency researchers but also to those interested in alcohol and drug use, educational difficulties, poverty and poor housing, unemployment, sexual behavior, aggression, other social problems, and human development generally.

A major aim in this survey, therefore, was to measure as many factors as possible that were alleged to be causes or correlates of offending. The males were interviewed and tested in their schools when they were aged about 8, 10, and 14, by male or female psychologists. They were interviewed in a research office at about 16, 18 and 21, and in their homes at about 25 and 32, by young male social science graduates. At all ages except 21 and 25, the aim was to interview the whole sample, and it was always possible to trace and interview a high proportion: 389 out of 410 still alive at age 18 (95%) and 378 out of 403 still alive at age 32 (94%), for example. The tests in schools measured individual characteristics such as intelligence, attainment, personality, and psychomotor impulsivity, while information was collected in the interviews about such topics as living circumstances, employment histories, relationships with females, leisure activities such as drinking and fighting, and offending behavior.

In addition to interviews and tests with the males, interviews with their parents were carried out by female social workers who visited their homes. These took place about once a year from when the male was about 8 until when he was aged 14–15 and was in his last year of compulsory education. The primary informant was the mother, although many fathers were also seen. The parents provided details about such matters as family income, family size, their employment histories, their child-rearing practices (including attitudes, discipline, and parental disharmony), their degree of supervision of the boy, and his temporary or permanent separations from them. Also, when the boy was aged 12, the parents completed questionnaires about their child-rearing attitudes and about his leisure activities.

The teachers completed questionnaires when the males were aged about 8, 10, 12 and 14. These furnished data about their troublesome and aggressive school behavior, their restlessness and attention difficulties, their school attainments and their truancy. Ratings were also obtained from their peers when they were in the primary schools, about such topics as their daring, dishonesty, troublesomeness and popularity.

Searches were also carried out in the central Criminal Record Office (National Identification Service) in London to try to locate findings of guilt of the males, of their parents, of their brothers and sisters, and (in recent years) of their wives and cohabitees. The minimum age of criminal responsibility in England is 10. The Criminal Record Office contains records of all relatively serious offenses committed in Great Britain or Ireland, and also acts as a repository for records of minor juvenile offenses committed in London. In the case of 18 males who had emigrated outside Great Britain and Ireland by age 32, applications were made to search their criminal records in the 8 countries where they had settled, and searches were actually carried out in four countries. Since most males did not emigrate until their twenties, and since the emigrants had rarely been convicted in England, it is likely that the criminal records are quite complete.

The latest search of conviction records took place in the summer of 1994, when most of the males were aged 40. Between ages 10 and 16 inclusive (the years of juvenile delinquency in England at that time), 85 males (21%) were convicted. Altogether, up to age 40, 164 males (40%) were convicted (Farrington *et al.*, 1996). In this chapter, the recorded age of offending is the age at which an offense was committed, not the age on conviction. There can be delays of several months or even more than a year between offenses and convictions, making conviction ages different from offending ages. Offenses are defined as acts leading to convictions, and only offenses committed on different days were counted. Where two or more offenses were committed on the same day, only the most serious one was counted. Most court appearances arose from only one offending day; the 760 recorded offenses up to age 40 corresponded to 686 separate occasions of conviction.

Convictions were only counted if they were for offenses normally recorded in the Criminal Record Office, thereby excluding minor crimes such as common assault, traffic infractions and drunkenness. The most common offenses included were thefts, burglaries and unauthorized takings of vehicles, although there were also quite a few crimes of violence, vandalism, fraud and drug abuse. In order not to rely on official records for information about offending, self-reports of offending were obtained from the males at every age from 14 to 32.

The Cambridge Study in Delinquent Development has a unique combination of features:

    a. Eight personal interviews with the males have been completed over a period of 24 years, from age 8 to age 32;

    b. The main focus of interest is on offending, which has been studied from age 10 to age 40;

    c. The sample size of about 400 is large enough for many statistical analyses but small enough to permit detailed case histories of the boys and their families;

    d. There has been a very low attrition rate, since 94% of the males still alive provided information at age 32;

    e. Information has been obtained from multiple sources: the males, their parents, teachers, peers, and official records;

    f. Information has been obtained about a wide variety of theoretical constructs, including intelligence, personality, parental child-rearing methods, peer delinquency, school behavior, employment success, marital stability, and so on.

Very few biological variables were measured in the Cambridge Study. The most important was resting heart rate, which was only measured at age 18. Height and weight were measured at ages 8, 10, 14, and 18, and grip strength at ages 10 and 14. Obviously, it would have been better to measure heart rate during all interviews. How far the results might be generalized beyond White inner-city working-class British boys is an empirical question.

3. MEASURES OF VIOLENCE

Up to age 40, 65 of the males were convicted for violence: 16% of 404 at risk, excluding 7 not convicted for violence who died up to age 32. These males were convicted for 117 violent offenses: 81 involving physical assault or threatening behavior, 18 robberies, and 18 offenses of possessing an offensive weapon, which often arose from violent incidents. (Half of those convicted for possessing an offensive weapon had a conviction for some other violent offense.) All of these offenses were relatively serious, since minor violence (common assault) was not routinely recorded in the Criminal Record Office. About 40% of convicted offenders were violent, but only 15% of offenses were violent.

Between ages 10 and 18 inclusive, 26 males (6%) were convicted for violence, while 53 (13%) were convicted for violence between ages 19 and 40. The majority of those convicted for violence between ages 10 and 18 were convicted again between ages 19 and 40 (56% of 25 at risk, compared with only 10% of 378 not convicted for violence up to 18: Odds Ratio or OR = 11.1). The peak ages for violence convictions were between 16 and 21; 52 of the 117 offenses were committed between these ages.

The vast majority of males with convictions for violence (55 out of 65) also had convictions for nonviolent offenses. In the Cambridge Study, violent offenses seemed almost to be committed at random in criminal careers, and the characteristics of violent offenders were very similar to those of nonviolent but equally frequent offenders (Farrington, 1991b). These results were replicated in the Oregon Youth Study (Capaldi & Patterson, 1996). Offenders seemed to be versatile rather than specialized, and most persistent offenders sooner or later committed a violent offense. Table 1 shows how the probability of being a violent offender increased with the total number of offenses committed. For example, 82% of those who committed 12 or more offenses had committed at least one violent offense.

While there is a great deal of versatility in offending, there is also some degree of specialization in violence. For example, in a Copenhagen birth cohort study of over 28,000 males followed up to age 27, Moffitt *et al.* (1989, p.20) concluded that a first-time violent offender was 1.9 times as likely to commit a violent act among his future offenses as a first-time property offender. Specialization is also seen in transition matrices showing the probability of one type of offense being followed by another (Farrington *et al.*, 1988, p.478; Stander *et al.*, 1989, p. 320; Tracy *et al.*, 1990, p. 124). Hence, it should not be concluded that violent offenders are indistinguishable from frequent offenders.

In the Cambridge Study, the main self-reports of violence were collected at ages 14, 18, and 32 (Farrington, 1989b). The time periods covered were up to age 14, between ages 15 and 18, and between ages 27 and 32. The numbers interviewed were 405, 389, and 378, respectively. At age 14, the self-reported violence measure was a score of the number of violent acts admitted out of seven (group fighting, fighting strangers in the street, carrying a weapon, attacking someone, using a weapon in a fight, fighting to get away from a police officer, attacking a police officer: see West & Farrington, 1973, p. 170). The 61 males (15%) who admitted three or more of these acts were contrasted with the remaining 344.

At age 18, the self-reported violence measure was the sum of four items, each scored 1–4: number of fights in the previous three years, number of fights started, number of times carried a weapon in case it was needed in a fight, and number of times used a weapon in a fight. The 79 boys (20%) with the highest scores (10 or more out of 16) were contrasted with the remaining 310 (West & Farrington, 1977, p.82). At age 32, the self-reported violent males were the 61 (16% of 377) who had either (a) been involved in four or more fights in the previous five years or (b) hit their wife or cohabitee without her hitting them (Farrington, 1989a, pp. 230 and 235).

The final measure of violence was a teacher rating of the boy's aggressive behavior at age 14. This was the sum of six items, all scored 1–3: frequently disobedient, frequently difficult to discipline, unduly rough during playtime (recess), over-competitive, quarrelsome and aggressive, and unduly resentful to criticism or punishment. The 89 boys (24%) with the highest scores (12 or more out of 18) were contrasted with the remaining 289 rated (West & Farrington, 1973, p.181).

Table 1. Total number of offenses versus probability of committing a violent offense

Total number of offenses	Number of offenders	Per cent of offenders violent
1	49	18
2	34	29
3 - 4	25	44
5 - 7	20	55
8 - 11	14	71
12 or more	17	82
Total	159	41

Note: The figures exclude offenders who died and who were not violent

Table 2. Inter-relationships among violence measures (odds ratio)

Measure (% violent)	Convicted 19 - 40	SR 14	SR 18	SR 32	TR 14
Convicted 10-40 (16)	X	5.3	5.0	4.7	4.5
Convicted 10-18 (6)	11.1	6.9	4.5	3.3	4.7
Convicted 19-40 (13)	X	5.9	4.4	4.0	4.7
Self-report 14 (15)		X	5.9	2.7	4.5
Self-report 18 (20)			X	2.3	4.4
Self-report 32 (16)				X	2.5
Teacher report 14 (24)					X

Notes: All odds ratios significant at p =.05
SR = Self-Report
TR = Teacher Report

Table 2 shows the inter-relationships among the various violence measures, using the odds ratio as the main measure of strength of association. All were significantly inter-related, but the weakest associations were with self-reported violence at age 32.

4. HEART RATE VERSUS VIOLENCE

Resting heart rate was measured during the interview at age 18 (West & Farrington, 1977, p.71). The pulse rate was measured using a pulsimeter, which included a pressure cup which was fitted over the right middle finger. The pulse was made visible by a needle movement across a dial, and this was counted using a stop-watch. The readings were taken towards the end of the interview (which lasted two hours on average), with the youth sitting quietly resting his arm on a desk. The cumulative number of beats was recorded after 30 and 60 seconds. If the youth moved, the procedure was recommenced.

About half of the males (48% of 389) smoked during the interview at age 18. It is known that smoking causes an increase in the heart rate. In agreement with this, the males who smoked had a significantly higher heart rate than those who did not (74 beats per minute compared with 68; $F = 32.6$, df = 1, 385, p <.0001). In order to correct for the effect of smoking during the interview, the heart rate of those who smoked was reduced by 6 beats per minute.

Table 3 divides heart rate basically into 5-beat blocks and shows the relationships with the violence measures. Low heart rates were significantly associated with violence. For example, the percentage convicted for violence at any age was high among those with heart rates of 55 or less (21%), 56–60 (20%), and 61–65 (31%), and low among those with heart rates of 66–70 (14%), 71–75 (6%), 76–80 (15%), and 81 or more (8%). Dichotomizing heart rate into 66 or more versus 65 or less, 25% of low heart rate males were convicted for violence, compared with 11% of high heart rate males (OR = 2.8).

A low heart rate was significantly related both to convictions for violence at ages 10–18 (OR = 4.9) and to later convictions at ages 19–40 (OR = 2.4). A low heart rate was also significantly associated with self-reported violence at age 18 (OR = 2.5) and with teacher-reported violence at age 14 (OR = 2.3). It was not significantly associated with self-reported violence at ages 14 or 32.

How far does the correction for smoking during the interview contribute to the relationship between a low heart rate and violence? Smoking during the interview was only marginally related to convictions for violence (OR = 1.7), but was more strongly related to self-reported violence at age 18 (OR = 4.3) and teacher-reported violence at age 14 (OR =

Table 3. Heart rate versus violence measures

Heart rate (N)	% Conv.10-40	% Conv.10-18	% Conv.19-40	% SR14	% SR18	%SR32	% TR 14
55 or less (30)	21	7	17	17	23	11	36
56 - 60 (60)	20	10	17	15	30	20	29
61 - 65 (76)	31	16	23	18	30	21	33
66 - 70 (82)	14	5	11	20	16	15	20
71 - 75 (67)	6	0	6	6	10	14	15
76 - 80 (34)	15	0	15	18	15	9	24
81 or more (38)	8	5	5	13	16	18	8
p value (chi-squared)	.004	.004	.056	NS	.025	NS	.020
66 or more (221)	11	3	9	14	14	14	17
65 or less (166)	25	12	20	17	29	19	32
Odds Ratio	2.8	4.9	2.4	1.2*	2.5	1.4*	2.3

Notes: Odds Ratios significant except for *
Conv. = Conviction
SR = Self-Report
TR = Teacher Report
NS = Not Significant

2.5). The uncorrected heart rate was slightly more strongly related to convictions for violence than the corrected heart rate; without the correction for smoking, 23% of boys with heart rates of 70 or less were convicted, compared with 9% of those with heart rates of 71 or more (OR = 2.9). However, the uncorrected heart rate was less strongly, although still significantly, related to self-reported violence at age 18 (OR = 2.1) and teacher-reported violence at age 14 (OR = 1.8). Clearly, the significant relationships between low heart rate and violence were not created by the correction for smoking. The corrected heart rate will be used in the remainder of this chapter.

5. CORRELATES OF VIOLENCE

For the present analyses, explanatory variables were dichotomized, at the median where possible. Because most variables were originally classified into a small number of categories (typically 3 or 4), and because fine distinctions between categories could not be made very accurately, this dichotomizing did not usually involve a great loss of information. Variables were not included in the analysis if more than about 10% of the sample were missing (not known) on them, or if less than about 20% of the boys were in the risk category.

There are many advantages of using dichotomized variables. First, they permit a "risk factor" approach, and also make it possible to study the cumulative effects of several risk factors. Second, they make it easy to investigate interactions between variables (which are often neglected with continuous variables because of the difficulty of studying them). Hence, they encourage a focus on types of individuals as well as on variables, permitting the investigation of relationships within different subgroups of individuals. Information about individuals is more useful for interventions than information about variables. Third, they make it possible to compare all variables directly by equating sensitivity of measurement. Some variables are inherently dichotomous (e.g. a broken family). In many studies, it is difficult to know whether one variable is more closely related to an outcome than another because of differential sensitivity of measurement rather than differential causal influence.

Fourth, dichotomous data permit the use of the odds ratio as a measure of strength of relationship, which has many attractions (Fleiss, 1981). It is easily understandable as the increase in risk associated with a risk factor. It is a more realistic measure of predictive efficiency than the percentage of variance explained (Rosenthal & Rubin, 1982). For example, an odds ratio of 2, doubling the risk of delinquency, might correspond to a correlation of about .12, which translates into an apparently negligible 1.4% of the variance explained. The percentage of variance explained gives a misleading impression of weak relationships and low predictability. Unlike correlation-based measures, the odds ratio is independent of the frequency distribution or prevalence of variables and independent of the study design (retrospective or prospective). Nevertheless, because of the mathematical relationship between the logarithm of the odds ratio and the phi correlation (Agresti, 1990, p.54), conclusions about relative strengths of associations based on odds ratios and phi correlations are similar. Also, the partial odds ratio emerges in logistic regression analyses as the major measure of strength of effect while controlling for other variables.

A key issue is whether the association between low heart rate and violence holds up after controlling for other correlates of violence. Table 4 shows the relationship between important risk factors measured in the Cambridge Study, heart rate, and three violence measures significantly associated with heart rate: convictions for violence at any age (10–40), self-reported violence at age 18, and teacher-reported violence at age 14. Only risk factors that were not measures of antisocial behavior (i.e. excluding troublesome or dishonest school behavior, heavy drinking, drug use, etc.) were included as possible risk factors, so that all the predictors of violence could be explanatory. As Amdur (1989) pointed out, a common fault of researchers is to include measures of the outcome variable as predictors. If two variables basically measure the same underlying construct, using one as a predictor of the other will artifactually increase the percentage of variance explained, but this is of little use in the explanation of violence.

The risk factors were divided into five categories: personality, attainment, family, socioeconomic, and physical. Among the personality variables, low nervousness of the boy (rated by parents), daring or taking many risks (rated by parents and peers), a lack of concentration or restlessness at ages 8–10 and 12–14 (rated by teachers), extraversion (on the Eysenck Personality Inventory) and impulsiveness (based on attitude questionnaire items such as "I generally do and say things quickly without stopping to think"; see Farrington, 1991a) were all related to two or three violence measures but not to heart rate. Unpopularity (rated by peers) was marginally related to teacher-rated violence but not to heart rate. Impulsivity on psychomotor tests at age 8–10, nervousness rated by parents and teachers at age 14, extraversion at ages 10 and 14 (on the New Junior Maudsley Inventory), and neuroticism at ages 10, 14 and 16 were not related to any outcome measures. Hence, of 42 comparisons between 14 personality variables and three violence measures, 15 (36%) were statistically significant.

Among the attainment variables, low nonverbal IQ at ages 8–10 and 14 (on the Progressive Matrices), low verbal IQ at ages 8–10 and 14 (on the Mill Hill Vocabulary test), low junior school attainment, low school track or stream, leaving school at the earliest possible age of 15, and taking no examinations were all related to all violence measures. All 24 comparisons of 8 attainment variables and three violence measures were statistically significant. Only low nonverbal IQ at 8–10 was (marginally) associated with a low heart rate (OR = 1.6).

Among the family variables, a young mother (under age 20 at the time of her first child), poor parental supervision or monitoring, separation from a parent for reasons other than death or hospitalization, a convicted parent, and a poor relationship with a parent

Table 4. Correlates of violence and heart rate (odds ratios)

Risk Factor at Age (%)	Conv.	SR	TR	HR
Personality				
Low nervousness 8 (49)	—	1.7	1.7	—
Daring 8-10 (30)	2.8	3.9	3.5	—
Unpopular 8-10 (32)	—	—	1.7	—
Lacks concentration 8-10 (20)	2.8	—	2.3	—
Lacks concentration 12-14 (26)	3.9	3.1	10.2	—
Extraverted 16 (54)	2.0	2.9	—	—
Impulsive 18 (27)	—	2.0	2.0	—
Attainment				
Low nonverbal IQ 8-10 (49)	2.7	2.6	1.9	1.6
Low verbal IQ 8-10 (50)	3.0	1.8	2.3	—
Low junior attainment 11 (53)	2.4	1.8	2.4	—
Low school track 11 (58)	2.8	3.1	2.2	—
Low nonverbal IQ 14 (59)	2.8	2.0	1.7	—
Low verbal IQ 14 (51)	1.9	2.0	1.8	—
Left school 15 (61)	3.0	3.8	2.2	—
No exams taken 18 (51)	3.5	3.7	2.4	—
Family				
Young mother (22)	—	1.9	1.8	—
Poor child-rearing 8 (47)	2.1	—	—	—
Poor supervision 8 (19)	2.7	2.3	2.1	—
Separated from parent 10 (22)	2.5	1.9	2.1	1.8
Poor child-rearing 14 (29)	2.0	—	—	—
Convicted parent 18 (30)	3.4	2.3	2.1	1.6
Poor relation with parent 18 (22)	3.6	1.9	—	1.7
Socioeconomic				
Low income 8 (23)	2.6	2.6	—	—
Poor housing 8-10 (37)	1.9	2.1	—	—
Large family size 10 (40)	3.0	2.2	2.1	—
Large family size 14 (54)	2.2	3.1	1.9	—
Low status job 18 (55)	3.3	2.3	2.3	—
Unstable job record 18 (24)	3.7	3.3	2.8	2.5
Physical				
Low weight 10 (47)	—	—	2.0	—
Low weight 14 (50)	—	—	1.7	—
Low grip strength 14 (50)	—	2.0	—	—
Team games 16 (60)	2.3	—	—	2.2

Notes: All odds ratios shown are significant at $p = .05$, two-tailed. Nonsignificant odds ratios are not shown
Conv.= Conviction 10-40
SR= Self-report at 18
TR= Teacher report at 14
HR= Heart rate at 18

were all related to two or three violence measures. Poor child-rearing (erratic or harsh discipline or attitude) at ages 8 and 14 was only related to convictions for violence. Hence, 15 out of 21 comparisons (71%) of 7 family variables and three violence measures were statistically significant. Separation from a parent, a convicted parent, and a poor relationship with a parent were all significantly associated with a low heart rate, but only marginally (OR = 1.6 to 1.8).

Among the socioeconomic variables, low family income, poor housing, large family size at ages 10 and 14, a low status job, and an unstable job record were all related to two or three violence measures. However, low socioeconomic status of the family (based on the occupational prestige of the family breadwinner) at ages 8–10 and 14 was not related to any outcome measure. Hence, 16 out of 24 comparisons (67%) of 8 socioeconomic variables and three violence measures were statistically significant.

An unstable job record at age 18 was a combined variable based on the average number of jobs per year since leaving school, the average number of weeks unemployed per year since leaving school, the longest time in any one job, and the number of times fired from a job (West & Farrington, 1977, p.65). This variable had the strongest association with low heart rate at age 18. Of 92 boys with an unstable job record, 60% had a low heart rate, compared with 38% of 294 boys with a more stable job record (OR = 2.5).

Among the physical variables, low weight at ages 10 and 14 was related to teacher-reported violence, and low grip strength (measured by a dynamometer) at age 14 was related to self-reported violence. Hence, as also found for soccer violence (Farrington, 1994a, p. 235), and for other outcome measures (Farrington, 1993, p.15; Farrington & West, 1993, p. 508), the smaller boys tended to be more violent. Those who played team games (e.g. soccer, rugby, cricket) at age 16 tended to be convicted for violence and also tended to have low heart rates, possibly because of the effects of regular exercise. Of 227 boys who played team games, 51% had a low heart rate, compared with 31% of the remaining 156 (OR = 2.2). However, most physical variables were not significantly associated with any violence measures: weight at ages 8 and 18, height at ages 8, 10, 14, and 18, and grip strength at age 10. Only 4 out of 33 comparisons (12%) of 11 physical variables and three violence measures were statistically significant. In total, 74 out of 144 comparisons (51%) of 48 risk factors and three violence measures were significant, far in excess of chance expectation.

There are many theoretical reasons for predicting relationships between violence and personality, attainment, family, and socioeconomic variables, but these will not be reviewed in this chapter (see e.g. Farrington, 1997). There are few theoretical reasons for predicting relationships between violence and physical variables such as height, weight, and grip strength. These variables were included in the analyses primarily because it was expected that they would be related to resting heart rate, and hence might influence or interact with the relationship between heart rate and violence. For example, it is known that heart rate decreases with increasing body mass. However, heart rate was not significantly related to height, weight, or grip strength in this Study. As expected in light of the link between exercise and heart rate, boys who played team games had lower heart rates on average than the remainder. It was thought important to include a measure of exercise in the analysis.

Heart rate was not significantly related either to measures of nervousness or neuroticism or to measures of daring or impulsiveness. Hence, these analyses do not support either of the major theoretical links between heart rate and violence.

6. REGRESSION ANALYSES FOR VIOLENCE

Regression analyses were carried out to investigate whether low heart rate was related to violence measures after controlling for all other key risk factors, and especially after controlling for an unstable job record and playing team games. Two methods of investigating independent predictors were used, namely ordinary least squares (OLS) regression and logistic regression. In practice, the two methods tend to produce similar results with dichotomous data (e.g. Cleary & Angel, 1984), and indeed the results obtained by the two methods are mathematically related (Schlesselman, 1982, p.245). The main differences between them follow from the fact that missing cases can be deleted variable by variable in OLS regression (thereby using as much of the data as possible), whereas in logistic regression a case that is missing on any one variable has to be deleted from the whole analysis, causing a considerable loss of data.

The most reliable predictors are those identified by both methods. The OLS regression analyses were carried out first, beginning with all significant predictors. Because they were computationally more intensive, the logistic regression analyses were then carried out second with all independent predictors identified in the OLS regression together with the next few predictors that would have entered the OLS equation. One-tailed p-values were used because all the predictions were in a clear direction. In order to avoid multicollinearity problems, it was sometimes necessary to delete highly correlated variables. For example, low junior school attainment was deleted because its high correlation with low school track clearly caused problems in the regression analyses.

Table 5 shows the results of the regression analyses. The most important independent risk factors for convictions for violence at any age were poor concentration or restlessness at age 12–14, a poor relationship with a parent at age 18, a convicted parent at age 18, playing team games at age 16, low verbal IQ at age 8–10, and a low heart rate at age 18. These were the only variables that were independently related to convictions in both regression analyses. Low heart rate had a partial odds ratio (controlling for all other variables) of 1.9.

The most important independent risk factors for self-reported violence at age 18 were daring or taking many risks at age 8–10, an unstable job record at age 18, high extraversion at age 16, a low school stream or track at age 11, large family size at age 14, a low heart rate at age 18, a low grip strength at age 14, poor concentration at age 12–14, and leaving school at the earliest age of 15. Low heart rate had a partial odds ratio of 2.0.

The most important independent risk factors for teacher-reported violence at age 14 were poor concentration at age 12–14, daring at age 8–10, a low heart rate at age 18, and unpopularity at age 8–10. Low heart rate had a partial odds ratio of 2.2.

It was interesting that a low heart rate was related to violence independently of playing team games in all analyses. It might perhaps have been expected that playing (physically aggressive) team games would have been associated both with violence and with a low heart rate. Hence, conceivably, these relationships could have produced an artifactual link between a low heart rate and violence. Therefore, it was important to demonstrate that the link between low heart rate and violence held independently of playing team gates.

Only two risk factors were independently related to violence in all six analyses: poor concentration and a low heart rate. (Daring was independently important in five of the six analyses.) These results suggest that a low heart rate may be one of the most important explanatory factors for violence.

7. INTERACTION EFFECTS

Raine (1993, pp. 186–189) has discussed some of the possible interaction effects between psychophysiological and psychosocial factors in influencing crime and violence. The most obvious theory suggests a multiplying or amplifying effect, with a disproportionally high risk of violence among those in both a psychophysiological risk category and a psychosocial risk category (or conversely a disproportionally low risk of violence among those not in the risk category of either variable). A second theory is that a psychophysiological factor will predict violence more strongly among those from relatively fa-

Table 5. Regression analyses for violence

	OLS		Logistic		
	F change	p	LRCS change	p	Partial OR
Convictions					
Lacks concentration 12-14	25.41	.0001	22.32(1)	.0001	3.0
Poor relation parent 18	18.92	.0001	11.46(3)	.0004	3.1
Convicted parent 18	13.29	.0002	16.95(2)	.0001	2.3
Team games 16	9.81	.001	9.20(4)	.001	2.3
Unstable job record 18	6.12	.007	—	—	—
Daring 8-10	5.03	.013	—	—	—
Low verbal IQ 8-10	4.35	.019	2.88(7)	.045	1.8
Low heart rate 18	3.36	.034	3.94(6)	.024	1.9
Extraverted 16	3.50	.031	—	—	—
Low status job 18	—	—	5.22(5)	.011	1.9
Self-Report					
Daring 8-10	27.17	.0001	26.27(1)	.0001	2.5
Unstable job record 18	15.83	.0001	15.60(2)	.0001	2.2
Extraverted 16	10.59	.0007	13.90(3)	.0001	3.1
Low school track 11	12.29	.0003	13.36(4)	.0002	1.9
Large family size 14	8.47	.002	9.45(5)	.001	2.3
Low heart rate 18	5.36	.011	4.84(7)	.014	2.0
Low grip strength 14	4.14	.021	2.73(9)	.049	1.8
Lacks concentration 12-14	3.23	.037	3.70(8)	.027	1.7
Left school 15	2.85	.046	5.20(6)	.011	2.2
Teacher Report					
Lacks concentration 12-14	102.27	.0001	73.81(1)	.0001	8.6
Daring 10	13.87	.0001	11.25(2)	.0004	2.7
Low heart rate 18	8.49	.002	8.50(3)	.002	2.2
Impulsive 18	4.55	.017	—	—	—
Unpopular 8-10	3.38	.033	3.63(4)	.028	1.9
Low weight 10	3.73	.027	—	—	—
Low nervousness 8	2.85	.046	—	—	—
Unstable job record 18	—	—	3.23(5)	.036	1.9

OLS= Ordinary Least Squares
LRCS= Likelihood Ratio Chi-Squared
OR= Odds Ratio
p values one-tailed
Order of entry of variables in logistic regression shown in parentheses.

vored backgrounds (i.e. not in a psychosocial risk category). This is based on the idea that the influence of a psychophysiological factor can be seen more clearly where the "social push" towards violence is low. A third theory suggests that good psychophysiological functioning could act as a protective factor against a psychosocial risk factor, or conversely that a favorable psychosocial background could act as a protective factor against a psychophysiological risk factor.

All possible two-way interaction effects between heart rate and the 48 risk factors in predicting the three violence measures were investigated. The simplest way of calculating two-way interaction effects in OLS multiple regression is to carry out a two-way analysis of variance, since the two approaches are mathematically identical with dichotomous predictors. Since the calculation of interaction effects in logistic regression was more laborious, and since the significance of effects in the logistic analysis was almost invariably lower than in the anova analysis, logistic interaction effects were only calculated where the p value in anova was just over .10 or less (see also Farrington, 1994b).

Table 6 summarizes the results of the anova and logistic analyses, showing all p values of .10 or less. The number of significant (at p = .05) interaction effects was much greater than the chance expectation of 7 in the anova analyses (21) but not in the logistic analyses (11). There were few signs of significant interaction effects in predicting self-reported violence. Generally, a low heart rate interacted with attainment and physical factors in predicting convictions, and with family and socioeconomic factors in predicting teacher-rated violence. There were hardly any interactions between heart rate and personality factors.

Interestingly, there was no significant interaction between heart rate and playing team games. It might have been expected that playing aggressive team games would cause a decrease in heart rate and hence perhaps exaggerate the relationship between low heart rate and violence. However, a low heart rate was related to convictions for violence both among the players (27% as opposed to 14%; OR = 2.2) and among the nonplayers (19% as opposed to 7%; OR = 3.2).

The most replicable interaction effects (for all three violence measures) were between a low heart rate and large family size at age 10, a poor relationship with a parent at age 18, and low grip strength at age 10 (but see later). Interaction effects between a low heart rate and low nonverbal IQ at age 14, no examinations taken by age 18, a convicted parent at age 18, a low status job at age 18, and an unstable job record at age 18 were replicated over two violence measures.

Table 6. Significance of interactions with heart rate

Risk Factor at Age	Conv. A	Conv. L	SR A	SR L	TR A	TR L
Daring 8-10	—	—	—	—	.051	—
Lacks concentration 12-14	—	—	—	—	.033	—
Low nonverbal IQ 8-10	.012	—	—	—	—	—
Low verbal IQ 8-10	.036	—	—	—	—	—
Low school track 11	.019	—	—	—	—	—
Low nonverbal IQ 14	.001	.015	.089	—	—	—
Left school 15	.013	—	—	—	—	—
No exams taken 18	.007	—	—	—	.004	.027
Young mother	—	—	—	—	.003	.019
Separated from parent 10	—	—	—	—	.005	.022
Convicted parent 18	.098	—	—	—	.036	—
Poor relation with parent 18	.017	—	.102	—	.031	.038
Large family size 10	.0001	.008	.037	—	.026	—
Low SES family 8-10	—	—	—	—	.024	.018
Low status job 18	.087	—	—	—	.075	—
Unstable job record 18	—	—	.053	—	.029	—
Low weight 8	.054	.036	—	—	—	—
Low weight 10	.024	.036	—	—	—	—
Low weight 14	.028	.061	—	—	—	—
Low height 10	—	—	—	—	.062	—
Low grip strength 10	.071	.040	.068	—	.041	.032

p values (two-tailed) are shown.
Conv. = Conviction 10-40
SR = Self-report at 18
TR = Teacher report at 14
A = Anova.
L = Logistic.

Table 7 shows the interaction effects in more detail. This table shows all significant interaction effects in logistic analyses and all replicated interaction effects over two or three violence measures. With only one exception, all the interactions were amplifying effects, with the highest probability of violence when a low heart rate co-occurred with a risk category of another variable (and a stronger relationship between heart rate and violence in the high risk category than in the low risk category). The exception was low grip strength, which showed an amplifying effect for convictions and self-reports but an anomalous effect for teacher-reported violence: heart rate had a bigger effect in the absence of low grip strength than in the presence of low grip strength. In light of these inconsistent results (which may be due to chance), and in light of the uncertain theoretical basis for any relationship between low grip strength and violence, low grip strength will not be considered further.

In most cases, amplifying effects coincided with protective effects. For example, large family size at age 10 was not related to convictions for violence among those with high heart rates (12% as opposed to 10%), and conversely heart rate was not related to convictions for violence among those from smaller families (11% as opposed to 10%). Symmetrical protective effects of this kind were not always seen. For example, a poor re-

Table 7. Major interaction effects with heart rate

Risk factor at age	% Violent			
	Low risk		High risk	
	High HR	Low HR	High HR	Low HR
Convictions				
Low nonverbal IQ 14	10	8	11	35
No exams taken 18	7	11	14	38
Convicted parent 18	7	16	20	42
Poor relation with parent 18	9	17	18	47
Large family size 10	10	11	12	42
Low status job 18	7	13	16	36
Low weight 8	15	22	7	28
Low weight 10	13	19	8	31
Low weight 14	11	18	8	31
Low grip strength 10	15	23	6	28
Self-reports				
Low nonverbal IQ 14	12	17	16	36
Poor relation with parent 18	14	24	15	41
Large family size 10	13	20	16	41
Unstable job record 18	13	20	22	47
Low grip strength 10	14	22	14	37
Teacher reports				
No exams taken 18	15	27	19	46
Young mother	17	25	17	58
Separated from parent 10	17	24	14	51
Convicted parent 18	16	24	18	46
Poor relation with parent 18	18	28	10	43
Large family size 10	15	22	19	46
Low SES family 8-10	22	28	11	37
Low status job 18	14	19	21	43
Unstable job record 18	16	23	21	51
Low grip strength 10	11	37	21	28

Note: HR = Heart rate

lationship with a parent was not related to self-reported violence among those with high heart rates (15% as opposed to 14%), but heart rate was related to self-reported violence among those with a good relationship with parents (24% as opposed to 14%; OR = 2.0).

Few interaction effects were significantly related to violence measures in logistic regression analyses over and above all the main effects. The only heart rate interaction effect that predicted convictions for violence over and above all main effects was with low weight at age 8 (LRCS change = 6.44, p = .011). Heart rate was more strongly related to convictions for violence among boys with low weight: OR = 5.5, as opposed to 1.6 (NS) among the heavier boys. No interaction effects significantly predicted self-reported violence.

Two heart rate interaction effects were significantly related to teacher-reported violence in logistic regression analyses over and above all the main effects: with a low SES family (LRCS = 4.70, p = .030) and with a poor relationship with a parent (LRCS = 4.16, p = .042). Heart rate was more strongly related to teacher-reported violence among low SES boys (OR = 4.7, as opposed to 1.4) and among boys who got on badly with a parent (OR = 6.6, as opposed to 1.7).

8. CONCLUSIONS

Resting heart rate is one of the few biological variables that can easily be measured in community samples. In the Cambridge Study in Delinquent Development, a low heart rate at age 18 was significantly related to convictions for violence up to age 40, and also to self-reported violence at age 18 and teacher-reported violence at age 14. It was not significantly related to self-reported violence at ages 14 or 32. It was probably not related to self-reported violence at age 32 because of the long time interval between measuring heart rate and measuring violence; a large number of the violence convictions, for example, occurred around age 18. However, the lack of a relationship between heart rate and self-reported violence at age 14 is surprising.

A low resting heart rate was significantly associated with convictions for violence, self-reported violence at age 18, and teacher-reported violence at age 14 independently of all other measured risk factors (personality, attainment, family, socioeconomic, and physical). It was strongly related to only two risk factors: an unstable job record at age 18 and playing team games at age 16. The lack of any relationship between heart rate and measures of nervousness or risk-taking casts doubt on explanations of the results based on fearlessness or low arousal. In six regression analyses, the most important independent risk factors for violence were poor concentration or restlessness, a low heart rate and high daring.

Several important interaction effects involving heart rate were detected. In all cases, the combination of a low heart rate and a risk category of another variable was associated with a disproportionally high probability of violence. The most replicable interaction effects were with large family size and a poor relationship with a parent. Interaction effects that predicted violence independently of main effects were with a low SES family and low weight. There were indications that a high heart rate acted as a protective factor against the effects of these risk factors.

Clearly, a low heart rate is an important correlate of violence. The challenge to future researchers is to measure heart rate and violence more frequently, to specify changes and developmental sequences in both, and to explain precisely why heart rate and violence are associated.

REFERENCES

Agresti, A. (1990) *Categorical Data Analysis*. New York: Wiley

Amdur, R. L. (1989) Testing causal models of delinquency: A methodological critique. *Criminal Justice and Behavior*, 16, 35–62.

Capaldi, D. M. & Patterson, G. R. (1996) Can violent offenders be distinguished from frequent offenders? Prediction from childhood to adolescence. *Journal of Research in Crime and Delinquency*, 33, 206–231.

Cleary, P. D. & Angel, R. (1984) The analysis of relationships involving dichotomous dependent variables. *Journal of Health and Social Behavior*, 25, 334–348.

Farrington, D. P. (1987) Implications of biological findings for criminological research. In S. A. Mednick, T. E. Moffitt, & S. A. Stack (Eds.) *The Causes of Crime: New Biological Approaches* (pp. 42–64). Cambridge: Cambridge University Press.

Farrington, D. P. (1989a) Later adult life outcomes of offenders and non-offenders. In M. Brambring, F. Losel, & H. Skowronek (Eds.) *Children at Risk: Assessment, Longitudinal Research and Intervention* (pp. 220–244). Berlin: De Gruyter.

Farrington, D. P. (1989b) Self-reported and official offending from adolescence to adulthood. In M. W. Klein (Ed.) *Cross-National Research in Self-reported Crime and Delinquency* (pp. 399–423). Dordrecht, Netherlands: Kluwer.

Farrington, D. P. (1991a) Antisocial personality from childhood to adulthood. *The Psychologist*, 4, 389–394.

Farrington, D. P. (1991b) Childhood aggression and adult violence: Early precursors and later life outcomes. In D. J. Pepler & K. H. Rubin (Eds.) *The Development and Treatment of Childhood Aggression* (pp. 5–29). Hillsdale, N.J.: Lawrence Erlbaum.

Farrington, D. P. (1993) Childhood origins of teenage antisocial behavior and adult social dysfunction. *Journal of the Royal Society of Medicine*, 86, 13–17.

Farrington, D. P. (1994a) Childhood, adolescent, and adult features of violent males. In L. R. Huesmann (Ed.) *Aggressive Behavior: Current Perspectives* (pp. 215–240). New York: Plenum.

Farrington, D. P. (1994b) Interactions between individual and contextual factors in the development of offending. In R. K. Silbereisen & E. Todt (Eds.) *Adolescence in Context: The Interplay of Family, School, Peers and Work in Adjustment* (pp. 366–389). New York: Springer-Verlag.

Farrington, D. P. (1995) The development of offending and antisocial behavior from childhood: Key findings from the Cambridge Study in Delinquent Development. *Journal of Child Psychology and Psychiatry*, 36, 929–964.

Farrington, D. P. (1997) Predictors, causes, and correlates of male youth violence. In M. Tonry & M. H. Moore (Eds.) *Youth Violence* (Crime and Justice, vol. 24). Chicago: University of Chicago Press, in press.

Farrington, D. P., Barnes, G. & Lambert, S. (1996) The concentration of offending in families. *Legal and Criminological Psychology*, 1, 47–63.

Farrington, D. P., Snyder, H. N., & Finnegan, T. A. (1988) Specialization in juvenile court careers. *Criminology*, 26, 461–487.

Farrington, D. P. & West, D. J. (1990) The Cambridge study in delinquent development: A long-term follow-up of 411 London males. In H-J. Kerner & G. Kaiser (Eds.) *Kriminalitat: Personlichkeit, Lebensgeschichte und Verhalten* (Criminality: Personality, Behavior and Life History) (pp. 115–138). Berlin, Germany: Springer-Verlag.

Farrington, D. P. & West, D. J. (1993) Criminal, penal, and life histories of chronic offenders: Risk and protective factors and early identification. *Criminal Behavior and Mental Health*, 3, 492–523.

Fleiss, J. L. (1981) *Statistical Methods for Rates and Proportions* (2nd ed.). New York: Wiley.

Kagan, J. (1994) *Galen's Prophecy: Temperament in Human Nature*. New York: Basic Books.

Kindlon, D. J., Tremblay, R. E., Mezzacappa, E., Earls, F., Laurent, D. & Schaal, B. (1995) Longitudinal patterns of heart rate and fighting behavior in 9- through 12-year-old boys. *Journal of the American Academy of Child and Adolescent Psychiatry*, 34, 371–377.

Moffitt, T. E., Mednick, S. A. & Gabrielli, W. F. (1989) Predicting careers of criminal violence: Descriptive data and predispositional factors. In D. A. Brizer & M. Crowner (Eds.) *Current Approaches to the Prediction of Violence* (pp. 13–34). Washington, D.C.: American Psychiatric Press.

Raine, A. (1993) *The Psychopathology of Crime: Criminal Behavior as a Clinical Disorder*. San Diego: Academic Press.

Raine, A., Venables, P. H., & Williams, M. (1990) Relationships between central and autonomic measures of arousal at age 15 years and criminality at age 24 years. *Archives of General Psychiatry* 47, 1003–1007.

Rosenthal, R. & Rubin, D. B. (1982) A simple, general purpose display of magnitude of experimental effect. *Journal of Educational Psychology*, 74, 166–169.

Schlesselman, J. J. (1982) *Case-Control Studies*. New York: Oxford University Press.

Stander, J., Farrington, D. P., Hill, G., & Altham, P. M. E. (1989) Markov chain analysis and specialization in criminal careers. *British Journal of Criminology*, 29, 317–335.

Tracy, P. E., Wolfgang, M. E. & Figlio, R. M. (1990) *Delinquency Careers in Two Birth Cohorts*. New York: Plenum.

Wadsworth. M. E. J. (1976) Delinquency, pulse rates, and early emotional deprivation. *British Journal of Criminology*, 16, 245–256.

West, D. J. (1969) *Present Conduct and Future Delinquency*. London: Heinemann.

West. D. J. (1982) *Delinquency: Its Roots, Careers and Prospects*. London: Heinemann.

West, D. J. & Farrington, D. P. (1973) *Who Becomes Delinquent?* London: Heinemann.

West, D. J. & Farrington, D. P. (1977) *The Delinquent Way of Life*. London: Heinemann.

7

BIOSOCIAL BASES OF AGGRESSIVE BEHAVIOR IN CHILDHOOD

Resting Heart Rate, Skin Conductance Orienting, and Physique

Adrian Raine, Chandra Reynolds, Peter H. Venables, and Sarnoff A. Mednick

Department of Psychology
University of Southern California
Los Angeles, California 90089-1061

INTRODUCTION

This chapter aims to follow up some of the biosocial issues outlined in Chapter 1, and attempt an initial answer to some of the questions posed by the biosocial model developed in that chapter. In doing so, new results will be presented from the Mauritius study, a longitudinal psychophysiological study of child and adult psychopathology conducted on the island of Mauritius. A specific focus will be placed on two main risk factors for antisocial behavior which have been pursued by the first author over the past two decades, namely, low resting heart rate, and reduced electrodermal orienting. The longitudinal approach will be emphasized in this chapter because it is felt that clearer answers to the questions posed in chapter 1 will emerge from prospective studies. In particular, this chapter is concerned with links between psychophysiological and social measures taken at age 3, and aggressive behavior measured at age 11 years in a large sample of 1,795 males and females, Indian and Creole participants. Interactions with temperament will also be explored because this construct has been linked with both autonomic activity (Scarpa et al., 1996; Kagan, 1988; Kagan et al. 1989) and aggression (Caspi et al. 1995; Caspi and Silva, in press).

Some of the key questions to be posed (and possibly answered) are as follows:

1. Do heart rate and skin conductance interact with social and demographic factors in predisposing to aggression?
2. What is the nature of the biosocial effects? Do they indicate additive, multiplicative, or crossover effects?
3. Are any such interactions specific to aggression, or do they also hold for non-aggressive, antisocial behavior?

Biosocial Bases of Violence, edited by Raine *et al.*
Plenum Press, New York, 1997

4. Do effects interact with gender and ethnicity, or are biosocial effects common to all groups?
5. What psychosocial factors may help protect against aggression development in those with the biological risk factor of low resting heart rate?
6. Is the low resting heart rate - aggression link independent of psychosocial, environmental, and biological influences on aggression?

A brief review of autonomic activity and aggression will first be given to orient the reader to this field of enquiry. New findings from the Mauritius study will then be presented which address the above questions. Finally, results will be related back to the biosocial model of violence described in Chapter 1, and implications and directions for future research drawn.

BRIEF REVIEW OF AUTONOMIC ACTIVITY AND AGGRESSION

New analyses to be reported below will focus on resting heart rate and skin conductance orienting. Consequently, this section provides a brief rationale for the focus on these two variables. More detailed reviews and discussion of other psychophysiological variables may be found in Fowles (1993) and Raine (1993).

Low Resting Heart Rate

Perhaps the best replicated correlate of antisocial and aggressive behavior in childhood and adolescence is low resting heart rate. One review indicated that all 14 studies on non-institutionalized conduct disordered, delinquent, and antisocial children and adolescents showed significant effects in the predicted direction of lower resting heart rates in antisocial children (Raine, 1993). A more recent meta-analysis (Raine et al., 1996) of 21 studies yielding 25 independent samples showed significant effect sizes in 22 samples, mixed findings in 2 samples, non-significant effects in the predicted direction in one sample, and a significant effect in the opposite direction in one sample. The average effect size was 0.53. Zahn and Kruesi(1993) failed to find lower HRL in children with oppositional defiant disorder, but commented that this could be due to the selected nature of this clinic-referred sample who may have anxious parents who pass on high heart rates to their offspring.

Particularly striking are prospective studies which find a link between HRL and antisocial behavior. Wadsworth (1976) found that lower resting heart rate in unselected 11 year old schoolboys predicted delinquency measured from ages 8–21 years. The very lowest HRLs were found in those who committed violent criminal offenses as adults. Similarly, Farrington (1987) found that resting HR measured at age 18–19 years in non-institutionalized males predicted to violent criminal offending at age 25. Findings from a third prospective study are reported in the next section.

Low resting HR also characterizes young children with a disinhibited temperament. Low HR and increased vagal tone measured at age 4 was correlated with disinhibited behavior at age 5.5 years, while those with relatively low fetal heart rates had lower levels of motoric activity and crying at age 4 months (Kagan, 1989; Snidman et al. 1991). A confirmation and extension of the link between heart rate and disinhibition has been provided in the Mauritius longitudinal study (, Raine, Venables and Mednick, 1996). Lower heart rate and skin conductance measured at age 3 was associated with disinhibition at age 3. Importantly, group differences in heart rate could not be attributable to group differences in height, weight, or respiratory complaints.

Skin Conductance Orienting Deficits in Antisocials

Key findings from nine studies which have assessed SC orienting to neutral stimuli in antisocial groups has been reviewed by Raine (1993), while McBurnett and Lahey (1994) also review SC orienting specifically in conduct disordered and delinquent children. Five out of nine studies find evidence for an orienting deficit as indicated by reduced frequency of SCRs to orienting stimuli. Frequency measures of SC orienting appear to produce stronger findings in these studies, perhaps because frequency measures tend to be more reliable than amplitude measures (Raine and Venables, 1984b); this in turn may be because amplitude is more affected by non-ANS factors such as the number and size of sweat glands (Venables and Christie, 1973).

A nine-year prospective study of crime by Raine et al. (1990a) showed that low resting HRL measured at age 15 years in normal unselected schoolboys predicted criminal behavior at age 24 years (Raine et al. 1990a). Similarly, a reduced frequency of SC orienting responses at age 15 years predicted to crime at age 24 years (Raine et al. 1990b). Group differences in social class, academic ability, and area of residence were not found to mediate the link between under-arousal and antisocial behavior.

There is some suggestive evidence that reduced orienting and lower HRL may be more characteristic of violent criminal than non-violent criminal offenders. Five of the 17 who became criminal in the above study had convictions for violence (assault or wounding). These were compared to the other 12 non-violent offenders and 84 controls on SC orienting and resting HR. Results of these analyses are shown in Figure 1. Effect sizes between each of the two offender groups relative to normal controls are shown above the bars.

The violent group had the lowest heart rates of all, with the non-violent offender group being intermediate between violent offenders and controls. The averaged effect size of 1.0 for the violent group is slightly higher than that of 0.85 obtained for previous studies on HR reviewed above. Despite the lack of statistical power due to small N size, HR level for the violent offender group was significantly lower than the normal controls at the end of the rest period ($p < .05$, two-tailed), with a trend ($p < .08$) for the start of the rest period. Conversely, non-violent offender versus control comparisons were non-significant. Similar analyses for SC orienting did not produce statistically significant differences between violent offenders and controls, but in all cases the violent offenders had the lowest values of all, with the non-violent offenders intermediate between violent and control groups.

Autonomic Protective Factors against crime Development

All psychophysiological research to date has attempted to ask the question "what psychophysiological factors predispose to crime?", and consequently has focused exclusively on risk factors for crime development. A potentially more important question to be posed, however, is "what factors protect a child predisposed to crime from becoming criminal?" Understanding protective factors against crime development is of critical conceptual importance because it can more directly inform intervention and prevention of antisocial behavior.

HEART RATE

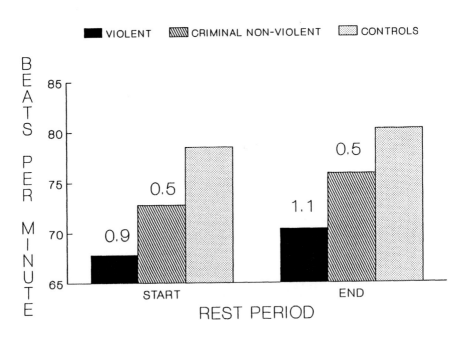

ORIENTING

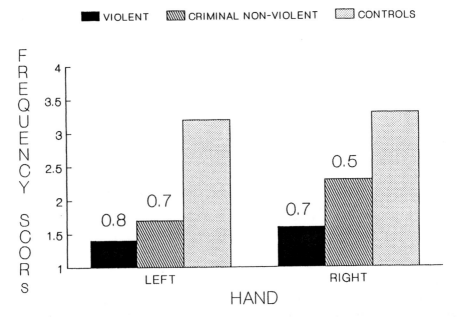

Figure 1. Resting heart rate and skin conductance orienting in violent criminals, non-violent criminals, and non-criminal controls.

The principle finding to emerge from the first work in this area is that *higher* autonomic activity during adolescence may act as a protective factor against crime development. Raine, Venables and Williams (1995, 1996) report on a 14 year prospective study in which autonomic and CNS measures of arousal, orienting, and conditioning were taken in 101 unselected 15-year-old male schoolchildren. Of these, seventeen adolescent antisocials who desisted from adult crime (Desistors) were matched on adolescent antisocial behavior and demographic variables with 17 adolescent antisocials who had became criminal by age 29 (Criminals), and 17 non-antisocial, non-criminals (Controls). Desistors had significantly higher heart rate levels, and higher SC orienting, higher SC conditioned orienting responses relative to Criminals. Findings suggest that individuals predisposed to adult crime by virtue of showing antisocial behavior in adolescence may be protected from crime by heightened levels of autonomic arousal and reactivity.

Interactions between Social and Psychophysiological Processes

Few people have investigated the interaction between psychophysiological and social factors, but the evidence available to date suggests that stronger and more consistent psychophysiology - antisocial behavior relationships may be found in those individuals who come from benign home backgrounds (Raine, 1988; Raine and Mednick, 1989). Evidence for this view is derived from both studies of resting HR and studies of SC activity.

With respect to resting HR, although HRL is generally lower in antisocials, it is a particularly strong characteristic of antisocials from higher social classes and those from intact homes. For example, Raine and Venables (1984a) found lower HRL to be associated with antisocial behavior in adolescents from high social classes, but not in those from lower social classes. Similarly, Wadsworth (1976) found lower HRL at age 11 years predicted criminal behavior in adulthood in those from intact homes, but not in those from broken homes. Maliphant et al. (1990) found particularly strong links between lower resting HRL and antisocial behavior in girls from privileged middle class backgrounds attending private schools in England.

Regarding skin conductance orienting activity, a similar pattern of findings have emerged. A smaller SC conditioned orienting response characterizes antisocial adolescents from high social classes (Raine and Venables, 1981). The reverse effect was observed in those from low social classes where antisocial individuals were characterized by higher conditioned orienting responses, indicating a crossover biosocial interaction (Raine et al. this volume). Schizoid criminals from intact home environments show reduced SC orienting, whereas schizoid criminals from broken homes do not (Raine, 1987). Similarly, criminals *without* a childhood history broken by parental absence and disharmony show poor SC conditioning (Hemming, 1981), while "privileged" (high socio-economic status) offenders who commit crimes of evasion show reduced SC arousal and reactivity (Buikhuisen et al., 1985). Again, reduced SC arousal and reactivity in these studies is particularly found in antisocials from benign home backgrounds.

One explanation for this pattern of results is that where the "social push" towards antisocial behavior is lower (high socio-economic status, intact homes), psychophysiological determinants of antisocial behavior assume greater importance (Raine and Venables, 1981). This will be referred to below as the "benign homes effect". Conversely, social causes of criminal behavior may be more important explanations of antisociality in those exposed to adverse early home conditions. The crossover effect observed in Raine and Venables (1981) adds to this by suggesting that in those from adverse rearing conditions, better information processing and arousal may actually facilitate antisocial behavior.

Eysenck (1977) argued that good conditioners from criminogenic homes may become conditioned into an antisocial way of life. This will be referred to below as the "antisocialization effect".

New Findings

The above review indicates that reduced resting HR and reduced SC orienting may be psychophysiological risk factors for antisocial and violent behavior. This section presents new findings which aim to address the questions posed at the beginning of the chapter. The central question is whether these two risk factors interact with social factors in predisposing to aggression, and what the nature of this interaction is.

New findings are derived from initial analyses of the Mauritius longitudinal study. In assessing these findings, it is important to bear in mind that they constitute provisional and initial analyses. In brief, subjects consist of 1,795 male and female children who were psychophysiologically tested at age 3 years (Venables 1978). 51% are male and 49% are female. The two main ethnic groups consist of Indian (69%) and Creole (29%). Although there are a few children of Chinese and European origin, these were excluded in analyses reported below which assessed for ethnic effects.

All 1,795 subjects were assessed on SC orienting and resting heart rate at age 3 years (see Venables, 1978 for full details). Inhibited versus disinhibited temperament was assessed at age 3 years (see Scarpa et al. in press for full details). 1,213 of the 1,795 were assessed by teachers at age 11 years on the Achenbach scale (Achenbach and Edelbrock, 1979). Analyses below focus on two key subscales of this checklist, Aggression and Delinquency. Two limitations of the Achenbach measures are that (a) scales for males and females are somewhat different (b) the "aggression" scales contains many items with no aggression component, while the "delinquency" scale contains aggression items. To provide purer indices of Aggression and Non-Aggressive Delinquency subscales common to both sexes, two new scales were constructed. Coefficient alpha for the Aggression scale were 0.72 (boys) and 0.72 (girls), with slightly lower reliabilities found for Delinquency (0.64 for boys, 0.68 for girls).

The key psychosocial variable to be considered in the analyses below was socioeconomic status (SES). This was taken at age 3, and consisted of a factor score based on a factor analysis of a variety of social variables which produced one major factor. Variables loading on this first principal component included number of years of education of the parents, parental occupation, additional educational training of the parents, appearance of the home, number of rooms per person, and number of rooms in the house. Data were available on 1,321 of the subjects.

Findings are divided into three sections: (a) biosocial analyses where the psychophysiological measures act as the dependent variable, (b) biosocial analyses where aggression and delinquency is the dependent variable, and © analyses of moderators and confounds in the HRL-aggression relationship.

Psychophysiology as Dependent Variables

The following analyses follow the more traditional design of using psychophysiology as the dependent variable, with aggression as the independent variable. Upper and lower quartile splits were used to divide subjects into high and low Aggression, Delinquency, total Antisocial, and Disinhibited groups. High and Low SES groups were then formed on the basis of a median split.

SC Orienting. The clearest biosocial effect was observed for SC orienting. A three-way ANOVA (Aggression x SES x Inhibition) on frequency of SC orienting responses produced a significant Aggression x SES interaction, $F(1,183) = 6.5$, $p < .01$. The interaction is illustrated in Figure 2. It can be seen that in the high SES group, Aggressives tend to give fewer orienting responses than Non-Aggressives, whereas this effect is reversed in the lower SES group where Aggressives showed greater orienting. As such, the effect is in the same direction as for the conditioning x SES interaction observed by Raine and Venables (1981). While the interaction is significant, the between-group comparison was only significant for the low SES group ($t = 2.1$, $df = 94$, $p < .04$).

This biosocial effect was not specific to aggressive behavior as the same interaction effect was observed for non-aggressive Delinquency ($F(1,125) = 4.3$, $p < .04$), with the same pattern of results emerging. No interactions were observed with either sex or ethnicity. The interaction effect was also specific to orienting, and was not observed for SC and HRL arousal measures. Furthermore, there were no three-way interactions involving Disinhibition.

Resting HRL. The above biosocial interaction was not observed for resting HRL. Nevertheless, there was a significant Aggression x Disinhibition interaction ($F(1,183) = 4.2$, $p < .04$). Results are illustrated in Figure 3. It can be seen that the main effect for Disinhibition indicates that Disinhibited children have lower HRLs than Inhibited children, while the trend for Aggression indicates that aggressives have lower HRLs than Non-Aggressives. The interaction effect indicates that amongst the Inhibited children, Aggressives had low HRL, whereas this effect is lost amongst the Disinhibited children. A trend towards a similar interaction was also observed for Delinquency ($F(1,125) = 2.9$, $p < .09$). However, no interactions were observed with either sex or ethnicity.

Interpretation. The biosocial effect for orienting is of some significance in that it mirrors the effect observed by Raine and Venables (1981) for SC conditioning. It takes

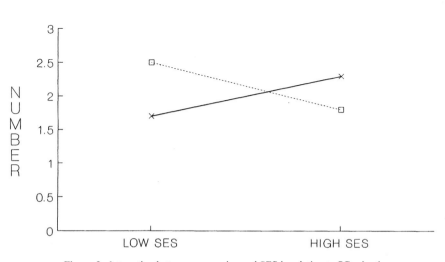

Figure 2. Interaction between aggression and SES in relation to SC orienting.

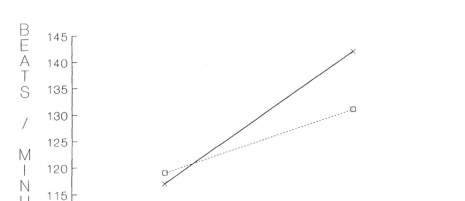

Figure 3. Interaction between aggression and temperament in relation to HRL.

that previous result four steps further by showing (a) orienting and SES prospectively collected at age 3 years predicts to aggression at age 11 years (b) the effects for males also hold for females ⊚ the effects generalize across ethnic groups (d) the effects appear to apply to delinquency as well as aggression.

These orienting results could be interpreted in the way that Raine and Venables (1981) interpreted their conditioning data along lines first suggested by Eysenck (1977), i.e, in terms of the benign homes effect (poor conditioning characterizes antisocials from benign home backgrounds) and the antisocialization effect (good orienting characterizes antisocials from poor home backgrounds). SC orienting is a sensitive measure of information processing (Dawson, 1990; Dawson et al. 1991). Poor orienting is thought to reflect a fundamental deficit in the ability to allocate attentional resources to environmental events. As such poor orienting in antisocials from benign homes may reflect an attentional deficit which retards classical conditioning and the ability to form associations between signals of punishment and the punishment itself. Good orienting in aggressives from poor homes may reflect good attention and more proficient learning of antisocial habits in more criminogenic homes.

High SC orienting has been associated with better prefrontal functioning (Hazlett et al. 1993) and larger structure of the prefrontal cortex (Raine et al. 1991). Aggressives from high SES backgrounds who have poor SC orienting may therefore have prefrontal deficits. It might be expected that the crimes these individuals go on to commit may be more impulsive, poorly planned, and non-adaptive, as prefrontal dysfunction has been associated with these traits (Raine et al. 1994). Conversely, aggressives who come from low SES environments show superior orienting, and presumably better prefrontal functioning. It may be that these individuals grow up to commit crimes that are relatively well-planned, adaptive, and functional, as better prefrontal functioning is associated with these traits. With respect to aggression, the low SES aggressives with good prefrontal functioning may be more likely to execute bank robberies, whereas the high SES aggressives with poor prefrontal functioning may be more likely to commit non-premeditated, impulsive assaults. Clearly, further research is needed to test this prediction. The more important point to emphasize is that aggressive behavior in different SES groups has different psychophysiological underpinnings, and that it may be worth examining whether such differences give rise to subtle differences in the form and type of aggressive outcome.

Analyses of the HRL data did not support a biosocial effect, as Aggression did not interact with SES. The interaction of Aggression with temperament, while not expressly biosocial in nature, is nevertheless of some interest. Low HRL only characterized Aggressives who were inhibited. If the Disinhibited group are predisposed to later crime and aggression as is indicated by Caspi et al. (1995) and Caspi and Silva (in press), this finding could be taken to indicate that low HRL relates to aggression only in a low risk (inhibited) group. Similarly, within criminal samples, psychopathic criminals do not differ to non-psychopathic criminals (Raine, 1993), whereas in community samples (low risk) the HRL - antisocial link is well replicated. This perspective may therefore have similarities with the biosocial approach outlined for the orienting data.

Psychophysiology as the Independent Variable

These analyses adopted the biological high risk design where subjects are grouped on the basis of the biological risk variable, with aggression as the dependent variable (Buchsbaum and Reider, 1979). Resting HRL at age 3 was divided above and below the median to form high and low HRL groups. Similarly, SES at age three years was divided above and below the median to form high and low SES groups. CBC variables formed the dependent variables in a three-way HRL x SES x Ethnicity ANOVA, with two levels of ethnicity, Indian and Creole.

Results. There were strong main effects for ethnicity on Aggression ($F(1,821) = 39.0$, $p < .0001$), Delinquency ($F(1,821) = 45.5, p < .0001$), and Total Antisocial ($F(1,821) = 50.4$, $p < .0001$) scores. Results are displayed in Figure 4. Creoles had much higher scores on age 11 aggression, delinquency, and antisocial scales than Indians, with effects sizes of 0.50, 0.51, and 0.55 respectively.

For CBC aggression scores there was a significant three-way HRL x SES x Ethnicity interaction, $F (1,82) = 3.9, p < .05$. A breakdown of this interaction shows that for Creoles, HRL interacts with SES such that the low HRL group are more aggressive than the high HRL group in those from high SES ($p < .05$, effect size = 0.30), whereas the reverse is the case for the low SES group. While the benign home effect was true for Creoles, it was not true for Indians.

The same interaction effect was also observed for Delinquency scores, although the effect was statistically more marginal ($p < .09$). For the overall Antisocial measure, the three-way interaction was again significant ($p < .5$). Again, high SES Creoles with low HRL had higher Antisocial scores than those with high HRL ($p < .05$, effect size = 0.33) (see Figure 5, lower half).

Exploratory analyses were conducted entering sex and temperament as a fourth factor. No interactions were observed with these variables. Analyses were also conducted on SC orienting, but no significant findings emerged.

Interpretation. The benign homes effect received some degree of support from analyses using the biological high risk approach with HRL. HRL interacts with SES in predisposing to aggression at age 11, such that the expected effect of low HRL predisposing to aggressive and antisocial behavior is observed in those from high SES backgrounds.

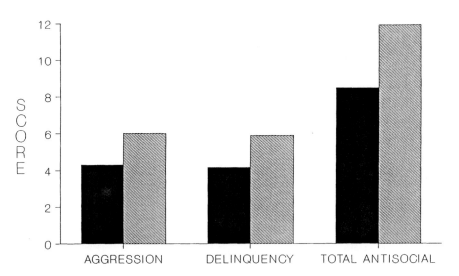

Figure 4. Ethnic differences in aggression, delinquency, and total antisocial behavior.

The fact that a similar effect was observed for non-aggressive delinquency suggests that the interaction effect is not specific to aggression per se, but generalizes to all forms of antisocial behavior. The failure to find interactions with sex is of importance because it demonstrates that the interaction effect holds for females as well as males. This indicates that, at least for age 11 aggression, the correlates of male aggression are also the correlates of female aggression.

The fact that this interaction was only true for Creoles and not Indians was unexpected. One clue as to the source of the three-way interaction involving ethnicity may be obtained from the main effect for ethnicity (see Figure 4). The interaction may only have been observed for Creoles because they have sufficiently high aggression scores for the interaction to emerge. Put another way, the low scores for Indians may produce a floor effect that precludes the interaction being observed.

HEART RATE LEVEL AND AGGRESSION: CONFOUNDS AND PROTECTIVE FACTORS

As discussed earlier, one of the best replicated psychophysiological correlates of antisocial behavior is low resting heart rate (Raine, 1993). Nevertheless, there are a number of possible confounds in the HRL- antisocial relationship that need addressing. In particular, because low heart rate is related to larger body size, and because delinquents and criminals are thought to differ to controls in terms of somatotype (Wilson and Herrnstein, 1985, Eysenck and Gudjonsson, 1989; Sampson, this volume), low HR in aggressives may be a function of body size. We have recently conducted analyses (Raine et al., under review) which were aimed to assess:

1. whether low HRL at age 3 years predicts to antisocial behavior at age 11 years

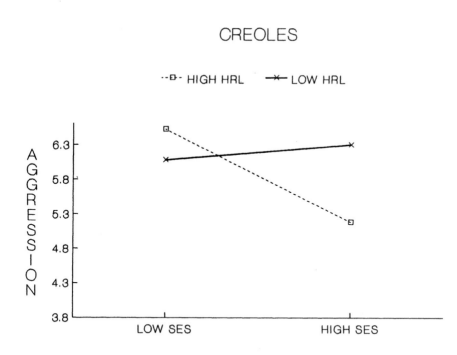

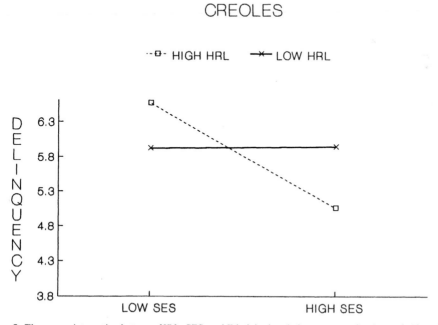

Figure 5. Three-way interaction between HRL, SES, and Ethnicity in relation to aggression (upper half) and delinquency (lower half).

2. whether effects are specific to aggression, or generalize to non-aggressive delinquency
3. whether interactions exists for sex and ethnicity
4. whether low SES mediates the link between HRL and aggression
5. whether large body size mediates the relationship between HRL and antisocial behavior

HRL and Aggression

High and low Aggression and Delinquency groups were again formed using quartile splits from the age 11 years CBC variables. For Aggression, a one-way ANOVA showed a significant effect of aggression $F(1,530) = 12.2$, $p < .001$. As can be seen in Figure 6, aggressives have lower HRLs relative to non-aggressives. For Delinquency however, there was only a trend ($p < .10$) towards lower HRL in high delinquency scorers.

In order to explore the possibility of two and three-way interactions, sex and ethnicity were entered as factors into the analyses. No significant interactions were observed for either Aggression ($p > .35$) or Delinquency ($p > .39$). As indicated in Figure 7, the low HRL - Aggression relationship was true for males as well as females, and for Indians as well as Creoles.

SES, HRL and Aggression

Is the link between HRL and aggression mediated by socio-economic status? That is, does low SES and a more negative, stressful home environment produce both alterations in ANS functioning and aggressive behavior? This question was answered by enter-

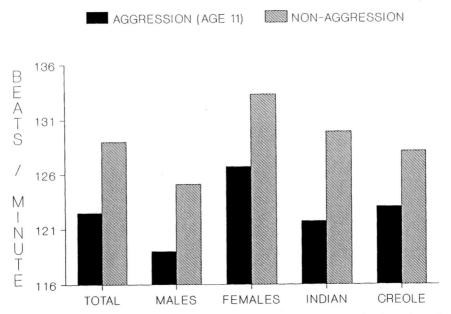

Figure 6. Differences between aggressives and non-aggressives in resting heart rate for the total sample, males and females, and Indians and Creoles.

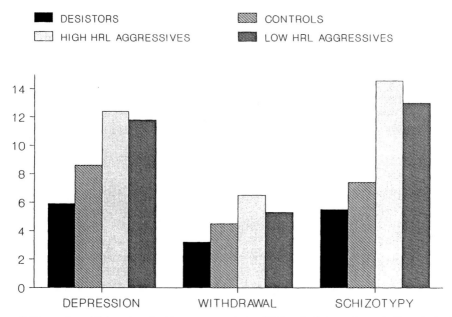

Figure 7. Depression, withdrawal, and schizotypy scores in low HR female desistors (upper half), and school performance in low HR male desistors (lower half).

ing SES as a covariate in the above analyses. Results remained unchanged. As such, the HRL - aggression relationship is independent of SES.

Body Size, HRL, and Aggression

To first assess whether there are any links between body size and antisocial behavior, height, weight, and chest (measured at age 10 years) were assessed in three separate one-way ANOVAs with both Aggression and Delinquency groups. Aggressives children were taller $(F(1,424) = 4.1, p < .05)$, weighed more $(F(1,424) = 4.4, p < .04)$, and had larger chests $(F(1,424) = 4.0, p < .05)$ relative to Non-Aggressives. These body size effects were relatively specific to aggression, in that Delinquents did not differ significantly to Non-Delinquents on height $(p > .28)$, weight $(p > .67)$, and chest $(p > .12)$. Furthermore, there were no significant interactions between Aggression and either sex or ethnicity $(p>.20)$.

The possibility arises that body size could mediate the link between HRL and aggression. To assess for this possibility, Aggressives were compared again to Non-Aggressives on HRL after covarying out the effects of all three body size indices. The effect for Aggression remained significant $(F(1,416) = 5.0, p < .03$, indicating that the relationship between HRL and Aggression is independent of the link between body size and Aggression. However, the margin effect previously observed between delinquency and HRL was lost after controlling for body size $(F(1,507) = 1.63, p > .20)$.

Signs of Malnutrition, HRL, and Aggression

To our knowledge, there are little or no data on links between malnutrition and aggression. Two commonly used indices of malnutrition based on body indices are stunting and wasting. Stunting refers to a reduction in observed height as a function of expected height, while wasting is a reduction in observed weight as a function of expected weight. To some extent therefore, stunting and wasting are indices that reflect a detrimental early environment, whereas the simple indices of height and weight, while influenced to some extent by the environment (Baker et al. 1992) may be viewed as relatively more reflective of genetic factors.

Two different predictions could be made of the links between stunting and wasting with aggression. It may be that poor nutrition may in some way predispose to aggression because it is conceivable that poor nutrition could in some way affect brain development, and thus predispose to aggression. Thus, it would be predicted that aggressives at age 11 are more stunting and wasted than controls. An alternative, opposite prediction is that aggressives have the reverse profile of being *less* stunted and wasted, having a particularly healthy body which provides them with the physical machinery to perpetrate violence. If the latter prediction is correct, then the next question to be asked is whether the link between HRL and aggression is mediated by a better physique.

To test these predictions, indices of stunting and wasting were developed, with norms for expected height and weight based on those children whose parents were at least 1 SD above the mean on SES. High scores on these indices reflect good physique, while low scores indicate stunting / wasting.

Aggressives has significantly higher scores on stunting than controls, $F(1,496) = 11.7$, $p < .001$, indicating less stunting (or better physique) in aggressives. There was also a trend towards aggressives being less wasted than controls, $F(1,496) = 3.5$, $p < .06$. No significant relationships with non-aggressive Delinquency wee observed for either stunting ($F(,621) = 12.$, $p > .27$) or wasting ($F(1,621) = 23.$, $p > .12$), indicating some specificity to aggression per se.

There was a main effect for Ethnicity on both stunting ($F(1,464) = 4.3$, $p < .04$) and wasting ($F(1,464) = 4.5$, $p < .04$), with Creoles have better physiques than Indians. However, there were no interactions between aggression and either sex or ethnicity. Is low HRL in aggressives mediated by physique ? This question was addressed by entering stunting and wasting together as covariates in an ANCOVA comparing aggressives with controls on HRL. The main group effect was still significant ($F(1,484) = 9.0$, $p < .003$) after controlling for the effects of physique.

Interpretation

Returning to the questions posed earlier, low HRL at age 3 years does characterize those who go on to become aggressive 8 years later at age 11 years. Effects seem somewhat stronger for aggressive than for non-aggressive delinquency. There are no interactions with sex and ethnicity, and the relationship is not mediated by either low SES, or larger body size. In addition, because HRL is measured at age 3 years, the HRL-aggression relationship is not an artifact of factors such as cigarette smoking, alcohol consumption, engagement in sporting activities and exercise leading to physical fitness.

The most important question that remains to be answered concerns why low HRL should predispose to aggression. There are two main theoretical interpretations of reduced arousal in antisocials. Fearlessness theory indicates that low levels of arousal are markers of low levels of fear (Raine, 1993; Raine in press). For example, particularly fearless individuals such as bomb disposal experts who have been decorated for their bravery have particularly low heart rate levels and reactivity (Cox et al. 1983; O'Connor et al. 1985), as do British paratroopers decorated in the Falklands war (McMillan and Rachman, 1987).

Lack of fear would predispose to antisocial and violent behavior since such behavior (e.g. fights and assaults) requires a degree of fearlessness to execute, while lack of fear, especially in childhood, would help explain poor socialization since low fear of punishment would reduce the effectiveness of conditioning. Fearlessness theory receives support from the fact that low HRL also provides the underpinning for a fearless or uninhibited temperament (see above).

A second theory explaining reduced arousal is stimulation-seeking theory (Eysenck, 1964; Quay, 1965; Raine, 1993; Raine, 1996). This theory argues that low arousal represents an aversive physiological state, and that antisocials seek out stimulation in order to increase their arousal levels back to an optimal or normal level. Antisocial behavior is viewed as a form of stimulation-seeking in that committing a burglary, assault, or robbery could be stimulating for some individuals.

A third possibility is that low HRL is a marker for some other biological process, such as brain dysfunction. For example, low HRL has been linked to enlargement of the third ventricle (Cannon et al. 1992). Such enlargement is thought to perhaps reflect tissue loss to the anterior hypothalamus, a structure involved in both the regulation of autonomic activity and aggressive behavior.

Low HRL has also been correlated with characteristics in children such as lack of empathy, (Zahn-Waxler et al., 1995), the witnessing of parental violence (Gottman and Katz, 1989), and coming from a home broken by divorce in the first four years of life (Wadsworth, 1976). Consequently, the possibility that low HRL is a marker for social and psychological processes should not be forgotten.

Protection against Low HRL Leading to Aggression

Although low HRL at age 3 predisposes to aggression at age 11, not all children with low HRL become aggressive. Why do some of these children desist from aggression? One possible explanation that we have speculated on for some time without data is that good school performance may protect against antisocial behavior because underaroused children may be able to obtain their arousal jag in life from academic stimulation and the consequent positive regard of teachers, parents, and peers, as opposed to having to engage in antisocial, risky behavior such as fighting. For example, there is some evidence that such factors may protect against delinquency (Stouthamer-Loeber et al., 1993). We also speculated that such desistors may be less likely to have comorbid conditions such as hyperactivity or depression.

Analyses. In order to attempt to identify possible protective factors, low and high HRL quartile groups were each divided above and below the median on aggression scores to form four groups: low HRL, high aggression (underaroused aggressives) low HRL, low aggression (underaroused desistors), high HRL, high aggression (overaroused aggressives), and high HRL, low aggression (controls). One-way ANOVAs were then used to compare groups on CBC scores, school performance measured at age 11 years, and IQ at age 11 years. Because it was not possible to develop an adequate common-item CBC scale for depression (due to the different nature of items defining depression in males and females), the two sexes were separately analyzed.

Results. For males, there was a group main effect for school performance, $F(3,212) = 3.4$, $p < .02$. Group means are displayed in Figure 7 (lower half) and show that Desistors had the highest scores of all groups. A breakdown of the group effect indicated that Desistors performed significantly better than both the overaroused aggressives (t = 3.9, df = 106, p < .004) and the controls (t = 2.1, df = 80, p < .04), but comparison with controls did not reach significance (p < .19) The group difference on school performance was surprising in the light that groups did not differ on verbal (p > .37) or performance IQ (p > .34) at age 11. Regarding CBC scores at age 11, no scales clearly differentiated Under-Aroused Desistors from both Controls and Under-Aroused Aggressives.

For females, groups did not differ on school performance (p > .76). However, a noteworthy finding from analysis of CBC scales was a strong main effect for Depression, F(3,177) = 20.3, p < .00001. Means are displayed in Figure 7 (upper half). A breakdown of the effect showed that Desistors had significantly lower scores than Controls (t = 3.8, df = 98, p < .0001), Underaroused Aggressives (t = 6.4, df = 70, p < .0001), and Overaroused Aggressives (t = 7.3, df = 87, p < .0001). The same effect was shown for Withdrawal, F(3,177) = 5.8, p < .0008, with Desistors being significantly lower than Controls (t = 2.0, df = 98, p < .06), Underaroused Aggressives (t = 2.6, df = 70, p < .01), and Overaroused Aggressives (t = 4.3, df = 87, p < .0001). Similarly, a main effect for Schizotypal Personality, F (3,177) = 26.4, p < .00001, showed that Desistors were less schizotypal than Controls (t = 2.3, df = 98, p < .03), Underaroused Aggressives (t = 6.5, df = 70, p < .0001), and Overaroused Aggressives (t = 7.8, df=87, p < .0001).

Interpretation. Different protective factors appeared for male and female Desistors. Male low HRL Desistors are characterized by good school performance, while female low HRL Desistors are characterized by less depression, withdrawal, and schizotypy.

The fact that female Desistors have lower scores on Depression than both Under- and Over-Aroused Aggressives is perhaps not surprising because Desistors are by definition low on aggression and as such might be expected to have less pathology than groups who are aggressive. The fact that they are also significantly lower than the other non-aggressive control group is more striking, and perhaps suggests that the lack of Depression, or conversely being happy, may protect against aggression.

Why are Desistors less depressed, or more happy, than other groups? One possibility is that they may have more friends and a better social network, as indicated by their low scores on Withdrawal and Schizotypal Personality. Greater social interaction may be a potent sources of stimulation for these underaroused Desistors, whereas underaroused aggressives may turn to aggression for stimulation because their high depression, withdrawal, and schizotypal tendencies forecloses this alternative.

For males, the fact that Desistors were the best school performers might suggest that they have particularly higher IQs, since high IQ has been found in one study to be a protective factor against crime (Kandel et al. 1988, Brennan et al. 1994). In the present study however, Desistors were not found to have higher IQs than others. It could be speculated that male Desistors have family environments that nurture, encourage and reward academic performance which may for these under-aroused males be a source of stimulation. Similarly, it is interesting to consider whether female Desistors have families which model social networking, or whether differences are more intrinsic to the child.

IMPLICATIONS FOR MODEL AND FUTURE RESEARCH

What implications do the findings presented in this chapter have for the biosocial model described in Chapter 1? Although no one study can hope to provide definitive answers to even a few of the questions posed in Chapter 1, a few initial pointers can be drawn out.

First, of the various types of biosocial interactions described in Chapter 1, the commonest finding to emerge was the crossover biosocial effect. This was somewhat surprising. In contrast, more intuitively plausible effects, the multiplicative, sequential, and additive interactions, were not found. The implication for future studies is that perhaps more effort should be placed on considering that the risk factors found to predispose to aggression in one relatively homogenous group (benign home background) can be reversed in another (e.g. criminogenic background). This is potentially important because ignoring the moderator can erroneously lead to null findings and wrong conclusions.

Second, Chapter 1 asked whether biosocial effects generalize to females as well as males, and ethnic minority groups as well as Caucasians. Findings generally support both suggestions. It seems that biosocial pathways to aggression are no respecter of sex of ethnic grouping. In addition, the fact that they are found in a very different culture (Mauritius) to western cultures provides some important cross-cultural generalizability to the findings. The important caveat to make however is that such findings still do not generalize to ethnic or gender groups in the United States. Furthermore, one biosocial crossover effect was observed for Creoles but not Indians.

Third, sequential biosocial effects were not observed, particularly with respect to the risk factor of low resting heart rate. The link between low HRL and aggression was not mediated by low SES or a detrimental early childhood environment leading to malnutrition. This does not mean that other environmental influences, not measured in this study, may still mediate the effect. Nevertheless, it seems more likely that low HRL is an important, non-artifactual risk factor for aggression which may have its roots in genetic or non-genetic, early biological influences.

Fourth, Chapter 1 raised the questions of the direction of "flow" from causal mechanisms to violence phenomena, and vice-versa. Results showed that while aggressives at age 11 have lower HRL at age 3 years, those with low HRL at age 3 were not strongly characterized by high aggression at age 11. This difference should alert other studies into the possibility that the two different approaches may result in different conclusions.

Finally, resting HR has been repeatedly linked to antisocial and aggressive behavior in childhood, and criminal behavior in adulthood. Given the technical ease with which it can be measured and its low cost, it is recommended that resting HR should be increasingly assessed in both laboratory and field studies of antisocial and aggressive behavior where it can be integrated with other macrosocial and microsocial variables. Such future studies will help address the outstanding question of how and why resting HRL predisposes to antisocial and violent behavior.

SUMMARY

This chapter focuses on how resting heart rate and skin conductance orienting interacts with socio-economic status measured at age 3 years in predisposing to aggression at age 11 years using data from the Mauritius longitudinal study. A crossover biosocial effect was observed for SC orienting, whereby aggressives from high SES backgrounds have low SC orienting, whereas aggressives from low SES backgrounds have high SC orienting. Similarly, aggressive behavior in Creoles was associated with low resting HR when subjects were from high SES backgrounds, whereas aggressive behavior was linked to high HR in those from low SES backgrounds. The link between low HRL at age 3 and high aggression at age 11 is not mediated by low SES, the larger body size observed in aggressives, or stunting and wasting indices of malnutrition. Good school performance protects low heart rate boys against aggression, while a lack of depression, withdrawal, and schizotypy protect low HRL girls from becoming aggressive. Overall, these findings suggest that the biological correlates of antisocial behavior may be moderated by social factors such as SES. Autonomic deficits may particularly characterize aggression in high SES environments where social predispositions to crime are minimized. Conversely, individuals from more criminogenic social backgrounds (low SES) who have advantageous autonomic features may more easily condition into and learn antisocial, aggressive habits. It is argued that low HRL is a well-replicated and potentially important autonomic predisposition to later aggressive, antisocial behavior. It is recommended that HR should be increasingly measured in both laboratory and field studies of antisocial and aggressive behavior where it can be integrated with other macrosocial and microsocial variables.

ACKNOWLEDGMENTS

This paper was written while the first author was supported by a NIMH grant (RO1 MH46435–02), and also a NIMH Research Scientist Development Award (1 KO2 MH01114–01). We wish to thank the Mauritius Government for their continued support of the Mauritius Child Health Project, David Farrington for consulting, Angela Scarpa for providing an index of temperament, and Brian Bell, Marie Claire Calambay, Meena Calinghen, Ramesh Cheeneebash, Cyril Dalais, Devi Jaganathen, Dr. A.C. Raman, and Dr. C. Yip Tong for assistance in data collection.

REFERENCES

Achenbach T.M. and Edelbrock C.S. (1979) The Child Behavior Profile: II. Boys aged 12–16 and girls aged 6–11 and 12–16. Journal of Consulting and Clinical Psychology 47 223–233.

Baker, L.A., Reynolds, C., and Phelps, E. (1992). Biometric analysis of individual growth curves. *Behavior Genetics 22* 253–264.

Brennan, P., Raine, A., Venables, P.H. and Mednick, S.A. (1994). Psychophysiological protective factors for children at high risk for antisocial outcome. *Psychophysiology 31* 30.

Buchsbaum, M.S., and Reider, R.O. (1979). Biologic heterogeneity and psychiatry research. *Archives of General Psychiatry 36* 1163–1169.

Buikhuisen, W., Bontekoe, E.H.M., C.D. Plas-Korenhoff, and S. Buuren (1985). Characteristics of criminals: The privileged offender. *International Journal of Law and Psychiatry 7* 301–313.

Cannon, T.D., Raine, A., Herman, T.M., Mednick, S.A., Schulsinger, F. and Moore, M. (1992). Third ventricle enlargement and lower heart rate levels in a high risk sample. *Psychophysiology 29* 294–301.

Caspi, A. Henry, B., McGee, R.O., Moffitt, T.E. and Silva, P.A. (1995). *Temperamental* origins of child and adolescent behavior problems: From age three to age fifteen. *Child Development 66* 55–68.

Caspi, A. And Silva, P.A. (in press). Temperamental qualities at age 3 predict personality traits in young adulthood: Longitudinal evidence from a birth cohort. *Child Development*

Cox, D., Hallam, R., O'Connor, K. and Rachman, S. (1983). An experimental study of fearlessness and courage. *British Journal of Psychology 74* 107–117.

Dawson, M.E. (1990). Psychophysiology at the interface of clinical science, cognitive science, and neuroscience: Presidential address, 1989. *Psychophysiology 27* 243–255.

Dawson, M.E., Schell, A.M. and Filion, D. (1991). The electrodermal system. In J.T., Cacioppo, L.G. Tassinary, and A. Fridlund (Eds). *Principles of psychophysiology: Physical, social and inferential elements.* (pp. 295–324). Cambridge: Cambridge University Press.

Eysenck, H.J (1964). *Crime and personality* (1st ed.) London: Methuen.

Eysenck, H.J. (1977). *Crime and personality* (3rd. ed.). St. Albans: Paladin.

Eysenck, H.J. and Gudjonsson, G.H. (1989). *The causes and cures of criminality*. New York: Plenum.

Farrington, D.P. (1987). Implications of biological findings for criminological research. In S.A. Mednick, T.E. Moffitt, & S.A. Stack (Eds.), *The causes of crime: New biological approaches* (pp. 42–64). New York: Cambridge University Press.

Fowles, D.C. (1993). Electrodermal activity and antisocial behavior. In J.C. Roy, W. Boucsein, D.C. Fowles and J. Gruzelier (Eds). *Electrodermal activity: From physiology to psychology*. New York : Plenum.

Gottman, J.M. and Katz, L.F. (1989). Effects of marital discord on young children's peer interaction and health. *Developmental Psychology 25* 373–381.

Hazlett, E., Dawson, M., Buchsbaum, M.S., and Nuechterlein, K. (1993). Reduced regional brain glucose metabolism assessed by PET in electrodermal nonresponder schizophrenics: A pilot study. *Journal of Abnormal Psychology 102* 39–46.

Hemming, J.H. (1981). Electrodermal indices in a selected prison sample and students. *Personality and Individual Differences 2* 37–46.

Kagan, J. (1989). Temperamental contributions to social behavior. *American Psychologist 44* 668–674.

Kandel, E., Mednick, S.A., Kirkegaard-Sorensen, L. and Hutchings, B. (1988). IQ as a protective factor for subjects at high risk for antisocial behavior. *Journal of Consulting and Clinical psychology 56* 224–226.

McBurnett, K. and Lahey, B.B. (1994). Biological correlates of conduct disorder and antisocial behavior in children and adolescents. In D.C. Fowles (Ed.). *Progress in experimental personality and psychopathology research*. New York: Springer.

Maliphant, R., Hume, F., and Furnham, A. (1990). Autonomic nervous system (ANS) activity, personality characteristics and disruptive behavior in girls. *Journal of Child Psychology and Psychiatry 31* 619–628.

McMillan,T.M. and Rachman-S-J. (1987). Fearlessness and courage: A laboratory study of paratrooper veterans of the Falklands War. *British Journal of Psychology, 1987 78* 375–383.

O'Connor, K., Hallam, R., and Rachman, S. (1985). Fearless and courage: A replication experiment. *British Journal of Psychology 76* 187–197.

Quay, H.C. (1965). Psychopathic personality as pathological stimulation-seeking. *American Journal of Psychiatry 122* 180–183.

Raine, A. (in press). Psychophysiology and antisocial behavior. In J.D. Master, J. Brieling, and D. Stoff (Eds). *Handbook of antisocial behavior*.

Raine, A. (1996). Autonomic nervous system activity and violence. In D.M. Stoff and R.F. Cairns (Eds.) *The Neurobiology of clinical aggression*. (pp. 145–168). Lawrence Erlbaum.

Raine, A. (1987). Effect of early environment on electrodermal and cognitive correlates of schizotypy and psychopathy in criminals. *International Journal of Psychophysiology 4* 277–287.

Raine, A. (1988). Antisocial behavior and social psychophysiology. In H. Wagner (Ed.) *Social psychophysiology and emotion: Theory and clinical application* (pp. 231–253). London: Wiley.

Raine, A. (1993). *The psychopathology of crime: Criminal behavior as a clinical disorder* San Diego: Academic Press.

Raine, A., Brennan, P., and Farrington, D.P. (in press). Conceptual and theoretical issues facing biosocial research on violence. In Raine, A., Brennan, P., Farrington, D.P. and Mednick, S.A. (Eds). *Biosocial bases of violence*. New York: Plenum.

Raine, A., Buchsbaum, M.S., Stanley, J., Lottenberg, S., Abel, L. and Stoddard, J. (1994). Selective reductions in pre-frontal glucose metabolism in murderers. *Biological Psychiatry. 36* 365–373.

Raine, A. and Mednick, S.A. (1989). Biosocial longitudinal research into antisocial behavior. *Review d'Epidemiologie et de Sante Publique 37* 515–524.

Raine, A., Reynolds, G.P. and Sheard, C. (1991). Neuroanatomical mediators of electrodermal activity in normal human subjects: A magnetic resonance imaging study. *Psychophysiology 28* 448–558.

Raine, A. and Venables, P.H. (1981). Classical conditioning and socialization - A biosocial interaction? *Personality and Individual Differences 2* 273–283.

Raine, A. and Venables, P.H. (1984 b). Tonic heart rate level, social class, and antisocial behavior. *Biological Psychology 18* 123–132.

Raine, A. and Venables, P.H. (1984 a). Electrodermal non-responding, schizoid tendencies, and antisocial behavior in adolescents. *Psychophysiology 21* 424–433.

Raine, A., Venables, P.H., and Mednick, S.A. (1996) *Reduced resting heart rate at age 3 years predicts to aggressive behavior at age 11 years: Findings from the Mauritius Child Health Study* (under review).

Raine, A., Venables, P.H. and Williams, M. (1990a). Relationships between CNS and ANS measures of arousal at age 15 and criminality at age 24. *Archives of General Psychiatry 47* 1003–1007.

Raine, A., Venables, P.H. and Williams, M. (1990b). Relationships between N1, P300 and CNV recorded at age 15 and criminal behavior at age 24. *Psychophysiology 27* 567–575.

Raine, A., Venables, P.H. and Williams, M. (1995). High autonomic arousal and electrodermal orienting at age 15 years as protective factors against criminal behavior at age 29 years. *American Journal of Psychiatry 152* 1595–1600.

Raine, A., Venables, P.H. and Williams, M. (1996). Better autonomic conditioning and faster electrodermal half-recovery times at age 15 as possible protective factors against crime at age 29 years. *Developmental Psychology 32* 624–630.

Sampson, R. And Laub, J.H. (In press). Unraveling the social context of physique and delinquency: A new long-term look at the Glueck's classic study. In Raine, A., Brennan, P., Farrington, D.P. and Mednick, S.A. (Eds). *Biosocial bases of violence.* New York: Plenum.

Scarpa, A.S., Raine, A., Venables, P.H. and Mednick, S.A. (in press). Heart rate and skin conductance in behaviorally inhibited Mauritian children. *Journal of Abnormal Psychology*

Snidman, N., Kagan, J. and McQuilkin, A. (1991). Fetal heart rate as a predictor of infant behavior. *Psychophysiology 28* 51.

Stouthamer-Loeber, M., Loeber, R., Farrington, D.P., Zhang, Q., van Kammen, W.B. and Maguin, E. (1993). The double edge of protective and risk factors for delinquency: Interrelations and developmental patterns. *Development and Psychopathology 5* 683–701.

Venables, P.H. (1978). Psychophysiology and psychometrics. *Psychophysiology 15* 30–315.

Venables, P.H. and Christie, M.J. (1973). Mechanisms, instrumentation, recording techniques, and quantification of responses. In W.F. Prokasy and D.C. Raskin (Eds). *Electrodermal activity in psychological research.* New York: Wiley.

Wadsworth, M.E.J. (1976). Delinquency, pulse rate and early emotional deprivation. *British Journal of Criminology 16* 245–256.

Wilson, J.Q. and Herrnstein. R.J. (1985). *Crime and human nature.* New York: Simon and Schuster.

Zahn, T.P. and Kruesi, M.J.P. (1993). Autonomic activity in boys with disruptive behavior disorders. *Psychophysiology 30* 605–614.

Zahn-Waxler, C., Cole, P.M., Welsh, J.D., and Fox, N.A. (1995). Psychophysiological correlates of empathy and prosocial behaviors in preschool children with behavior problems. *Development and Psychopathology 7* 27–48.

8

BIOSOCIAL RISK FACTORS FOR DOMESTIC VIOLENCE

Continuities with Criminality?

Neil S. Jacobson and Eric T. Gortner

Department of Psychology
University of Washington
Seattle, Washington 98105

INTRODUCTION

Domestic violence occupies an unusual position in psychological research. While being one of the most widespread and devastating of societal problems, domestic violence is also one of the areas with the least amount of quality empirical work. We need to look no further than the historical and cultural context of wife assault to recognize that violence against women has not only been largely ignored and minimized across time, but has also been sanctioned by our patriarchal culture (Bograd, 1988; Dobash & Dobash, 1979). Part of this apparent lack of research enthusiasm may be explained by the delicate ethical issues that are raised when science attempts to intervene in such a politically perilous area (Jacobson, 1994). The accurate dissemination of results from this type of research is particularly difficult given the propensity of the mass media to confuse and misinterpret issues of correlation, causality, and responsibility. Another possible deterrent to would-be researchers is that domestic violence is not part of any mainstream, but rather overlaps with many different disciplines, including psychology, public health, law, and social work domains.

Descriptive scientific work has illuminated some of the basic facts of domestic violence, and they are disturbing: each year over 1.5 million women in the United States are physically assaulted by their partners (Straus & Gelles, 1986,1990). Estimates are that up to one-third of all women will be physically assaulted by a partner during their adulthood (Browne, 1993). The impact of domestic violence on women is devastating, ranging from severe emotional trauma, including post-traumatic stress disorder and depression, to severe physical injury, including death (Holtzworth-Munroe, Smutzler & Sandin, in press). Assaults on women by male partners are more likely to result in injury or death in comparison to attacks by other types of assailants (Browne, 1993). The destructive impact of domestic violence is not limited to women. There is evidence that battering has significant

Biosocial Bases of Violence, edited by Raine *et al.*
Plenum Press, New York, 1997

detrimental impact on children, including increased risk for serious behavioral problems, such as conduct disorder, as well as emotional and cognitive difficulties (Davies & Cummings, 1994; Holtzworth-Munroe, Smutzler & Sandin, in press).

Although research work on the prevalence and effects of battering has grown in the past decade, we still know surprisingly little about the proximal causes of the domestic violence. We also know little that is definitive about the characteristics of batterers or the relationships of violent couples. There has been no shortage of theoretical speculation as to the causes of battering, but few theories have been subjected to empirical scrutiny. Much of what passes for knowledge comes from case studies and retrospective reports.

In this chapter, we will focus on a growing body of research examining potential biological, psychological, and social risk factors for domestic violence. We will also briefly examine pieces of the larger research base on biology, criminal behavior, and impulsivity in order to generate questions about possible continuities between general antisocial behavior and battering. By necessity, our focus will be on men. Battering is essentially a male phenomenon. Although bilateral physical aggression is quite common in couples (O'Leary et al., 1989; Straus, Gelles & Steinmetz, 1980), there is an important gender distinction in terms of the violence's impact and function. In contrast to violence directed from men toward men, violence from female partners rarely causes physical harm and is often defensive in nature (Cantos, Neidig & O'Leary, 1994). As a result, men's violence is unique in its capacity to create genuine fear and terror in women victims (Jacobson et al., 1994). It is this specific fear that facilitates the primary function of battering: to control, intimidate, and subjudcate the partner. In marriage, it is not easy to find wives who batter husbands.

RISK FACTORS FOR BATTERING

Battering occurs in a particular societal context, where patriarchal values predominate, and physical force by men against women is legitimized. Although patriarchy is the important overall context surrounding battering, it, in itself, is an incomplete explanation for the phenomenon of male battering. Not all men are physically violent in relationships. Also, there is a growing body of research that shows similar prevalence rates of battering in lesbian and gay male relationships, which can't be attributed to patriarchy (Dutton, 1994). The complexity of battering is likely to be best accounted for by considering how other psychological, social, and biological variables interact with patriarchy.

Many of the studies on domestic violence suffer from methodological limitations, including poor sample selection, reporter biases, and inadequate control groups (see LaTaillade & Jacobson, in press). Researchers are increasingly attending to these methodological problems, and, perhaps as a result, there is an emerging convergence of findings on possible risk factors for domestic violence. It is important to note that the term "risk factors" in this chapter simply denotes variables that have an association with increased rates of domestic violence. These variables may or may not have a causal role in domestic violence. The cross-sectional and correlational nature of most research on domestic violence prevents us, at this point, from making causal inferences. For instance, consider the established association between alcohol usage and battering. It is possible that alcohol usage has a direct causal relationship to battering through a disinhibition effect, for example. Alternatively, alcohol usage may simply tend to co-occur with other variables, such as antisocial personality or impulsivity, that have a direct causal relationship to battering. Since very few researchers have followed the course of violence longitudinally, we know very little about true predictive variables for domestic violence. The distinction between causal and marker variables blurs in an area where experimental procedure is difficult to uphold.

The majority of studies looking at risk factors for domestic violence have focused on the characteristics of men who batter. These variables can be roughly broken down into three categories: (1) sociodemographic factors, (2) psychopathology, personality, and couple factors, and (3) physical/biological factors (for more complete reviews see Holtzworth-Munroe, Bates, Smutzler, & Sandin, in press; Saunders, 1995).

Sociodemographic Factors

Age. Rates of domestic violence are consistently higher among younger couples. It appears that couples who are under the age of 30 have disproportionately higher rates than older couples (e.g., those over 40). Although there are notable exceptions, it may be that battering as a phenomenon peaks in the 20's and then decreases with age. For instance, Hampton and Gelles (1994) found that African-American couples under age 30 reported rates of severe violence that were three times greater than those of couples over age 40. There are several possible explanations for this trend. First, there is a general tendency for battering to reduce or extinguish over time. This is not a compelling explanation, though, because there is little evidence for an extinction curve that covaries with age. It may be that as men get older, peer group influence may decline, their abilities to use force may deteriorate, or the incorrigibles may become detained by the legal system. Unfortunately, to-date there has been no systematic exploration of what might account for this finding. Second, it may that physical violence is less necessary to maintain control in battering relationships once it is established. That is, across time, the threat of physical violence becomes a controlling factor, rendering actual violent acts less "necessary" and less frequent. In this sense, emotional or psychological abuse, such as threats, may become more effective controlling mechanisms than physical aggression. For instance, in our own research examining the longitudinal course of battering relationships, we found that even in cases where physical violence decreased dramatically over time, emotional abuse continued at high levels for virtually all batterers (Jacobson, Gottman, Gortner, Berns, & Shortt, in press). This suggests that it would be a mistake to consider, in isolation, reductions of physical violence over time as necessarily representing reductions in the overall abusiveness of battering relationships.

Socioeconomic Status. Couple's social class has consistently been negatively correlated with domestic violence. Couples with lower income levels (less than $10,000) have reported rates of husband violence that are more than twice as high as higher income couples (Hampton & Gelles, 1994). This correlation may be particularly illuminating with regard to severity of violence, with lower SES couples engaging in more severe forms of violence (Hotaling & Sugarman, 1990). Relatedly, unemployment is also typically associated with battering, especially among young men (Howell & Pugleisi, 1988). The overrepresentation of battering among men with lower SES may be attributable to the increased stressors facing those who are economically marginalized and the decreased resources to cope with typical life stresses (Gelles, 1993). It is also possible that there is a reporting artifact involved in this literature, where those from lower socioeconomic classes are more likely to report the presence of violence or are more likely to become involved in the criminal-justice system. To the extent that violence is more normative in one's environment, it stands to reason that one may be more willing to admit to such violence.

Ethnicity. Battering is related to ethnicity/race. Although battering occurs among couples of all ethnicities, rates of domestic violence are typically higher in minority couples, as compared to white couples. For instance, prevalence estimates of husband violence in African-American and Latino couples are approximately 17%, compared to estimates of 12% in White couples (Hampton, Gelles, & Harrop, 1989). This finding has held in some studies, even when socioeconomic factors were controlled for (McLaughlin et al., 1992). However, there are important exceptions to these findings. For instance, Centerwall (1995) compared rates of intraracial domestic homicides in African-American and White populations in New Orleans. When household crowding was unaccounted for in the two groups, African-American populations showed a sixfold greater relative risk of intraracial domestic violence. However, when household crowding was controlled for, this sixfold difference disappeared. This highlights the importance of considering the many different socioeconomic factors that may influence observed racial differences in domestic violence. Taken together, the data on ethnicity remain ambiguous and it is difficult to determine which, if any, facets of race independently influence battering prevalence, as compared to other concomitant factors associated with minority status, such as poverty, urban stress, and discrimination. As Holtzworth-Munroe, Smultzer, and Bates (in press) note, there is still a lack of sensitivity among researchers to important cultural issues surrounding domestic violence in communities of color.

Marital Status. Unmarried couples that cohabitate appear to be at a greater risk for domestic violence than those that are either married or dating couples. Since many of these sociodemographic variables are correlated, there are important confounds that need to be considered. For instance, couples that cohabitate are often younger and of a lower socioeconomic status than married couples. However, McLaughlin et al. (1992) found an association between cohabitation and violence even when controlling for age and occupational status. It may be that couples who live together, but remain unmarried, are less committed or involved in relationships, which may relate to risk for violence (Stets & Straus, 1989). At this point, it remains unclear how cohabitation relates to increased rates of battering.

Psychopathology, Personality, and Couple Factors

Depression. Men who batter generally report higher levels of depressive symptomatology than their nonviolent counterparts (e.g., Julian & McKenry, 1993; Pan, Neidig, & O'Leary, 1994). However, it is important to note that many of these men do not report enough depressive symptomatology to warrant a clinical diagnosis of depression. On self-report instruments, batterers often score in the moderate range of depression symptoms. One possibility is that these reports of low level depressive symptoms are reflective of underlying low self-esteem issues or the result of arrest or separation (Saunders, 1995). Unfortunately, since many of the studies looking at battering and depression fail to include a nonviolent, but equally maritally distressed control group, it is difficult to determine whether depressive symptoms are related to battering or marital distress.

Alcohol Use. One of the more consistent findings is the connection between alcohol use and battering (e.g., Leonard & Blane, 1992; McKenry, Julian, & Gavazzi, 1995). There is often a direct correlation between a male partner's overall level of alcohol consumption and his use of physical violence in the relationship. It seems that alcohol use uniquely affects the propensity for batterers to use more severe, often life-threatening, forms of physical violence, such as hitting with an objects, using a knife or a gun, or prolonged beating (Saunders, 1995). For instance, Pan et al. (1994) found that the presence of alcohol difficulties increased the probability of severe, as opposed to mild, physical violence by 70%.

Anger/ Hostility. Not surprisingly, men who batter consistently report higher levels of anger and hostility than their nonviolent counterparts (e.g., Leonard & Blane, 1992; O'Leary, Malone, & Tyree, 1994). The few observational studies of actual marital interactions also support the notion that violent men interact with their partner in more hostile, aggressive, and contemptuous ways than their nonviolent counterparts (Jacobson et al., 1994; Margolin et al., 1988). This finding has held even when controlling for relationship distress (Jacobson et al., 1994). An important differentiating feature may be the form of anger that batters use. Jacobson et al. (1994) examined anger among batterers in a more fine-tuned manner, by differentiating between provocative forms of verbal aggression (e.g., belligerence or contempt) and other, less provocative, forms (e.g., criticism, disagreement, or defensiveness). Overall, both violent wives and husbands showed more anger than did nonviolent partners. However, violent husbands were clearly differentiated from their nonviolent counterparts by their higher usage of provocative anger, such as contempt and belligerence. Taken together, these findings suggest that anger, especially provocative forms of anger, and hostility play a central role in the lives of battering men. At this point, it is unclear whether men who batter are characterologically angry, or whether their anger is relationship-specific.

General Personality Features. Researchers and theorists alike have also attended to more general personality characteristics of batterers. Personality questionnaires that include DSM-IV Axis II-type personality disorders, such as the Millon Clinical Multiaxial Inventory (MCMI; Millon, 1983), have offered decidedly mixed results. In general, batterers tend to show more dysfunction than nonviolent men on both mood and personality scales (e.g., Hamberger & Hastings. 1991; Murphy, Meyer, & O'Leary, 1993). However, such questionnaire studies have yet to show a particular constellation of personality features that reliably discriminates between violent and nonviolent men. The results point toward diffuse pathology, with alcohol/drug and antisocial measures typically being the most consistently elevated scales. The most promising findings in the personality area are from Dutton's work on borderline personality organization (BPO) (Dutton, 1995). Individuals with borderline personality organization are thought to be intolerant of being alone and suffer from abandonment anxiety. This excessive dependence on partners, coupled with intense abandonment anxiety and anger, sets the stage for violent proclivities in these men. Dutton connects the formation of the attachment-rage components of BPO to early development, especially paternal rejection and insecure attachment. In general, research on BPO has yielded promising results, with numerous studies finding aspects of BPO overrepresented among battering men (see Dutton, 1995 for review).

Abuse History. Severe batterers appear more likely to report childhood physical abuse history than nonviolent men (e.g., Dutton & Hart, 1992). A more important variable may be childhood exposure to violent incidents, with violent men more likely to report witnessing incidents of physical violence between their parents than nonviolent men (Doumas, Margolin, & John, 1994; Gottman et al., 1995). Although these studies are often plagued by methodological problems, including retrospective reports and incomplete definitions of parental violence, it may be that violent or abusive experiences predispose a child to increased aggressiveness and violence later in life. As with many of the aforementioned variables, childhood abuse may be a risk factor for general antisocial behavior, which in turn relates to a higher incidence of battering.

Relationship Factors

Battering occurs within the context of an intimate relationship. Some investigators have sought to look at the characteristics of violent couples in order to understand how their interactions might be different than nonviolent couples. From an advocacy perspective, this type of work is troubling, since it has the potential to be misinterpreted in such a way that places responsibility for violence on the female victim. For instance, if certain wife behaviors are found to be related to escalation of husband violence, some might jump to the conclusion that his violence is being "caused" by her behavior. However, from a contextual point of view, understanding the covariation between behaviors in their natural context is categorically different than ascribing causality or responsibility (see Jacobson, 1994). With regard to male battering, it is clear that men are responsible for their violent acts and must be held accountable for these acts. But even battering men do not batter twenty-four hours a day, and we need to explain when these acts occur, as well as what predicts their onset, offset, increase, and decrease. Research that attends to couple dynamics seeks only to understand the context of such violence and how couple interactions might affect its course.

Couple Distress. It is clear that couples in battering relationships report more distress than their nonviolent counterparts. They also appear to be more distressed than nonviolent, but unhappy couples. This latter comparison is important in that it suggests that there is a level of unhappiness in violent couples that supersedes "normal" unhappy, distressed couples. However, a substantial percentage of couples (approximately 60%) who report repeated physical violence do not report themselves as maritally dissatisfied (O'Leary et al., 1989; Bauserman & Arias, 1992). These cases do not appear to be restricted to just newlyweds. So, there are significant numbers of couples where men batter and couple distress is not reported. As O'Leary et al. (1989) note, these findings raise important issues about the meaning of physical violence in relationships, particularly with regard to partners minimizing or justifying physical aggression. It is also not clear what, if any, causal role couple distress plays in the occurrence of battering. Is couple distress and unhappiness a precursor to physical violence, or is it a reflection of the existence of violence in a relationship? Alternatively, is couple distress a marker of other troubling features in the relationship, such as alcohol abuse, that have a more direct relation to violence?

Communication Patterns. Investigators have also compared the manner in which violent couples communicate as compared to nonviolent couples. The particular focus has been on the sorts of exchanges that violent couples engage in when discussing problems in their relationships. As noted above, violent couples generally show higher levels of anger, contempt, and belligerence than nonviolent couples, even when distress is controlled for. Another unique feature of violent couples may be that they are more likely to engage in what is termed "negative reciprocity" communication patterns (Margolin et al., 1988; Cordova et al., 1993). Negative reciprocity represents an interaction pattern where angry behavior on the part of one partner directly increases the likelihood of an angry response from the other partner. In this manner, angry or hostile affect is likely to have an escalating effect in violent couples. Interestingly, it appears that negative reciprocity interaction patterns are not contained to just violent husbands reciprocating with negative behavior. Wives of violent husbands also tend to display and reciprocate negative behaviors in similar rates to their husbands in laboratory interactions. In sum, violent couples do engage in repetitive, rigid, and escalating patterns of negative communication, characterized by high levels of belligerence and contempt. These communication patterns may provide a basis for understanding how even ordinary conflict may escalate to physical violence in battering couples.

Biological/ Medical Factors

To-date there have been few studies that have examined biological factors in battering. The vast majority of empirical work has focused on what are traditionally considered psychological or social variables. However, researchers and theorists alike are beginning to attend to similarities between general antisocial behavior and domestic violence. There is a substantial literature on biological risk factors for antisocial behavior and criminality (Brennan & Mednick, in press). As a result, it is likely that theoretical and empirical models of battering will further incorporate biological variables. The following section outlines some of the possible biological mechanisms that may underlie and influence male propensities toward violence.

Physiological Reactivity. An important exception to this tendency to dichotomize research into psychological or physiological variables is the work on domestic violence by Jacobson, Gottman, and their associates (Jacobson et al., 1994; Gottman et al., 1995). In this study, the investigators collected data on physiological, psychological, and social variables in order to examine the marital interactions of violent and nonviolent couples. A central component of this study was a 15-minute laboratory interaction where the couples discussed two problem areas in their relationship. During this interaction, each partner's physiological arousal was monitored. The investigators measured cardiac interbeat interval, pulse transmission time to the finger, finger pulse amplitude, skin conductance level, and general somatic activity level . Perhaps the most provocative finding from this study related to male batterers' heart rate reactivity during the conflict discussion. It is most common for heart rate to rise at the beginning of an interaction, compared to a baseline, especially if the topic of discussion is an area of conflict. However, for a subset of male batterers, roughly 20% in this sample, heart rate actually decreased during the interaction, suggesting that they were becoming more physiologically calm during the conflict discussion. This subsample of men were called Type 1 batterers, while those who became physiologically aroused during the interaction were called Type 2 batterers. The investigators discovered that Type 1 batterers were quite different from Type 2 batterers. Type 1 batterers were more belligerent and contemptuous toward their wives than Type 2 men. The violence toward their wives was more severe, and these men were generally more violent outside their marriage, toward friends, strangers, and coworkers or bosses. Type 1 batterers were also more likely to have witnessed physical violence between their parents. With regard to psychopathology, Type 1 men were more likely to be assessed as antisocial, drug dependent, and aggressive-sadistic. Taken together, the Type 1 batterers were clearly the most severe batterers in this sample and were categorically distinct from the other men in the study. Most interesting, a physiological variable proved to be a powerful discriminator between these two types of batterers.

We are left with how to interpret these heart rate reactivity findings and how they relate to more severe batterers. Gottman et al. (1995) and Jacobson et al. (1995) provided one possible explanation. They noted that heart rate deceleration, perhaps through changes in vagal tone, might be related to focused attention. They suggest that these men may be narrowing their attentional focus onto their partner, particularly at the beginning of the interaction, in order to maximize the impact of their verbal aggression. This focusing allows for a more controlled and "effective" display of aggression.

Additional questions revolve around any potential causal role that this physiological variable may play. Is this a dispositional characteristic for these men? For instance, is this lowered heart reactivity correlated in some fashion to underarousability, which has been connected with criminality in the past? Is this variable related to a physiological process which serves as precursor to domestic violence? Might lowered heart rate reactivity interact with social and psychological stressors? Alternatively, perhaps this lowered physiological arousal is an artifact of exposure to longer-term and more-severe violence. It may be a habituation response to conflict and severe violence, and hence the laboratory interaction environment was not stressful or physiologically evocative for these men. Type 1 men reported being exposed to higher levels of parental violence as children. They also reported significantly higher levels of bilateral or unilateral mother-to-father violence in their families. Although it appears that children, in general, do not typically habituate to intraparental conflict, there may some element in the nature of the violence that Type 1 children are exposed to that results in them learning to cope with a stressful family environment by not responding physiologically (see Jacobson et al., 1995).

Margolin and colleagues (1988) observed physically aggressive, verbally aggressive, withdrawing, and nondistressed/nonaggressive couples during two 10-minute problem-oriented discussions. Part of their study included subjects' self-reports on physiological and emotional arousal during the interaction. Margolin and colleagues discovered that physically aggressive men reported greater physiological arousal during conflict discussions than nonviolent men. This finding could suggest that part of the cycle of violence in battering couples relates to the men's increasing anger and physiological arousal, which likely escalates into out-of-control behavior in true physical confrontations. The finding that physically aggressive men display more negative affect and concurrently experience greater physiological arousal than nonviolent men supports the anger-management model that underlies most treatment programs for batterers.

Margolin et al.'s (1988) findings with regard to physiology may appear, on the surface, to be somewhat at odds with the Gottman et al.'s (1995) findings on heart rate reactivity noted above. However, there are important methodological differences between the studies which may explain the pattern of results. First, the physically aggressive couples in Margolin et al.'s (1988) study were more distressed than the other comparison couples in the study. Therefore, interpretation is difficult with regard to this group, since they differed on two variables: presence of violence and marital distress. Second, a substantial percentage of the nondistressed nonviolent couples in the Margolin et al. (1988) study did report some history of physical abuse. Third, in general, the physically aggressive group in this study was of a lesser severity, in terms of violence, than the couples in the Jacobson et al. (1994) study (see Cordova et al.(1993)). Finally, Margolin et al. (1988) relied on self-report for physiological data, while Jacobson et al. (1994) used direct measurements of physiological process. Self-report accounts of physiological process tend to correlate poorly with direct physiological measures. However, even in the face of these methodological differences, the findings from these two studies may still be consistent. In Jacobson et al's (1994) study, only a subset of severe batterers showed reduced physiological arousal during the interaction. The majority of the batterers showed increased physiological arousal. It is likely that Margolin et al's (1988) sample was mainly composed from this latter population of less-severe batterers. In other words, the Type 1 batterers that Gottman et al. (1995) are discussing may be categorically different men than those discussed in the Margolin et al. (1988) study.

Head Injury. Aggression is a common correlate of traumatic head injuries (e.g., Miller, 1994). Several studies by Rosenbaum, Warnken, and their colleagues have found a relationship between head injury and male battering. In one study, Rosenbaum et al. (1994) found that 53% of violent men in the sample had a history of head injury, compared to 25% of nonviolent sample. In another study, Rosenbaum and Hoge (1989) found that 61% of male batterers had histories of significant head injury. Warnken et al. (1994) compared a group of head-injured men with a comparison group of orthopedically-injured men. Although these two groups did not significantly differ with regard to either husband or wife reports of physical violence, female partners of head-injured men reported more post-injury verbal abuse by the men. Head-injured men also reported more post-injury problems with impulse control, such as losing their temper, arguing, and yelling more. These investigators have hypothesized that head injury can have both direct and indirect effects on battering. First, head injury can result in frontotemporal lobe dysfunction, which is often implicated in lowered impulse control and aggression (Warnken et al., 1994). Second, head injury can result in profound personality change, which can, in turn, influence violent behavior. Although the data so far seem to implicate head injury with battering, it is important to note that not all batterers have suffered head injury. Furthermore, the relationship between frontal lobe dysfunction and antisocial social behavior is still under question (Kandel & Freed, 1989; Raine & Scerbo, 1991).

Testosterone. Investigators have also begun to look at androgen levels in battering men as a possible biological mechanism related to increased aggression, with the expectation that high testosterone levels are connected to aggressive behavior. Booth & Dabbs (1993) measured testosterone levels in a normative sample of over 4000 former US military service men. Results revealed that men whose testosterone level was above the mean were more likely to have experienced trouble marital relations and were more likely to hit or throw things at their partner. Dabbs et al. (1995) examined testosterone levels, crime, and prison behavior in male prison inmates. They found that those prisoners that had committed crimes involving sex and violence, and those who evidenced confrontive behavior while in prison, had higher testosterone levels than those who were involved with property or drug crimes. McKenry et al. (1995) also examined testosterone levels, among other physiological and social variables, in battering men. When physiological variables were examined independently of psychosocial variables, higher testosterone level was a significant individual predictor of male violence toward partner. However, when all physiological and psychosocial variables were considered conjointly, testosterone level showed a trend toward significant prediction ($p < .10$). Although the data are preliminary and the effects are generally weak, there may be a connection between elevated testosterone levels and violence in battering men.

Biological Factors in Antisocial Behavior and Criminality: Possible Connections with Domestic Violence

As noted above, Gottman et al.(1995) discovered a physiological variable that powerfully distinguished between two types of batterers. Type I batterers were characterized by lowered heart rate reactivity during conflict interactions. These individuals were certainly the more severe batterers within an already quite violent sample. Interestingly, these men were also more likely to engage in violence and general antisocial behavior outside of their intimate relationship. It appears that these Type 1 batterers are the men that often interface with the criminal-justice system, and are therefore more likely to represent the prototypical batterer encountered by police, women's shelters, and court-ordered treatment programs. These men also tend to have concurrent substance abuse problems. With these men, physical violence and antisocial behavior are not confined to their marriage or intimate relationships, but rather hostility and violence are characteristic of how these individuals interact with the world. Most importantly, the use of violence by these men does not appear to emanate from anger-management problems. The opposite is true: these men seem to finely focus and control their aggression for maximal effect.

Domestic violence becomes more complicated once we adequately consider these Type 1 men. What emerges is an individual with a general pattern of aggression, who comes from a chaotic family background that features more severe physical and emotional abuse. This cluster of Type 1 batterers suggests that there may be some form of continuity between battering and more generalized antisocial behavior. This opens the realm of domestic violence to the literature on general antisocial behavior, impulsivity, and criminality. In the following section, we will briefly consider a fraction of the research on biological factors underlying impulsivity and antisocial behavior. We will focus on those pieces that seem to best connect with batterers. As there has been virtually no systematic research examining these variables and battering, we introduce these pieces as interesting speculation and possible future research ideas.

Underarousability and Sensation-Seeking. There is a large literature suggesting that physiology is related to criminality (Moffitt & Mednick, 1988; although see Raine, 1993). Specifically, individuals at risk for criminal behavior are proposed to be chronically underaroused, evidenced by low levels of physiological reactivity, such as lowered cardiovascular and cortical arousal. Individuals are thought to find this state of hypophysiologic reactivity aversive and stimulation is sought (sensation seeking), which, in turn may predispose to risk seeking and criminality (Schalling, Edman, & Asberg, 1983). In addition to lower physiological reactivity, criminal individuals are thought to have attentional deficits that may reflect underlying cognitive impairments that predispose to criminal behavior (Raine & Venables, 1984). In an important prospective study, Raine, Venables, & Williams (1990) examined cortical and autonomic activity, which were measured during adolescence (age 15), for 101 male school children in order to predict later criminality at age 24. They found that later-criminals had lower resting heart rate, lower skin conductance, and slower electroencephalographic (EEG) activity than noncriminals at adolescence. Raine, Venables & Williams (1995) also found that antisocial adolescents who desist from criminal behavior by age 29 show higher levels of physiological arousal and orienting response than those antisocial adolescents that become adult criminals. This led the investigators to suggest that high levels of autonomic arousal and orienting, which is thought to be an index of attention, may serve as protective factors against continued and future criminal behavior.

Robert Hare and colleagues have written and done extensive research on physiological arousal and its association with psychopathic criminal behavior (Hare, 1978; Hare, Frazelle, & Cox, 1978). Hare describes psychopathy as a personality disorder and form of chronic mental disorder. The psychopath is conceptualized as an individual lacking in guilt or anxiety, who suffers from impulsivity and an inability to form close interpersonal bonds. The psychopathic individual is thought to have difficulties in anticipating negative consequences for himself or others and this is proposed to be related to lowered autonomic responding. Psychopaths are also proposed to lack normal fear and startle responses to aversive stimuli (Patrick, 1994). However, Hare and others' work has also shown contradictions with regard to psychopaths' heart rate reactivity in the anticipation of aversive stimuli (see Dolan, 1994 for a review; also Raine, 1993). In addition, there continue to be serious problems with assessment of psychopathy (see Lilienfeld, 1994).

On the surface, there are certain similarities between the clinical descriptions of Hare's psychopaths and the Type 1 batterers identified by Gottman et al. (1995). In particular, Type 1 batterers seem to engage in the serious and wide ranging types of violence related to psychopathy, as well as more generalized antisocial behaviors. Interpersonally, Type 1 batterers certainly appear to be callous and lacking in guilt or anxiety. In terms of similarity of physiological reactivity between general criminality and battering, the picture is less clear. Although both batterers and those at-risk for criminality show signs of hypophysiologic reactivity, the kinds of reactivity may be different. While criminals seem to be underaroused at resting state, this may not be true for Type 1 batterers: their resting heart rates were not different from Type 2 batterers. Instead, it appears that Type 1 men became calmer as the conflict and stress increase. It may be that a sensation-seeking hypothesis is inappropriate for what is occurring with Type 1 men, and the hypophysiological process we observed is more of a reflection of attentional focus in the service of increased aggression (see Gottman et al., 1995; Jacobson et al., 1995). Clearly, more research examining physiological reactivity in different types of batterers is needed in order to draw firmer conclusions.

Serotonin. There is an accumulating research base examining neurochemical functioning and problems of aggression, violent suicidality, and impulsivity. Although nearly every identifiable neurotransmitter system has been implicated in aggressive behavior, there is growing empirical consensus that serotonin (5-hydroxytryptamine) functioning is related to aggression and impulsivity (Brown & Linnoila, 1990; Linnoila & Virkkunen, 1992). Specifically, it is hypothesized that low levels of serotonin are related to increased impulsivity, aggression, and suicidality. Linnoila and Virkkunen (1992) have proposed a "low serotonin syndrome" where deficits in serotonin activity at the level of the raphe nuclei are the driving force behind reductions in serotonergic output and resultant disturbances in both glucose metabolism and diurnal rhythm regulation. They postulate that the hypoglycemia is directly related to lower thresholds for impulsive violent behavior, and the disturbances in diurnal rhythm regulation are conducive to chronic low-grade dysphoria. Numerous studies have shown that serotonergic functioning may be altered in individuals with aggressive behaviors. Brown and Linnoila (1990) have concluded that these results "constitute one of the most highly replicated findings in biologic psychiatry." (p.37). Additionally, preliminary work suggests that selective serotonin reuptake intake inhibitors may be effective treatments for impulsive aggressive behavior in personality disordered patients (Kavoussi, Liu, & Coccaro, 1994). However, as with much of the research in this area, there are important methodological problems. For instance, the vast majority of these studies have inferred serotonin functioning through cerebrospinal fluid levels of 5-hydroxyindoleacetic acid (CSF 5-HIAA), which is the major metabolite of serotonin. CSF 5-HIAA is affected by many factors that are difficult to control for, including age, sex, body height, circadian and seasonal rhythms, drug intake, medical conditions, diet, physical activity, as well as amount and handling of CSF. In short, CSF 5-HIAA is not necessarily a "pure" marker of serotonergic activity.

To-date, there has only been one study examining serotonin functioning in batterers. McKenry, Julian, & Gavazzi (1995) examined prolactin levels in violent and nonviolent men. Prolactin is polypeptide produced by the anterior pituitary and thought to index overall CNS serotonin activity. Prolactin was not related to male violence in this study. In retrospect, McKenry et al. (1995) noted the low variability of prolactin levels in this study and potential inadequacy of prolactin as a proxy for serotonin as potential methodological problems. Nonetheless, research has been promising with regard to a serotonergic factor in aggressive and impulsive behavior and future domestic violence studies should examine the role of serotonin in battering.

Pre/Perinatal Factors and Medical Histories. Brennan and Mednick (in press) have recently reviewed the literature on medical and physical factors in antisocial behavior. In general, research has been mixed with regard to prenatal (conception through seventh month of gestation) and perinatal (seventh month of pregnancy to 28 days after birth) factors. Some studies have discovered that behavior disordered children have had higher rates of pregnancy complications than controls, although there have been notable exceptions. There is some preliminary evidence suggesting that minor physical anomalies (MPAs) and violent offending are related (Kandel, Brennan, Mednick & Michelsen, 1989). MPAs relate to prenatal complications and suggest interference with neurological development. Interestingly, studies are beginning to examine and find stronger results when considering interaction effects between pre/perinatal factors and social environments. For instance, Werner (1987) discovered that the effects of perinatal stress have differing effects on delinquency depending on whether children were exposed to a disruptive family environment. Other researchers have also examined dietary factors and mineral toxicity, as they relate to violence behavior. However, there are important methodological problems with extending this research to violent offenders, much less to battering populations. First, a significant number of these studies have looked at the association between perinatal factors and antisocial outcomes for children only. It is not clear to what extent these factors predict antisocial outcomes in adults, as there is not a one-to-one continuity between childhood antisocial behavior and adult criminality. Second, the outcome variables in childhood have been significantly variable from study to study, with some examining aggressive behavior and others hyperactive behavior, or some considering property damage crimes while others address physical assault.

SYNTHESIS: BIOSOCIAL MODELS AND INTERACTIONS

Raine, Brennan, and Farrington (in preparation) have discussed the emerging importance of biosocial models of violence. These models attempt to incorporate, in a comprehensive manner, both biological and social factors of violence, with the focus being on the interaction between these factors. Social variables in this framework include those that have traditionally been defined as psychological, such as individual difference variables like intelligence. Raine and colleagues describe various biosocial approaches in the study of violence. These include: additive and multiplicative biosocial effect models, where the presence of *both* biological and social deficits proves to be more predictive of violent behavior than either deficit in isolation, and a sequential biosocial effect model, where, for instance, a biological factor leads to changes in the social environment that directly influence violent behavior.

Although there have not been many studies in the field of general violence and criminality that have examined biosocial interactions and violence, the few studies that have provide very interesting results (see Raine, Brennan, & Farrington, in press, for a review; also Brennan & Mednick, in press). Investigators are increasingly discovering that the combination of biological and social factors uniquely improves prediction of violent or aggressive behavior. For instance, in a study examining the biosocial interaction of delivery complications and maternal rejection on adult violent crime, Raine, Brennan, and Mednick (1994) found that the combination of both delivery complications, which were interpreted as representing perinatal neurological damage, and maternal rejection was related to a disproportionate increase in adult violent crime. Maternal rejection was operationalized as either unwanted pregnancy, attempts to abort pregnancy, and/or public institutional care of child. Similarly, Mednick and Kandel (1988) discovered that the combination of minor physical anomalies, which, as noted above, are thought to relate to neurological damage, and unstable family environment predicted later violent crime. Moffitt (1990) examined how neuropsychological deficits and family environment related to adolescent aggression. She found that mean aggressive scores were four times greater for those children who experienced both neuropsychological deficits and family adversity. In a prospective study examining delinquent outcomes, Werner (1987) discovered that the impact of perinatal stress, and therefore presumably neurological damage, on delinquent outcome was strongest for those subjects who were exposed to a disruptive family environment. This included such stressors as marital discord, separation from parents, or family mental health problems. In sum, there is a growing body of evidence implicating biosocial interactions as important predictive factors.

Studies incorporating multivariate prediction models or examining complex interactions among numerous variables are a recent development in the field of domestic violence (Holtzworth-Munroe, Bates, Smutzler, & Sandin, in press). Even more unusual are studies that incorporate both variables from different domains, such as biological and social variables. However, researchers are increasingly examining multivariate, multimethod models, as it is becoming clear than the complexity of domestic violence is unlikely to be captured at the univariate level. One example is McKenry, Julian, & Gavazzi's (1995) study of 102 married men and their attempt to formulate a biosocial model of male violence. This was one of the few studies to include biological, social, and psychological variables, as well as include specific interactional predictions. The social and psychological factors that McKenry and colleagues measured included social stress, marital quality, socioeconomic status, and hostility. With regard to biological factors, they obtained measures or indices of serotonin functioning, alcohol usage, and testosterone levels. McKenry and colleagues found that *both* the physiological and social variables had a significant effect on violence. Specifically, alcohol usage, marital quality, and family income had the strongest impact, with higher testosterone levels showing a trend. However, none of the hypothesized interaction effects between biological and social variables reached statistical significance. As McKenry et al. (1995) note, this lack of interaction effects may have been due to the combination of both low sample size and restricted variance in some of the measures. Nonetheless, this study represents an important advance in attempts to conceptualize and study domestic violence in a multifaceted manner.

The work by Jacobson, Gottman, and colleagues also represents an initial attempt to incorporate biological variables into the study of domestic violence (Jacobson et al., 1994; Gottman et al., 1995). As noted earlier, Jacobson et al. (1994) measured both social and physiological variables in their investigation of domestically violent couples. Some of the results incorporate the sequential and additive biosocial effect approaches described above. For instance, the finding of lowered heart rate reactivity among more severe batterers can be interpreted as a sequential biosocial effect, where history of family violence (social factors) may influence heart rate reactivity through vagal tone (biological factor), which facilitates greater attentional focus (psychological factor) and, in turn, results in more intense violence. This is would represent but one pathway to violent outcomes. It is likely that the social factor of family violence exerts a direct influence on violence, in addition to any indirect pathway via physiology. Additionally, it is likely that biological factors, such as a heart rate reactivity or frontal lobe dysfunction, also exert direct and indirect pathways to violence and should be explored further.

Another finding from Jacobson and Gottman's work that relates to the biosocial approach involves attempts at predicting which women are more likely to leave abusive relationships. Jacobson and colleagues (1994) followed violent relationships for a two-year period. Of those violent couples available at two-year follow-up, nearly 40% were separated or divorced . Most interestingly, all separations involved couples were the batterers were Type 2 battering men: there were no divorces or separations among the Type 1 men. Jacobson and colleagues examined both social and physiological variables in order to better understand which variables at initial assessment predicted separation or divorce two years later. In line with the biosocial approach, the combination of physiological and psychological variables provided the best classificatory prediction with regard to which couples had separated/divorced or remained together (Jacobson et al., in press). Specifically, high levels of physiological arousal in combination with high levels of negative affect (in both partners) related to likelihood of separation or divorce.

The interpretation of this finding is complicated with regard to the role of biological factors. Finger pulse amplitude (FPA) is an estimate of the relative volume of blood reaching the finger on each heart beat (see Gottman et al., 1995). The sympathetic nervous system can change the distribution of central versus peripheral blood flow by constricting or dilating the peripheral blood vessels. Unfortunately, this is also a measure that is hard to calibrate, given individual differences in finger size and density. The direction of FPA change in battering men whose relationships were likely to end at two-year follow-up suggests a blood flow pattern away from the periphery and toward the trunk and central organs. This blood flow pattern is consistent with an alarm response, where blood is drawn toward central organs and main muscles (and also away from the periphery where blood loss might be greatest in the event of injuries.) It is possible that these batterers were experiencing alarm over being challenged in these interactions by their assertive and defensive partners. An early signpost of battered womens' likelihood of leaving an abusive relationship may be her willingness to "stand up" and provide aversive consequences in the face of anger and violence from her partner (Follingstad et al., 1992). What we were witnessing in those couples that were more likely to divorce may have been this phenomenon in action. In this sense, our physiological variable was a marker for what was occurring in the social interaction: the husbands were alarmed over losing control of the interaction. Alternatively, it may be that more unstable battering relationships are those where the male batterer indeed has the high arousability and intimacy-anger surrounding female independence that Dutton(1995) and others have proposed. The observed FPA changes may in some fashion relate to this physiological arousability. As we noted above, no battered women left any of the men who became physiologically calmer during the conflict discussions.

CONCLUDING COMMENTS AND FUTURE DIRECTIONS

Research in domestic violence continues to grow. Over the past decade, there has been a dramatic increase in scientific inquiry into the issue of battering. In this chapter, we have reviewed some of the major risk factors for domestic violence. However, our understanding of how, or even if, these factors function as causal mechanisms for male battering continues to be hampered by overly simplistic research models. We cannot give justice to the complexity of battering by singly focusing on univariate, cross-sectional designs. Until we consistently advance the field of domestic violence research through multivariate and longitudinal models that incorporate both biological and social variables, and their interactions, we will continue to be unable to distinguish between variables that are a cause, a marker, or a result of male battering. Although our analyses need to begin with the understanding that violence against women is embedded in, and sanctioned by, our patriarchal culture, we must further refine which psychological, social, and biological variables, both independently and conjointly, help explain why particular men in particular contexts are more likely to engage in physical violence toward women. Through our review of the existing literature, this chapter has highlighted some of those variables that hold promise in explaining battering behavior. Given the devastating consequences of domestic violence for women, children, and families, it is imperative that this research actively continue forward.

REFERENCES

Bauserman, S.K. & Arias, I. (1992). Relationships among marital investment, marital satisfaction, and marital commitment in domestically victimized and nonvictimized wives. *Violence and Victims, 7*, 287–296.

Bograd, M. (1988). Feminist perspectives on wife abuse: An introduction. In K. Yllo and M. Bograd (Eds.), *Feminist perspectives on wife abuse.* Newbury Park, CA:Sage.

Booth, A. & Dabbs, J.M. (1993). Testosterone and men's marriages. *Social Forces, 72*, 463–477.

Brennan, P.A. & Mednick, S.A. (in press). Perinatal and medical histories of antisocial individuals. In J.D. Maser, D.M. Stoff, & J. Breiling, (Eds.), *Handbook of Antisocial Behavior,*

Brown, G.L., & Linnoila, M.I. (1990). CSF serotonin metabolite (5-HIAA) studies in depression, impulsivity, and violence. *Journal of Clinical Psychiatry. 51*, 31–41.

Browne, A. (1993). Violence against women by male partners: Prevalences, outcomes, and policy implications. *American Psychologist, 48*, 1077–1087.

Cantos, A.L., Neidig, P.H., & O'Leary, K.D. (1994). Injuries of women and men in a treatment program for domestic violence. *Journal of Family Violence, 8*, 113–124.

Centerwall, B.S. (1995). Race, socioeconomic status, and domestic homicide. *Journal of American Medical Association, 273*, 1755–1758.

Cordova, J.V., Jacobson, N.S., Gottman, J.M., Rushe, R., & Cox, G. (1993). Negative reciprocity and communication in couples with a violent husband. *Journal of Abnormal Psychology, 102, 559–564.*

Dabbs, J.M., Carr, T.S., Frady, R.L., & Riad, J.K. (1995). Testosterone, crime, and misbehavior among 692 male prison inmates. *Personality and Individual Differences, 18*, 627–633.

Davies, P.T., & Cummings, E.M. (1994). Marital conflict and child adjustment: An emotional security hypothesis. *Psychological Bulletin, 116(3)*, 387–411.

Dobash, R.E., & Dobash, R.P. (1979). *Violence against wives: A case against patriarchy.* New York: The Free Press.

Dolan, M. (1994). Psychopathy: A neurobiological perspective. *British Journal of Psychiatry, 165*, 151–159.

Doumas, D., Margolin, G, & John, R.S. (1994). The intergenerational transmission of aggression across three generations. *Journal of Family Violence, 9*, 157–175.

Dutton, D.G. (1995). Male abusiveness in intimate relationships. *Clinical Psychology Review, 15(6)*, 567–581.

Dutton, D.G. (1994). Partriarchy and wife assault: The ecological fallacy. *Violence and Victims, 9*, 167–182.

Dutton, D.G. & Hart, S.D. (1992). Evidence for long-term, specific effects of childhood abuse and neglect on criminal behavior in men. *International Journal of Offender Therapy and Comparative Criminology, 36,* 129–137.

Follingstad, D.R., Rutledge, L.L., Berg, B.J., Hause, E.S., & Polek, D.S. (1990). The role of emotional abuse in physically abusive relationships. *Journal of Family Violence, 5,* 107–120.

Gelles, R.J. (1993). Through a sociological lense: Social structure and family violence. In R.J. Gelles & D.R. Loseke (Eds.), *Current Controversies on Family Violence,* 102–196. Newbury Park. Ca. Sage.

Gottman, J.M., Jacobson, N.S., Rushe, R., Shortt, J., Babcock, J., LaTaillade, J., & Waltz, J. (1995). The relationship between heart rate reactivity, emotionally aggressive behavior, and general violence in batterers. *Journal of Family Psychology, 9,* 227–248.

Hamberger, K. & Hastings, J. (1991). Personality correlates of men who batter and nonviolent men: Some continuities and discontinuities. *Journal of Family Violence, 6,* 131–147.

Hampton, R.L. & Gelles, R.J. (1994). Violence toward black women in a nationally representative sample of black women. *Journal of Comparative Family Studies, 25,* 105–119.

Hampton, R.L., Gelles, R.J., & Harrop, J.W. (1989). Is violence in black families increasing? A comparison of 1975 and 1985 national survey rates. *Journal of Marriage and the Family, 51,* 969–980.

Hare, R.D. (1978). Electrodermal and cardiovascular correlates of psychopathy. In R.D. Hare & D. Schalling (Eds.). *Psychopathic Behavior: Approaches to Research* (pp.107–144). New York: John Wiley and Sons.

Hare, R.D., Frazelle, J., & Cox, D.N. (1978). Psychopathy and physiological responses to threat of an aversive stimulus. *Psychophysiology, 15,* 165–172.

Holtzworth-Munroe, A., Bates, L., Smutzler, N., & Sandin, B. (in press). A brief review of the research on husband violence. Part I: Maritally violent versus nonviolent men. *Aggression and Violent Behavior.*

Holtzworth-Munroe, A., Smutzler, N., & Sandin, B. (in press). A brief review of the research on husband violence. Part II: The psychological effects of husband violence on battered women and their children. *Aggression and Violent Behavior.*

Holtzworth-Munroe, A., Smutzler, N., & Bates, L. (in press). A brief review of the research on husband violence. Part III: Sociodemographic factors, relationship factors, and differing consequences of husband and wife violence. *Aggression and Violent Behavior.*

Hotaling, G. & Sugarman, D. (1990). A risk marker analysis of assaulted wives. *Journal of Family Violence, 5,* 1–13.

Howell, M. & Pugliesi, K. (1988). Husbands who harm: Predicting spousal violence by men. *Journal of Family Violence, 3,* 15–27.

Jacobson, N.S., Gottman, J.M., Gortner, E.T., Berns, S, & Shortt, J. (In press). Psychological factors in the longitudinal course of battering. *Violence and Victims.*

Jacobson, N.S., Gottman, J.M., & Shortt, J. (1995). The distinction between type 1 and type2 batterers-further consideration: Reply to Ordnuff et al. (1995), Margolin et al. (1995), and Walker (1995). *Journal of Family Psychology, 9,* 272–279.

Jacobson, N.S. (1994). Rewards and dangers in researching domestic violence. *Family Process, 33,* 81–85.

Jacobson, N.S., Gottman, J.M., Waltz, J., Rushe, R., & Babcock, J. (1994). Affect, verbal content, and psychophysiology in the arguments of couples with a violent husband. *Journal of Consulting and Clinical Psychology, 62,* 982–988.

Julian, T. & McHenry, P. (1993). Mediator of male violence towards female intimates. *Journal of Family Violence, 8,* 39–55.

Kandel, E. & Freed, D. (1989). Frontal-lobe dysfunction and antisocial behavior: A review. *Journal of Clinical Psychology, 445,* 404–413.

Kandel, E., Brennan, P.A., Mednick, S.A., & Michelsen, N.M. (1989). Minor physical anomalies and recidivistic adult violent offending. *Acta Psychiatrica Scandinavica, 79,* 103–107.

Kavoussi, R.J., Liu, J., & Coccaro, E.F. (1994). An open trial of sertraline in personality disordered patients with impulsive aggression. *Journal of Clinical Psychiatry, 55,* 137–141.

La Taillade, J. & Jacobson, N.S. (In press). Domestic violence: Antisocial behavior in the family. In J.D. Maser, D.M. Stoff, J. Breiling (Eds.), *Handbook of Antisocial Behavior,*

Leonard, K.E. & Blane, H.T. (1992). Alcohol and marital aggression in a national sample of young men. *Journal of Interpersonal Violence, 7,* 19–30.

Lilienfeld, S. (1994). Conceptual problems with the assessment of psychopathy. *Clinical Psychology Review, 14,* 17–38.

Linnoila, V.M., & Virkkunen, M. (1992). Aggression, suicidality, and serotonin. *Journal of Clinical Psychiatry, 53,* 46–51.

McLaughlin, I., Leonard, K., Senchak, M. (1992). Prevalence and distribution of premarital aggression among couples applying for a marriage license. *Journal of Family Violence, 7,* 109–113.

McKenry, P.C., Julian, T.W., & Gavazzi, S.M. (1995). Toward a biopsychosocial model of domestic violence. *Journal of Marriage and the Family, 57*, 307–320.

Margolin, G., John, R.S., & Gleberman, L. (1988). Affective responses to conflictual discussion in violent and nonviolent couples. *Journal of Consulting and Clinical Psychology, 56*, 24–33.

Mednick, S.A., & Kandel, E. (1988). Genetic and perinatal factors in violence. In S.A. Moffitt and T. Moffit (Eds.) *Biological contributions to crime causation* (pp. 121–134). Dordrecht, Holland: Martinus Nijhoff.

Miller, L. (1994). Traumatic brain injury and aggression. *The Psychobiology of Aggression*, 91–103.

Millon, T. (1983). Millon Clinical Multiaxial Inventory, manual. Minneapolis, MN: Interpretive Scoring Systems.

Moffitt, T.E. (1990). The neuropsychology of juvenile delinquency. In M. Tonry and N. Morris (Eds.). *Crime and justice: A review of research* (Volume 12). Chicago: University of Chicago Press.

Moffitt, T.E., & Mednick, S.A. (Eds.). (1988). *Biological contributions to crime causation*. Boston: Martinus Nijhoff (published in cooperation with NATO Scientific Affairs Division).

Murphy, C.M., Meyer, S., & O'Leary, K.D. (1993). Family of origin violence and MCMI-II psychopathology among partner assaultive men. *Violence and Victims, 8(2)*, 165–176.

O'Leary, K.D., Barling, J., Arias, I., Rosenbaum, A., Malone, J., & Tyree, A. (1989). Prevalence and stability of physical aggression between spouses: A longitudinal analysis. *Journal of Consutlin and Clinical Psychology, 57*, 263–268.

O'Leary, K.D., Malone, J., & Tyree, A. (1994). Physical aggression in early marriage: Prerelationship and relationship effects. *Journal of Consulting and Clinical Psychology, 62*, 594–602.

Pan, H., Neidig, P., O'Leary. D. (1994). Predicting mild and severe husband to wife physical aggression. *Journal of Consulting and Clinical Psychology, 62*, 975–981.

Patrick, C.J. (1994). Emotion and psychopathy: Startling new insights. *Psychophysiology, 31*, 319–330.

Raine, A. (1993). *The Psychopathology of Crime: Criminal Behavior as a Clinical Disorder*. San Diego: Academic Press.

Raine, A., Brennan, P., & Farrington, D. (in preparation). Biosocial bases of violence: Conceptual and theoretical issues. In A. Raine, P. Brennan, D.P. Farrington, & S.A. Mednick (Eds.). *Biosocial bases of violence*. New York: Plenum.

Raine, A., Brennan, P, & Mednick, S. (1994). Birth complications combined with early maternal rejection at age 1 year predispose to violent crime at age 18 years. *Archives of General Psychiatry, 51*, 984–988.

Raine, A. & Scerbo, A. (1991). Biological theories of violence. In J.S. Milner (Ed.) *Neuropsychology of aggression* (pp. 1–25). Boston: Kluwer Academic.

Raine, A., & Venables, P.H. (1984). Electordermal non-responding, schizoid tendencies, and antisocial behavior in adolescents. *Psychophysiology, 21*, 424–433.

Raine, A., Venables, P.H., & Williams, M. (1995). High autonomic arousal and electrodermal orienting at age 15 years as protective factors against criminal behavior at age 29 years. *American Journal of Psychiatry, 152*, 1595–1600.

Raine, A., Venables, P.H., & Williams, M. (1990). Relationships between central and autonomic measures of arousal at age 15 years and criminality at age 24 years. *Archives of General Psychiatry, 47*, 1003–1007.

Rosenbaum, A., & Hoge, S.K. (1989). Head injury and male aggression. *American Journal of Psychiatry, 146*, 1048–1051.

Rosenbaum, A., Hoge, S.K., Adelman, S.A., Warnken, W.J., Fletcher, K.E., & Kane, R.L. (1994). Head injury in partner-abusive men. *Journal of Consulting and Clinical Psychology, 62*, 1187–1193.

Saunders, D.G. (1995). Prediction of wife assault. In J.C. Campbell (Ed.), *Assessing Dangerousness*. (pp. 68–95). Sage.

Schalling, D., Edman, G, & Asberg, M. (1983). Impulsive cognitive style and ability to tolerate boredom: Psychobiological studies of temperamental vulnerability. In M. Zuckerman (Ed.). *Biological bases of sensation seeking, impulsivity, and anxiety* (pp. 110–137). Hillsdale, NJ: Erlbaum.

Stets, J.E., & Straus, M.A. (1989). The marriage license as a hitting license: A comparison of assaults in dating cohabitating and married couples. *Journal of Family Violence, 4*, 161–180.

Straus, M.A. & Gelles, R.J. (1990). *Physical violence in American families: Risk factors and adaption to violence in 8,145 families*. New Brunswick, NJ: Transaction.

Straus, M.A. & Gelles, R.J. (1986). Societal change and change in family violence from 1975 to 1985 as revealed by two national surveys. *Journal of Marriage and the Family, 48*, 465–479.

Straus, M.A., Gelles, R.J., & Steinmetz, S.K. (1980). *Behind closed doors: Violence in the American Family*. Garden City, NY: Doubleday.

Warnken, W.J., Rosenbaum, A., Fletcher, K.E., Hoge, S.K., & Adelman, S.A. (1994). Head-injured males: A population at risk for relationship aggression? *Violence and Victims, 9*, 153–165.

Werner, E.E. (1987). Vulnerability and resiliency in children at risk for delinquency: A longitudinal study from birth to adulthood. In J.D. Burchard & S.N. Burchard (Eds.). *Primary Prevention of Psychopathology*, (pp. 16–43). Newbury Park, CA: Sage.

9

EMOTIONALITY AND VIOLENT BEHAVIOR IN PSYCHOPATHS

A Biosocial Analysis

Christopher J. Patrick, Kristin A. Zempolich, and Gary K. Levenston

Department of Psychology
Florida State University
Tallahassee, Florida 32306-1051

1. INTRODUCTION

According to the classic clinical description of psychopathy offered by Hervey Cleckley, violence and persistent criminality are not essential aspects of the disorder. He theorized that the primary features of psychopathy derive from a constitutional deficit in affectivity that actually diminished the likelihood of intense emotional displays, vengeful grudges, and angry aggression:

"It is my opinion that when the typical psychopath, in the sense with which the term is here used, occasionally commits a major deed of violence, it is usually a casual act done not from tremendous passion or as a result of plans persistently followed with earnest compelling fervor. There is less to indicate excessively violent rage than a relatively weak emotion breaking through even weaker restraints. The psychopath is not volcanically explosive, at the mercy of irresistible drives and overwhelming rages of temper. Often he seems scarcely wholehearted, even in wrath or wickedness." (Cleckley, 1976, p. 263).

However, empirical research within the past 15 years has demonstrated a strong relationship between psychopathy and violent behavior among incarcerated male offenders. Psychopathic criminal offenders are more likely to participate in a wider range and higher rate of violent acts than nonpsychopaths (Hare, 1981; Hare & McPherson, 1984). These findings stand in apparent opposition to Cleckley's conceptualization of the psychopath as emotionally detached and imperturbable.

It will be argued here that in exploring links between psychopathy and violence it is necessary to consider subtypes of aggressive behavior, and the role that emotionality plays in each. For instance, nonpsychopaths are less likely than psychopaths to commit assaults or routinely carry weapons, but they are more likely to commit murder (Hare & McPherson, 1984). It also appears that nonpsychopaths are more likely to commit acts of violence against friends or loved ones whereas psychopaths more often aggress against strangers.

Biosocial Bases of Violence, edited by Raine *et al.*
Plenum Press, New York, 1997

On the basis of these empirical findings, Hare (1981) has argued that the violent behavior of psychopaths is likely to be instrumental and cold-blooded rather than passionate.

Another important consideration pertains to the specific diagnostic features that mediate relationships between psychopathy and violence in criminal populations. Recent research indicates that *criminal* psychopathy involves two distinct facets, an emotional/interpersonal facet encapsulating the core symptoms emphasized by Cleckley, and an antisocial deviance component (Harpur, Hakstian, & Hare, 1988, 1989). Cleckley's patients were largely noncriminals from upper-middle class homes. Thus, it is possible that the heightened prevalence of violence among incarcerated psychopaths is related primarily to antisociality, and not to the core emotional symptoms of psychopathy.

This paper explores links between psychopathy and violence from the standpoint of affect- and temperament-related constructs. In the first section, recent research on emotional response in psychopaths is described, and an effort is made to characterize the two components of criminal psychopathy in terms of temperament traits. The second section explores correlations between the two factors of psychopathy and violent behavior. In the concluding section, links between the two psychopathy factors and social variables are examined. It is argued that an understanding of relationships between psychopathy and violent offending requires consideration of both constitutional and social variables.

2. EMOTION, TEMPERAMENT, AND PSYCHOPATHY

2.1. Conceptualization of Emotion

A prominent theme in modern theories of emotion is the notion of affects as response dispositions, or states of readiness for adaptive behavior (Izard, 1993; Lang, 1995; Plutchik, 1984). The term "motive" implies motion or movement: The affectively aroused organism is tuned for strategic, self-preservative action (Lang, Bradley, & Cuthbert, 1990). This perspective assumes that at a basic level, emotional reactions reflect the operation of two brain motive systems: an aversive system governing defensive reactions, and an appetitive system governing approach and consummatory behaviors (Gray, 1987; Konorski, 1967; Lang, Bradley, Cuthbert, & Patrick, 1993). These systems are primitive and subcortical (Lang, 1994; LeDoux, 1995), and ubiquitous among mammals (Schneirla, 1959). They underlie elementary conditioning phenomena (Konorski, 1967), exert a broad governing influence on emotional activation and expression, and account for the valence (pleasantness) and arousal dimensions of affective self-report (Lang, 1995; Russell, 1978).

However, these systems also interact with other areas of the brain, including higher declarative memory systems (LeDoux, 1995). Consequently, emotional reactions can be influenced by prior learning, by ongoing information processing, and by the unique constraints of a situational context. For example, a negative emotional reaction can be elicited by a simple sensory cue (e.g., a light signalling shock) or by a complex symbolic stimulus (e.g., a verbal description of a frightening event), and the reaction may lead to different behavioral expressions (e.g., freezing, flight, attack) depending upon the surrounding context.

Emotional processing is thus hierarchic (Stritzke, Lang, & Patrick, in press), involving the interaction of primitive, action-mobilization centers and other brain regions, including higher information processing systems. This hierarchic interplay accounts for the diversity that is evident in affective report and expression, and potentially for discrete emotions (Ekman, 1992; Izard, 1993) and their variants. Nevertheless, at the most basic level, affective reactions are presumed to reflect either appetitive or defensive motivation (Lang et al., 1990).

2.2. The Psychopathic Personality

The classic clinical description of the psychopathic personality was put forth by Cleckley (1976). His diagnostic criteria for the syndrome highlighted affective and inter-personal characteristics, including: general poverty of affect, deficient insight, lack of nervousness, absence of remorse or shame, superficial charm, pathological lying, egocentricity and incapacity for love, and failure to form intimate or even close relationships. Cleckley believed that these attributes were symptomatic of a deep-rooted emotional deficit, and that the behavioral features of psychopathy (i.e., irresponsibility, impulsive antisocial acts, failure to learn from experience, reckless behavior under the influence of alcohol, and a lack of long-term goals) were a byproduct of this deviation. His concept of psychopathy did not include explicit reference to aggressive behavior.

Hare (1980, 1991) devised the Psychopathy Checklist (PCL) as a means of identifying Cleckley psychopaths in prison settings. The recently revised version of the Checklist (PCL-R) comprises 20 items, each rated on a 0–2 scale (absent, equivocal, or present) on the basis of information obtained from a semi-structured interview and from prison files. The constituent item scores are summed to provide an overall score, with totals of 30 or more leading to a diagnosis of psychopathy.

Factor analyses of the Psychopathy Checklist (Harpur, Hakstian, & Hare, 1988; Harpur, Hare, & Hakstian, 1989) have revealed two oblique dimensions, labeled emotional detachment and antisocial behavior by Patrick, Bradley, & Lang (1993). PCL Factor 1 is marked by items reflecting the core affective and interpersonal symptoms of psychopathy that Cleckley emphasized. PCL Factor 2 is marked by items describing a chronic antisocial lifestyle, including child behavior problems, impulsiveness, irresponsibility, and absence of long-term goals. The PCL includes an item that deals specifically with anger and aggressiveness ("poor behavioral controls"), and this item loads on the antisocial behavior factor.

The two PCL factors show divergent relationships with independent indices of personality and behavior. Scores on the emotional detachment factor are negatively correlated with self-report anxiety scales and positively related to measures of social dominance, narcissistic personality, and Machiavellianism (Harpur et al., 1989; Hare, 1991), reflecting traits of shallow affectivity and self-serving exploitation of others. Ratings on the antisocial behavior factor are positively correlated with impulsivity, sensation seeking, and frequency of criminal offending (Harpur et al, 1989; Hare, 1991), and also substance abuse (Smith & Newman, 1990).

The diagnostic category of antisocial personality disorder (APD) described in DSM-IV (American Psychiatric Association, 1994) is not equivalent to psychopathy as defined by Hare's Checklist. The criteria for APD consist primarily of behavioral symptoms exhibited during childhood and adulthood, and empirically, DSM-IV APD is linked to the behavioral but not the emotional component of the PCL (Hare, Hart, & Harpur, 1991). As a result, many individuals who meet the DSM criteria for antisocial personality would not qualify as psychopaths according to Cleckley's definition. In prison populations, the prevalence of APD can be as high as 80% (Hare, 1991), in contrast to a base rate of only 25–30% for PCL-defined psychopathy (Hare et al., 1991).

Table 1. Items loading on the two factors of Hare's (1991)
Psychopathy Checklist-Revised (PCL-R)

1. Emotional detachment	2. Antisocial behavior
Glibness/superficial charm	Proneness to boredom
Grandiose sense of self-worth	Parasitic lifestyle
Pathological lying	Poor behavior controls
Conning/manipulative	Early behavior problems
Lack of remorse or guilt	Lack of realistic, long-term goals
Shallow affect	Impulsivity
Callous/lack of empathy	Irresponsibility
Failure to accept responsibility	Juvenile delinquency
	Revocation of conditional release

Note. The above is derived from the factor analytic research of Harpur et al. (1988,
1989) and Hare et al. (1990). The remaining PCL-R items (promiscuous sexual behavior, numerous short-term marital relationships, and criminal versatility) do not load on
either factor (Hare, 1991).

2.3. Affective Reactivity in Criminal Psychopaths

Cleckley (1976) theorized that the diagnostic features of psychopathy resulted from
a core affective deficit. He maintained that the true psychopath is incapable of intense
emotional reactions, either positive or negative. Other theorists (e.g., Fowles, 1980; Hare,
1970; Lykken 1957, 1995) have postulated that psychopathy involves a selective deficit in
negative emotional reactivity (i.e., fear). Much of the empirical research on affect and
psychopathy has been directed toward testing this low fear hypothesis, and considerable
support has been obtained for it. The most reliable findings are that psychopathic individuals show diminished electrodermal arousal during anticipation of noxious stimulation
(e.g., loud noise or electric shock), and impaired acquisition of a passive avoidance response (Hare, 1978, 1986; Lykken, 1995; Siddle & Trasler, 1981).

Patrick, Bradley, and Lang (1993) recently investigated this hypothesis using the
startle probe reflex as an index of fear. Eyeblink startle reactions elicited by unwarned
noise probes were recorded while criminal offenders, grouped using the PCL-R, viewed
pleasant, neutral, and unpleasant slides. Offenders with low or moderate PCL-R ratings
showed normal *linear* startle modulation, with blink responses larger during unpleasant
slides and smaller during pleasant slides, relative to neutral. In contrast, psychopaths
showed an abnormal *quadratic* startle pattern, with blink responses diminished during
both pleasant and unpleasant slides relative to neutral.

In view of evidence that startle reflex potentiation is mediated by subcortical defensive systems in the brain (Davis, 1989; Patrick, Berthot, & Moore, 1996), this finding provided strong support for the hypothesis that psychopathy involves a deficit in fear
reactivity. Furthermore, Patrick et al. (1993) found that this deficit was specific to the
emotional detachment features of the PCL-R: When subjects with high ratings on the antisocial behavior factor were subdivided into those low and high in emotional detachment,
only the latter group showed a deviant pattern of startle modulation. Subjects with high
scores on the behavioral component of psychopathy only (the majority of whom, along
with psychopathic subjects, met criteria for APD; Patrick, Cuthbert, & Lang, 1994)
showed robust startle potentiation during aversive slide viewing.

In a followup experiment (see Patrick, 1994, 1995), 75 federal prisoners were tested
in a procedure involving anticipation of an aversive noise blast. Acoustic startle reactions
were recorded during a warning cue preceding the stressor, and during intertrial intervals

(ITIs). Higher PCL-R emotional detachment predicted diminished startle potentiation, defined as the difference in blink reflex magnitude during warning cue periods as compared to ITIs. This negative relationship was strongest during the first block of the experiment, when the stressor was most potent. Higher emotional detachment was also associated with smaller heart rate and skin conductance orienting responses to the warning cue in block 1 of the experiment.

Figure 1 depicts startle findings for four subgroups selected a priori from this sample: (a) *nonpsychopaths*, defined by low scores on both psychopathy factors (*n* = 18); (b) *detached* offenders, scoring high on the Emotional Detachment factor but low on the Antisocial Behavior factor (*n* = 14); (c) *antisocial* offenders, scoring high on the Antisocial Behavior factor only (*n* = 8); and (d) *psychopaths*, obtaining high scores on both factors (*n* = 18). It can be seen that both the psychopathic and the detached groups showed reduced startle potentiation during noise anticipation relative to the nonpsychopathic and antisocial groups. Together with the findings of Patrick et al. (1993), these data suggest that psychopathy is characterized by diminished defensive reactivity, and that this deficiency is tied to the core affective features of the syndrome.

Concerning positive (appetitive) emotional reactivity, Lykken (1995) has argued that true psychopaths are normal in this realm. In the study by Patrick et al. (1993), psychopaths did not differ from nonpsychopaths or antisocial subjects in their physiological responses to pleasant affective stimuli: The normal pattern of startle reflex inhibition for positive as compared to neutral slides was observed in all prisoner groups. Consistent with this, Schmauk (1970) reported that psychopaths exhibited normal passive avoidance learning when a material reward was at stake. On the basis of findings from related studies of

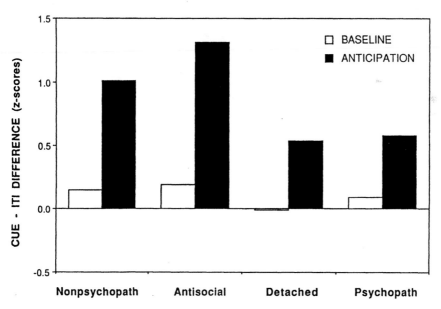

Figure 1. Mean blink magnitude difference scores (visual cue minus intertrial interval [ITI]) during baseline and during Block 1 of noise anticipation for four prisoner groups, defined according to scores on the two factors of Hare's (1991) Psychopathy Checklist-Revised. (From "Emotion and Psychopathy: Startling New Insights" by C. J. Patrick, 1994, *Psychophysiology*, 31, p. 328. Copyright 1994 by the Society for Psychophysiological Research. Reprinted with the permission of Cambridge University Press.)

passive avoidance learning in psychopaths, Newman and his colleagues (Gorenstein & Newman, 1980; Newman & Kosson, 1986; Newman, Widom, & Nathan, 1985) concluded that psychopathy may actually involve heightened reward sensitivity. In summary, the weight of the available evidence suggests that psychopathy is characterized by diminished defensive reactivity, but at least normal responsivity to appetitive cues (cf. Fowles, 1983).

2.4. Psychopathy and Temperament

Individual differences in emotionality and behavioral restraint are central to conceptions of temperament. Although temperament has been defined in a variety of ways, the term is typically used in reference to stable, enduring tendencies, particularly those of an affective nature. For example, emotional characteristics occupy a central role in in Buss and Plomin's (1975, 1984) theory of temperament. These investigators identified emotionality, activity, sociability, and impulsivity as basic dimensions of temperament, with emotionality linked to variability in thresholds for elicitation of defensive reactions, and activity and sociability related to positive emotional tendencies.

Patrick (1994) administered the self-report Emotionality-Activity-Sociability (EAS) Temperament Survey (Buss & Plomin, 1984), together with Buss and Plomin's (1975) Impulsivity scale, to a representative subset of the larger sample of inmates rated on the PCL-R, including the majority of those who participated in the noise anticipation procedure described in the preceding section. Correlations between PCL-R ratings and temperament scales scores were computed; in exploring relationships for the two oblique PCL-R factors (Harpur et al., 1988), partial correlational analyses were performed to control for the effects of the other factor.

Total PCL-R ratings were positively related to self-reported Impulsivity and EAS Anger, the latter a facet of Emotionality reflecting a tendency to respond to threat with aggressive attack (Buss & Plomin, 1984). Divergent relationships was noted for the two psychopathy factors, especially with the Emotionality subscales: PCL-R emotional detachment was negatively correlated with Fear and Distress, whereas antisocial behavior was positively related to these scales. Furthermore, it was the antisociality factor that accounted for the observed relationship of overall psychopathy to Impulsivity and Anger; these scales were uncorrelated with emotional detachment ratings after controlling for antisociality. These results coincide with the psychophysiological findings described earlier, in which offenders high in Emotional Detachment showed deficient startle potentiation during anticipation of noxious noise, whereas purely antisocial offenders exhibited normal or exaggerated startle potentiation.

In a more recent paper, Patrick (1995) reported correlations between the PCL-R and Tellegen's (1982) Multidimensional Personality Questionnaire (MPQ) for a sample of 174 male prisoners. The MPQ assesses a range of traits linked to three broad, temperament-oriented dimensions: Positive Emotionality (reward-seeking and proneness to pleasant mood), Negative Emotionality (propensity toward negative mood states), and Constraint (behaviorally cautious and inhibited versus impulsive and unrestrained). Overall psychopathy was associated with high scores on the MPQ Social Potency and Aggression subscales, with low Social Closeness, and with low scores on the higher-order CON factor. The emotional detachment factor of the PCL-R was related to high overall PEM (as a function of heightened Social Potency and Achievement scores), and with low Stress Reaction. The antisocial behavior factor, on the other hand, was associated with high overall NEM (including elevated Stress Reaction, Alienation, and Aggression), with low overall CON, and with low Achievement and Wellbeing.

Independently, Krueger, Schmutte, Caspi, Moffitt, Campbell, and Silva (1994) reported correlations between the MPQ and delinquent involvement as indexed by self- and informant-reports, police contacts, and court convictions in a sample of 18-year old men and women ($N = 862$). Higher rates of delinquency were associated with higher scores on the NEM factor of the MPQ and its component scales, and lower scores on the CON factor and constituent scales. In view of the fact that delinquency in this study was defined in terms of illegal behaviors and deviant behaviors, it is likely that the construct assessed was most closely related to factor 2 of the Psychopathy Checklist (antisocial behavior). In this regard, the MPQ correlations for delinquency are markedly similar to those reported by Patrick (1995) for the PCL antisociality factor.

The observed relationships between the MPQ and the PCL-R have implications for an understanding of links between psychopathy and aggressive behavior. The contrasting relationships between the two psychopathy factors and temperament-related personality traits are of particular interest. Antisocial deviance is associated with heightened stress, aggression, and behavioral disinhibition. This suggests that if there is a connection between psychopathy and defensive ("angry" or "reactive"; see, e.g., Buss, 1961, and Dodge, 1991) aggression, it is mediated by the antisocial deviance component. On the other hand, PCL-R emotional detachment, after controlling for antisociality, is linked to high social dominance and ambitiousness, and low anxiety. This suggests that aggression in the "true", Cleckley psychopath is more likely to be appetitively-oriented (i.e., "instrumental" or "proactive"; cf. Buss [1961], Dodge [1991]) than defensively-motivated.

High overall scores on the Psychopathy Checklist were reliably correlated with MPQ dominance and aggressiveness, low social affiliativeness, and low overall Constraint. The relationship of overall psychopathy to low Constraint, and to Buss and Plomin's (1975) Impulsivity scale, implies that inhibitions (including those against aggression; Megargee, 1982) are likely to be weaker among very high PCL-R scorers.

3. PSYCHOPATHY AND VIOLENCE

As noted earlier, several studies have demonstrated a link between criminal psychopathy and violent behavior, and some of the more recent reports have employed Hare's Checklist as a basis for diagnosis. There are no published studies examining relationships between the two factors of the PCL-R and violence. However, in an unpublished paper, Harpur and Hare (1991) reported that high psychopathy was associated with higher rates of instrumental violence as exemplified by weapons possession and use.

The present paper focuses on new data from the author's ongoing program of research on emotion and personality in criminal offenders. Participants in this study were adult male residents of the Federal Correctional Institution in Tallahassee Florida, a low medium security prison housing individuals convicted of federal offenses. Data reported here are from a sample of 228 offenders, with a mean age of 32 (range = 19 to 45). The sample included 95 Black inmates (41.7%), 27 White-Hispanic inmates (11.8%), 105 White inmates (46.1%), and 1 Asian inmate (0.4%). Measures discussed below were collected from varying proportions of the sample. The MPQ data described in the preceding section (Patrick, 1995) were obtained from this sample.

All participants in the study were assessed for psychopathy using Hare's (1991) PCL-R. The scatterplot in Figure 2 shows scores on the two PCL-R factors for all subjects in the sample. The primary effects reported below are correlations examining relationships between overall PCL-R and factor scores and dependent variables of interest in the sample as a whole;

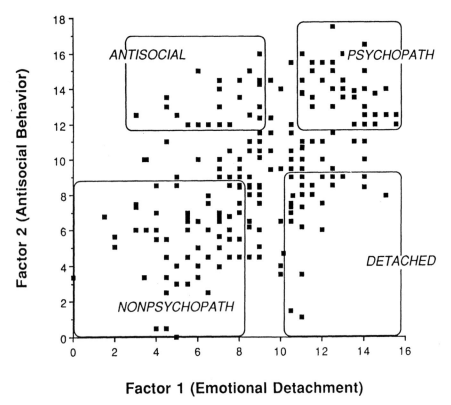

Figure 2. Scatterplot of Revised Psychopathy Checklist (Hare, 1991) factor scores for a sample of 225 male federal prison inmates, depicting diagnostic grouping based on levels of emotional detachment and antisocial behavior. Some subjects had scores that matched another subject's; these points overlap in the figure.

correlations for the two factors are partial correlations, controlling for the influence of the other factor. All correlations were evaluated at a two-tailed, .05 significance level. In some cases, effects were clarified by examining dependent variable means for subgroups defined on the basis of PCL-R total and factor scores: Individuals with low scores on both factors and with a total PCL-R score of 20 or less were classified as nonpsychopaths; those low in emotional detachment but high in antisocial behavior were defined as antisocial offenders; those high in emotional detachment but low in antisocial behavior were defined as detached offenders; and those with high scores on both factors and total PCL-R scores of 30 or greater were classified as psychopaths. These subgroups are also depicted in Figure 2.

The following dependent variables were coded from the PCL-R structured interview or from prison files.

3.1. Violent Offenses

Federal crime categories in the United States comprise a select range of offenses, including drug crimes (i.e., possession for purpose of trafficking, distribution, importation, conspiracy), bank robbery, certain weapons offenses (e.g., use of a weapon in the commission of a drug crime, transporting explosives), some types of fraud and forgery, and military crimes.

The modal current offense in the present sample involved drugs. A substantial proportion of offenders in the sample had prior criminal records, including violence-related offenses.

For each inmate, the number and types of violent charges or convictions listed in official criminal records contained in prison files were coded. Crimes coded as violent included robbery, assault, weapons offenses (e.g., possession of weapon, carrying concealed weapon, use of weapon in commission of a felony), murder or attempted murder, sexual offenses, kidnapping and false imprisonment, and arson. A highly significant relationship was found between overall PCL-R scores and total number of violent crimes committed, $r = .31$. The emotional detachment and antisocial behavior factors of the PCL-R showed lesser but significant correlations with violent crime, r's $= .14$ and $.16$, respectively.

Three of the individual violent crime categories showed sufficient base rates to permit separate analyses: assault, weapons offenses, and robbery. Robbery rates were not significantly related to PCL-R total or factor scores. For assaults, a significant relationship was found with overall psychopathy scores, $r = .22$, and with the antisocial behavior factor, $r = .24$; the relationship with the emotional detachment factor was negligible, $r = -.04$. If assaultiveness reflects a propensity for angry, impulsive aggression, it appears that this propensity is linked to the antisociality component of psychopathy, and not the core personality features.

A different pattern of relationships was found for weapons offenses: Possession of weapons was correlated significantly with overall psychopathy and with the emotional detachment factor, r's $= .27$ and $.17$, respectively, but not with the antisocial deviance factor, $r = .08$. If the use of weapons reflects calculated self-empowerment, it appears that the emotional detachment component of psychopathy may be more linked to deliberate, instrumental aggression (cf. Hare, 1981; Harpur & Hare, 1991).

Figure 3 depicts mean frequency of assault and weapons-related offenses for nonpsychopathic, antisocial, detached, and psychopathic subgroups, defined according the criteria mentioned earlier (see Figure 2). These group results are largely consistent with the correlational data incorporating the entire sample. Both detached and psychopathic subjects showed a higher rate of weapons charges than either nonpsychopaths or antisocial subjects. Antisocial subjects showed the highest frequency of assault charges. Although overall psychopathy was positively related to frequency of assaults, it can be seen from Figure 3 that individuals defined as psychopathic on the basis of high scores on both PCL-R factors were actually less likely to be assaultive than purely antisocial subjects, although they were more frequently assaultive than nonpsychopaths.

3.2. Other Aggression

In addition to violent offenses, the following indices of aggressive behavior were coded from structured interviews and prison files: number of aggression-related DSM Conduct Disorder criteria met (e.g., bullying or threatening others; initiating physical fights; physical cruelty to people or animals; forcible sexual activity); reported frequency of fights as a child; reported frequency of fights as an adult; and history of physical abuse of a spouse or relationship partner. All of these variables showed significant positive correlations with overall psychopathy and with PCL-R antisocial behavior scores, but negligible relationships with PCL-emotional detachment (see Table 2). These results indicate that psychopathy is linked to various other forms of physical aggression, and that the antisocial deviance factor accounts primarily for this relationship.

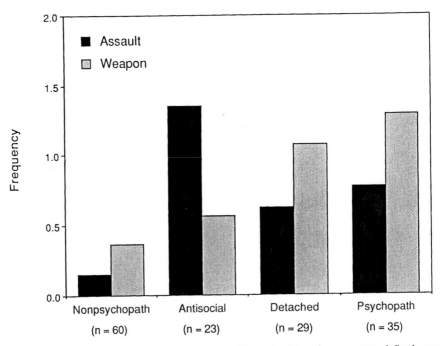

Figure 3. Mean number of assault and weapons-related offenses for four prisoner groups, defined according to scores on the two factors of Hare's (1991) PCL-R.

3.3. Other Deviant Behaviors

Table 2 also depicts correlational relationships between PCL-R total and factor scores and the following indices of deviant behavior derived from structured interview information and prison files: number of DSM Conduct Disorder criteria, excluding aggression-related criteria; age of first official arrest or conviction; number of adult criteria met for DSM APD, excluding the criterion of physical aggressiveness; and total number of nonviolent charges or convictions listed in official criminal records.

Predictably, overall psychopathy and the antisocial behavior factor of the PCL-R were strongly and positively related to Conduct Disorder symptoms, DSM adult antisociality, and number of nonviolent offenses. PCL-R emotional detachment, after controlling for antisociality, was positively related to DSM adult antisocial behavior and to nonviolent offense history; however, emotional detachment in itself was unrelated to child Conduct Disorder, and it showed a *positive* relationship with age at first offense. The implication is that the criminal deviance of detached offenders in our sample (i.e., meeting the Cleckley criteria for psychopathy, but scoring relatively low in PCL-R antisocial deviance) was restricted primarily to adulthood.

4. RISK AND PROTECTIVE FACTORS IN PSYCHOPATHY AND ANTISOCIALITY

The evidence reviewed thus far suggests that the core personality traits of psychopathy are associated with a deficit in emotional reactivity that may represent a constitutional risk factor for criminal deviance (Hare, 1970; Lykken, 1995). However, the data also suggest that some individuals possessing these core traits are not impulsively deviant and that they successfully avoid encounters with the law for much of their lives. These individuals do not show a heightened propensity for most forms of violent behavior, although they may engage in instrumentally aggressive behavior (e.g., employment of weapons). On the other hand, it appears that a significant proportion of persistently antisocial individuals are not psychopathic in the classic (Cleckley) sense. These individuals appear to be prone to impulsive, retaliatory forms of aggression such as fighting, spousal abuse, and assault.

It has been suggested that a variety of psychosocial risk factors exist that increase the probability of criminality and violent behavior in individuals who are not constitutionally predisposed toward antisociality, and that there are protective factors that may decrease the rate of antisocial behavior in individuals who are so predisposed (Lykken, 1995; Raine, 1993). This section examines relationships between the two factors of psychopathy and potential protective and risk factors. The main prediction is that detached individuals showing the core features of psychopathy in the absence of severe antisociality will show greater psychosocial advantages relative to individuals who are persistently antisocial but not psychopathic.

4.1. Social and Cognitive Variables

Figure 4 depicts relationships between PCL-R scores and two indices of social and cognitive functioning for inmates in the study sample: socioeconomic status, computed from educational and occupational data using the Hollingshead and Redlich criteria (1958) criteria, and Verbal IQ as measured by the vocabulary subtest of the Shipley Institute of Living Scale (Shipley, 1940). The latter measure was available only for a small subset of the overall sample (N = 91).

Consistent with prediction, the data reveal a significant positive, relationship between SES and emotional detachment (after controlling for antisociality), indicating advantaged educational and occupational status among primarily detached offenders. The

Table 2. Correlations between psychopathy scores and behavioral deviance indices

Deviance index	Overall psychopathy	Emotional detachment	Antisocial behavior
Aggression-related			
CDO-agg	.43*	−.05	.48*
Fights as child	.30*	−.13	.40*
Fights as adult	.24*	−.03	.26*
Domestic assault	.31*	−.05	.33*
Other			
CDO-nonagg	.61*	.06	.59*
Age at 1st offense	−.31*	.21*	−.47*
Adult APD	.76*	.38*	.60*
Nonviolent offenses	.46*	.18*	.28*

Note. Overall Psychopathy = total score on Hare's (1991) Revised Psychopathy Checklist. Emotional Detachment and Antisocial Behavior = scores on Psychopathy Checklist factors 1 and 2, respectively (Harpur et al., 1988, 1989; Hare et al., 1990). CDO-agg = number of aggression-related DSM-IV conduct disorder criteria met; CDO-nonagg = number of nonaggression-related conduct disorder criteria met. Adult APD = number of DSM-IV adult antisocial personality disorder criteria met, excluding the criterion of aggressiveness. Correlations for each of the two psychopathy factors are partial correlations, controlling for the influence of the other factor.
*$p<.05$.

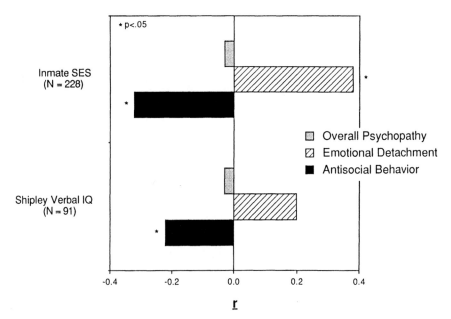

Figure 4. Correlations between ratings of criminal offenders on the two PCL-R factors and indices of social and cognitive functioning. Asterisks denote correlations significant at the .05 level.

relationship between SES and antisocial behavior (after controlling for emotional detachment) was negative, indicating diminished educational and occupational achievement among primarily antisocial offenders. The relationship between SES and overall psychopathy scores was not significant.

A similar pattern of results was found for the Verbal IQ measure: The correlation between antisociality and Shipley verbal scale scores was negative, whereas the relationship between emotional detachment and Verbal IQ was positive, although the latter correlation only approached significance. The relationship between overall PCL-R scores and Verbal IQ was not significant.

4.2. Family Variables

Further analyses were conducted to examine relationships between PCL-R scores and three family history variables: (a) whether or not the inmate grew up in a single parent home; (b) father's occupational status, coded using the Hollinghead and Redlich (1958) criteria; and (c) whether or not the inmate was abused as a child. The first variable was coded on the basis of interview information, and thus was available for most of the sample. The second variable was coded on the basis of file information, and thus was available for only a subsample of individuals ($N = 127$). The latter variable was coded on the basis of both file and interview information; if either source indicated a history of abuse, the variable was coded positively.

Overall psychopathy and emotional detachment scores were negatively associated with single parent status, r's = -.17 and -.21, respectively, suggesting that psychopaths and emotionally detached individuals were *less* likely to come from single-parent homes. Figure 5 (upper panel) depicts the proportion of subjects reporting single parent rearing

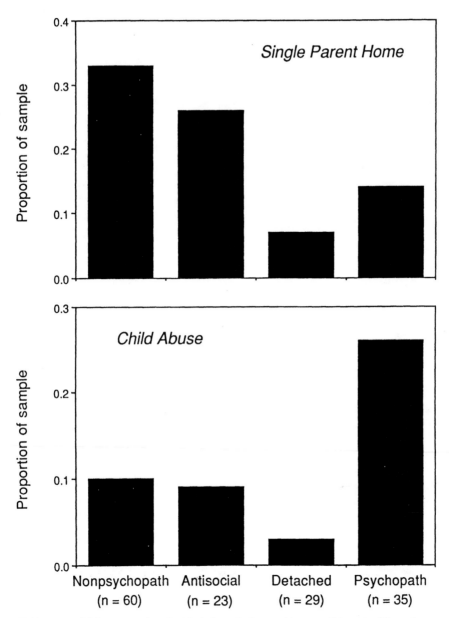

Figure 5. Upper panel: Mean proportion of subjects from single parent homes within each of four prisoner groups, defined according to scores on the two factors of Hare's (1991) PCL-R. Lower panel: Mean proportion of subjects exposed to abuse during childhood, according to interview and file records, within each of four prisoner groups defined according to scores on the two PCL-R factors.

within the nonpsychopathic, antisocial, detached, and psychopathic inmate subgroups, de-fined according to PCL-R cutoffs (see Figure 2). Consistent with prediction, detached of-fenders reported the lowest rate of single parent rearing. Interestingly, both nonpsychopaths and antisocial offenders showed relatively high rates of single parent rearing in relation to both detached and psychopathic offenders, accounting for the absence of a predicted relationship between antisociality and single parent status. One interpretation might be that individuals who lack the core features of psychopathy but who nevertheless become incarcerated are more likely to hail from broken homes than psychopathic individuals.

The data for father's occupational status were entirely consistent with prediction. A significant positive relationship was observed for the emotional detachment factor (after controlling for antisociality), r = .26, indicating more advantaged familial backgrounds among primarily detached offenders. The relationship between father's occupation and antisocial behavior (after controlling for emotional detachment) was reliably negative, r = -.19, indicating less privileged familial status among primarily antisocial offenders. The relationship between SES and overall psychopathy scores was not significant, *r* = .04.

For the child abuse variable, the only significant relationship was with overall psychopathy scores, r = .14, indicating that high PCL-R scorers were more likely to show a history of victimization during childhood than low PCL-R scorers. This correlation was evident in separate analyses of the data coded from prison files as well as the information obtained from diagnostic interviews, although the most robust relationship was found when both sources were combined. Figure 5 (lower panel) depicts the proportion of subjects with child abuse histories in the four PCL-R defined diagnostic groups. As expected, detached offenders as a group were most "protected" from child abuse. By far the greatest incidence of reported child abuse was found among offenders with high scores on both factors of the PCL-R (i.e., psychopaths).

5. CONCLUSIONS

On the basis of the findings reviewed in this paper, the following conclusions are advanced:

i. The core personality traits of psychopathy ("emotional detachment") are associated with diminished response to emotional cues, and specifically cues that are negative or aversive. This deficit is evidenced by a diminution or absence of normal startle reflex potentiation during exposure to fearful or threatening stimuli. In terms of temperament traits, the core features of psychopathy are linked to high dominance and low anxiousness.

ii. The classic personality features of psychopathy are dissociable from persistent antisocial deviance: Not all severely antisocial individuals are psychopaths, and not all psychopathic individuals are chronically antisocial. Antisociality per se is linked to high negative emotionality (including heightened stress reactivity and aggression) and low Constraint (high impulsivity).

iii. The two factors of criminal psychopathy are differentially related to violent behavior. The antisocial behavior component, consistent with its observed temperamental correlates, is associated with impulsive, angry aggression (e.g., fighting, spousal abuse, assault). The emotional detachment component is associated more with calculated, instrumental aggression (e.g., weapons use; cf. Harpur & Hare, 1991).

An important caveat to the violence data is that in this study the dependent measures were not entirely independent of the diagnostic predictors, since aggressiveness ("poor behavioral controls") is one criterion for psychopathy. Although it is possible to remove this item from the scoring of the PCL-R (cf. Harpur & Hare, 1991), it is quite likely that the scoring of other PCL-R items (e.g., absence of remorse, low empathy, impulsivity) is influenced by the presence or absence of violent behavior. Thus, at this stage, the reported relationships between psychopathy and violence should be regarded as primarily descriptive.

In future research, it should be possible to address this issue by having diagnosticians rate offenders based on clinical data, excluding information about violence. In the meantime, the fact that contrasting relationships between the two psychopathy factors and violence were mirrored by meaningful, contrasting relatioships in the independently derived temperament data are encouraging.

iv. Individuals who are criminal but not psychopathic appear more likely to come from single parent homes. Those who are persistently antisocial but who lack the core personality features of psychopathy—individuals especially prone to assaultive violence—appear more likely to come from economically disadvantaged backgrounds, to be lower in verbal ability, and to achieve lower socioeconomic status themselves.

On the other hand, individuals who possess the personality features of psychopathy but who avoid chronic criminal careers showed the least evidence of family impoverishment. They were most likely to come from economically advantaged, dual parent homes, and to have avoided abuse as a child. As adults, they showed greater socioeconomic achievement and evidence of higher verbal intelligence.

These data suggest an interaction between temperament traits and social influences in determining trajectories toward violence and criminal deviance.

v. In terms of social risk factors, the only variable that distinguished criminal psychopaths (i.e., chronically antisocial individuals possessing the core personality traits of psychopathy) from other offenders was a heightened incidence of abuse as a child. This finding must be viewed with caution as the data were not derived from official child abuse registries.

Assuming this effect is real, however, several interpretations are possible. One is that among individuals who are temperamentally predisposed to psychopathy, child abuse contributes to a criminally deviant expression of these tendencies. Another possibility is that children who are temperamentally deviant are more likely to elicit physical punishment from caretakers in the absence of counteracting protective factors (e.g., educational and economic advantage on the part of parents). A third possibility is that individuals who are temperamentally predisposed toward psychopathy are more likely to have parents with similar traits who rely upon coercive and punitive methods of control (Lykken, 1995). All of these possibilities are consistent with a biosocial perspective on crime causation.

ACKNOWLEDGMENTS

Preparation of this chapter was supported by Grants MH48657 and MH52384 from the National Institute of Mental Health.

REFERENCES

American Psychiatric Association. (1994). *Diagnostic and statistical manual of mental disorders* (4th ed.). Washington, DC: Author.

Buss, A. (1961). *The psychology of aggression*. New York: Wiley.

Buss, A.H., & Plomin, R. (1975). *A temperament theory of personality development*. New York: Wiley.

Buss, A.H., & Plomin, R. (1984). *Temperament: Early developing personality traits*. Hillsdale, NJ: Erlbaum.

Cleckley, H. (1976). *The mask of sanity* (5th ed.). St. Louis, Mo.: Mosby.

Davis, M. (1989). Neural systems involved in fear-potentiated startle. In M. Davis, B.L. Jacobs, & R.I. Schoenfeld (Eds.), *Annals of the New York Academy of Sciences, vol. 563: Modulation of defined neural vertebrate circuits* (pp. 165–183). New York: Author.

Dodge, K. A. (1991) The structure and function of reactive and proactive aggression. In D. J. Pepler & K. H. Rubin (Eds.), *The development and treatment of childhood aggression* (pp. 201–218). Hillsdale, NJ: Lawrence Erlbaum.

Ekman, P. (1992). An argument for basic emotions. *Cognition and Emotion, 6*, 169–200.

Fowles, D.C. (1980). The three arousal model: Implications of Gray's two-factor learning theory for heart rate, electrodermal activity, and psychopathy. *Psychophysiology, 17*, 87–104.

Fowles, D.C. (1983). Motivational effects on heart rate and electrodermal activity: Implications for research on personality and psychopathology. *Journal of Research in Personality, 17*, 87–104.

Gorenstein, E.E., & Newman, J.P. (1980). Disinhibitory psychopathology: A new perspective and a model for research. *Psychological Review, 87*, 301–315.

Gray, J.A. (1987). *The psychology of fear and stress* (2nd ed.). Cambridge: University of Cambridge Press.

Hare, R.D. (1970). *Psychopathy: Theory and research*. New York: Wiley.

Hare, R.D. (1978). Electrodermal and cardiovascular correlates of psychopathy. In R.D. Hare & D. Schalling (Eds.), *Psychopathic behavior: Approaches to research* (pp. 107–143). Chichester: Wiley.

Hare, R.D. (1980). A research scale for the assessment of psychopathy in criminal populations. *Personality and Individual Differences, 1*, 111–119.

Hare, R. D. (1981). Psychopathy and violence. In J. R. Hays, T. K. Roberts, & K. S. Solway (Eds.), *Violence and the violent individual* (pp. 53–74). New York: Spectrum Publications.

Hare, R. D. (1986). Twenty years of experience with the Cleckley psychopath. In W. H. Reid, D. Dorr, J. I. Walker, & J. W. Bonner (Eds.). *Unmasking the psychopath* (pp. 3–27). New York: W. W. Norton & Co.

Hare, R.D. (1991). *The Hare Psychopathy Checklist-Revised*. Toronto: Multi-Health Systems.

Hare, R.D., Hart, S.D., & Harpur, T.J. (1991). Psychopathy and the proposed DSM-IV criteria for antisocial personality disorder. *Journal of Abnormal Psychology, 100*, 391–398.

Hare, R. D., & McPherson, L. M. (1984). Violent and aggressive behavior by criminal psychopaths. *International Journal of Law and Psychiatry, 7*, 35–50.

Harpur, T.J., Hakstian, A.R., & Hare, R.D. (1988). Factor structure of the psychopathy checklist. *Journal of Consulting and Clinical Psychology, 56*, 741–747.

Harpur, T.J., Hare, R.D., & Hakstian, A.R. (1989). Two-factor conceptualization of psychopathy: Construct validity and assessment implications. *Psychological Assessment: A Journal of Consulting and Clinical Psychology, 1*, 6–17.

Harpur, T. J., & Hare, R. D. (1991, August). *Psychopathy and violent behavior: Two factors are better than one*. Paper presented at the 99th Annual Meeting of the American Psychological Association, San Francisco.

Hollingshead, A. B., & Redlich, F. C. (1958). *Social class and mental illness*. New York: Wiley.

Izard, C. E. (1993). Four systems for emotion activation: Cognitive and noncognitive processes. *Psychological Review, 100*, 68–90.

Konorski, J. (1967). *Integrative activity of the brain: An interdisciplinary approach*. Chicago: University of Chicago Press.

Krueger, R. F., Schmutte, P. S., Caspi, A., Moffitt, T. E., Campbell, K., & Silva, P. A. (1994). Personality traits are linked to crime among men and women: Evidence from a birth cohort. *Journal of Abnormal Psychology, 103*, 328–338.

Lang, P. J. (1994). The motivational organization of emotion: Affect-reflex connections. In S. Van Goozen, N. E. Van de Poll, & J. A. Sergeant (Eds)., *The emotions: Essays on emotion theory* (pp. 61–93). Hillsdale, NJ: Lawrence Erlbaum.

Lang, P. J. (1995). The emotion probe: Studies of motivation and attention. *American Psychologist, 50*, 372–385.

Lang, P.J., Bradley, M.M., & Cuthbert, B.N. (1990). Emotion, attention, and the startle reflex. *Psychological Review, 97*, 377–398.

Lang, P.J., Bradley, M.M, Cuthbert, B.N., & Patrick, C.J. (1993). Emotion and psychopathology: A startle probe analysis. In L. Chapman & D. Fowles (Eds.), *Progress in experimental personality and psychopathology research, vol. 16* (pp. 163–199). New York: Springer.

LeDoux, J. E. (1995). Emotion: Clues from the brain. *Annual Review of Psychology, 46*, 209–235.

Lykken, D.T. (1957). A study of anxiety in the sociopathic personality. *Journal of Abnormal and Clinical Psychology, 55*, 6–10.

Lykken, D. T. (1995). *The antisocial personalities*. Hillsdale, NJ: Erlbaum.

Megargee, E. I. (1982). Psychological determinants and correlates of criminal violence. In M. E. Wolfgang & N. A. Weiner (Eds.), *Criminal violence* (pp. 81–170). London: Sage Publications.

Newman, J. P., & Kosson, D. S. (1986). Passive avoidance learning in psychopathic and nonpsychopathic offenders. *Journal of Abnormal Psychology, 95*, 252–256.

Newman, J.P., Widom, C.S., & Nathan, S. (1985). Passive avoidance in syndromes of disinhibition: Psychopathy and extraversion. *Journal of Personality and Social Psychology, 48*, 1316–1327.

Patrick, C. J. (1995, Fall). Emotion and temperament in psychopathy. *Clinical Science*, 5–8.

Patrick, C. J. (1994). Emotion and psychopathy: Startling new insights. *Psychophysiology, 31*, 319–330.

Patrick, C. J., Berthot, B. D., & Moore, J. D. (1996). Diazepam blocks fear-potentiated startle in humans. *Journal of Abnormal Psychology, 105*, 89–96.

Patrick, C.J., Bradley, M.M., & Lang, P.J. (1993). Emotion in the criminal psychopath: Startle reflex modulation. *Journal of Abnormal Psychology, 102*, 82–92.

Patrick, C. J., Cuthbert, B. N., & Lang, P. J. (1994) Emotion in the criminal psychopath: Fear image processing. *Journal of Abnormal Psychology, 103*, 523–534.

Plutchik, R. (1984). Emotions: A general psychoevolutionary theory. In K. Scherer & P. Ekman (Eds.), *Approaches to emotion* (pp. 197–219). Hillsdale, NJ: Erlbaum.

Raine, A. (1993). *The psychopathology of crime*. San Diego: Academic Press.

Russell, J. A. (1978). Evidence of convergent validity on the dimensions of affect. *Journal of Personality and Social Psychology, 36*, 1152–1168.

Schmauk, F. J. (1970). Punishment, arousal, and avoidance learning in sociopaths. *Journal of Abnormal Psychology, 76*, 325–335.

Schneirla, T. C. (1959). An evolutionary and developmental theory of biphasic processes underlying approach and withdrawal. In *Nebraska Symposium on Motivation: 1959* (pp. 1–42). Lincoln: University of Nebraska Press.

Shipley, W. C. (1940). A self-administering scale for measuring intellectual impairment and deterioration. *Journal of Personality, 9*, 371–377.

Siddle, D.A.T., & Trasler, G.B. (1981). The psychophysiology of psychopathic behavior. In M.J. Christie & P.G. Mellett (Eds.), *Foundations of psychosomatics* (pp. 283–303). New York: Wiley.

Smith, S.S., & Newman, J.P. (1990). Alcohol and drug abuse-dependence disorders in psychopathic and nonpsychopathic criminal offenders. *Journal of Abnormal Psychology, 99*, 430–439.

Stritzke, W. G. K., Lang, A. R., & Patrick, C. J. (in press). Beyond stress and arousal: A reconceptualization of alcohol-emotion relations with special reference to psychophysiological methods. *Psychological Bulletin.*

Tellegen, A. (1982). *Brief manual for the Multidimensional Personality Questionnaire.* Unpublished manuscript, University of Minnesota.

10

BIOSOCIAL INTERACTIONS AND VIOLENCE

A Focus on Perinatal Factors

Patricia A. Brennan,[1] Sarnoff A. Mednick,[2] and Adrian Raine[2]

[1]Department of Psychology
Emory University
Atlanta, Georgia 30322
[2]Department of Psychology
University of Southern California
Los Angeles, California 90089-1061

1. INTRODUCTION

In this chapter, we will be examining data within the framework of the biosocial model developed by Adrian Raine and presented earlier in this volume. In this model, the differential effects of biological and social factors on violence are considered in a systemic fashion. As outlined in the model, biological and social factors can act alone or in a variety of combinations to increase the risk for violent outcome. One of the neglected areas of biosocial research on violence is the examination of statistical interactions between biological and social factors (Brennan & Raine, in press). This chapter will focus on such interactions and on the role of perinatal factors in the outcome of early-onset aggression and persistent criminal violence.

1.1. Developmental Psychopathology Perspective

Examination of the persistent or violent offender fits in well with a developmental psychopathology perspective. Current theories suggest that the persistent or violent criminal offender may be distinct from other offenders (Loeber, Wung, Keenan, Giroux, Stouthamer-Loeber, Van Kammen & Maughan, 1993; Moffitt, 1993). Our past research has shown that individuals arrested previously for a violent offense are significantly more likely than other criminal offenders to be arrested for violence in the future (Brennan, Mednick & John, 1989). These persistent and violent offenders may be behaviorally distinct from other individuals at a young age. Aggressive behavior has been found to be quite stable from age three to adulthood (Olweus, 1979). Persistent offenders have a young age of onset for crime—they are aggressive as children, delinquent in adolescence and continue to commit criminal offenses, including violence, in adulthood (Farrington, 1991).

Biosocial Bases of Violence, edited by Raine *et al.*
Plenum Press, New York, 1997

In developmental psychopathology, one focuses on cumulative risk and protective factors that result in negative behavioral or emotional outcomes. When focusing on persistent or violent offending, it makes sense to look at risk factors that occur in early childhood—prior to the onset of this developmental process. Perinatal factors are one such type of early potential risk. Perinatal factors include low birthweight, pregnancy complications such as mother's poor nutrition and viral infections, and delivery complications such as hypoxia or lack of oxygen to the fetus during labor. These medical risk factors have been hypothesized to lead to central nervous system damage, which in turn, is thought to increase the risk for persistent criminal offending (Moffitt, 1993). Central nervous system damage as a mediating factor fits well with the literature on violence and aggression, as neurological and neuropsychological deficits have been found to be especially related to persistent or violent offending (Gorenstein, 1990).

1.2. Perinatal Factors and Crime

Very little research has been carried out on the relationship between perinatal factors and criminal behavior. The research that exists suggests that perinatal factors may be more related to violence, rather than delinquency or property offending. For example, Lewis and her colleagues have noted this pattern in their examination of the health histories of delinquents. Violent delinquents are found to have more perinatal problems noted in their hospital records than other delinquents or controls (Lewis, Shanok & Balla, 1979). In a separate longitudinal study, Kandel and Mednick (1991) found that delivery complications were correlated with later arrests for violence, but were not correlated with later arrests for property offenses.

Existing research also suggests that perinatal factors may be especially related to aggressive and criminal outcomes for individuals with high social risk. For example, in a prospective longitudinal study in Kauai, Werner (1987) found that the effects of perinatal stress on delinquent outcome were strongest for children exposed to a disruptive family environment. A disruptive family environment was defined by Werner as separation from the mother, marital discord, absence of the father, legitimacy of the child or parental mental health problems. These social risk factors, in combination with the biological risk resulting from perinatal complications, increased the likelihood for delinquent outcome in her Kauai sample. In our research in Denmark, we have found similar biosocial interactions predicting to the outcomes of violent and persistent criminal behavior.

The research on perinatal factors and crime has been heavily focused on male offenders. Our empirical results presented in this chapter are also focused on males. Unfortunately, it is difficult to determine the predictors of violence in females as the low base rate of female violence hinders much of this research. Even in very large samples, the low rates of female violence preclude reliable statistical analyses. Because our analyses are restricted to males, our results cannot be generalized to female violence.

2. DANISH COHORT STUDIES ON PERINATAL FACTORS AND VIOLENT CRIME

We have examined the relationship between perinatal factors, social risk, and violent crime in the context of several longitudinal cohort studies in Denmark. In one study, we assessed the combined effects of delivery complications and child institutionalization on violent criminal outcome. Early institutionalization has been found to be a risk factor for psychopathology in many contexts. For example, Elaine Walker (1981) found that early institutionalization predicted to more severe symptomatology in the Copenhagen high risk sample, a cohort of children with schizophrenic parents.

2.1. Institutionalization

We examined the combined effects of institutionalization and delivery complications on the outcome of criminal violence in males. Our sample included all males born at Rigshospitalet in Copenhagen between September 1, 1959 and December 31, 1961—a total N of 4,269. Delivery complications data were recorded at the time of the birth, information concerning early childhood institutionalization was obtained in an interview with the mother at age one, and violent arrest histories were checked in the National Police Register at age 32.

Our delivery complications measure was a frequency score of items developed by Danish and American obstetricians and pediatric neurologists. It is important to note that these data were collected at the time of the birth, rather than through a follow-back study of hospital records. Very detailed information was therefore available. Examples of items on the delivery complications scale include: forceps extraction, breech position, umbilical cord prolapse and preeclampsia. For the purposes of analysis, individuals with two or more delivery complications were categorized as having *high* delivery complications, and individuals with one or no delivery complications were categorized as *low* in delivery complications.

Early institutionalization was defined as four or more months of institutional care of the infant during the first year of life. Data for this measure were obtained from an interview with the mother when the child was one year of age. It should be noted that individuals with this early risk factor also suffer from later social ills (such as continued family instability) as well.

Criminal violence was defined as present or absent based on arrest records obtained from the Danish National Police Register. All arrests are recorded in a central location in Denmark and the criminal register is one of the most accurate in the western world. The following offenses are defined as violent—murder, attempted murder, robbery, rape, assault (including domestic assault) and illegal possession of a weapon. The last item points out the attention of methodology to cultural context. Illegal possession of a weapon has an associated threat of violence in a country, like Denmark, where guns and many knives are forbidden. It should be noted that this type of arrest is rare, and the exclusion of this category from our definition of violence does not change our results.

The top portion of Figure 1 presents the relationship between delivery complications, institutionalization and violent offending in our cohort of Danish males. Logistic regression results reveal a significant interaction such that males with high delivery complications and a history of early institutionalization have higher rates of violence than the other groups in the sample ($\chi 2$(df=1, N=4269) = 6.36, $p<.05$).

Note this pattern of results—this is a consistent pattern that we have found in examining perinatal factors and criminal outcome. This result concerns violent offending as an outcome. Most violent offenders are also persistent offenders. We examined whether the same result would hold for persistent criminal offenders in this cohort. Individuals were categorized as criminal on the basis of arrest records for index offenses (such as property crimes, violent crimes, sex offenses, and narcotics offenses). Individuals were categorized as *persistent* criminal offenders if they had been arrested for at least one index offense prior to the age of 18 and at least one index offense after the age of 18. This definition is

Violent Offending

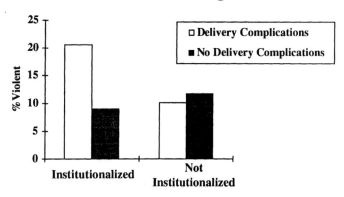

Persistent Criminal Offending

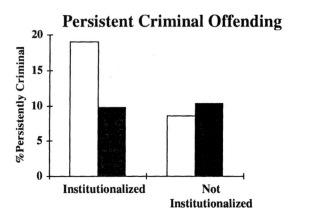

Nonviolent Criminal Offending

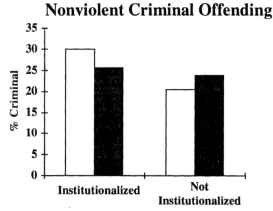

Figure 1. Delivery complications, early institutionalization and criminal outcomes.

in keeping with Moffitt's distinction (1993) of life-course persistent vs. adolescence-limited offenders.

The middle portion of Figure 1 reveals that the results for persistent criminal offending were highly similar to those found for violent offending. Individuals with both high delivery complications and early institutionalization were found to have higher rates of persistent criminal offending than other groups in the sample ($\chi 2$ (df=1, N=4269) = 5.07, $p<.05$).

Next we examined whether this result was specific to violent or persistent offending. We examined whether delivery complications and early rejection predicted to nonviolent criminal offending in this same sample. Nonviolent criminal index offenses included property offenses, narcotics offenses, and sex offenses. In the case of these nonviolent offenders, we found that the statistical interaction was not apparent ($\chi 2$ (df=1, N=4269) = 1.16, p=ns). The combination of delivery complications and early institutionalization does not appear to be related to the outcome of nonviolent offending (see bottom portion of Figure 1). This result is consistent with results obtained by previous researchers (Kandel & Mednick, 1991; Lewis et al., 1979).

2.2. Maternal Schizophrenia

In the context of the Rigshospitalet sample, we have examined delivery complications in combination with other vulnerabilities or social risks in the prediction of violent outcome. We will present several examples of these findings to point out the consistency of our findings. For example, we have examined the combined effects of maternal schizophrenia (defined by a record of admission to a psychiatric hospital with a diagnosis of schizophrenia) and delivery complications on violent criminal outcome. Again, a similar interaction was found ($\chi 2$ (df=1, N=4269) = 4.76, $p<.05$). Children with a schizophrenic mother and high numbers of delivery complications evidence higher rates of violence than other groups in the sample (see Figure 2).

2.3. Maternal Rejection

In the Rigshospitalet sample, we also examined the combined effects of early maternal rejection and delivery complications in the prediction of violent offending in adolescence (Raine, Brennan & Mednick, 1994). Again, we found a similar interaction ($\chi 2$

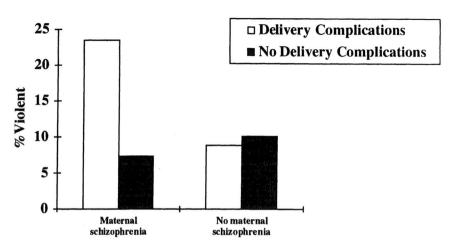

Figure 2. Delivery complications, maternal schizophrenia and violent offending.

(df=1, N=4269) = 14.4, *p*<.001) such that individuals with high maternal rejection and high delivery complications evidence the highest rates of violence (see Figure 3).

The findings presented thus far have been focused on delivery complications in particular. As stated before, in our studies of delivery complications we assume that central nervous system damage or dysfunction is a likely mediating factor between delivery complications and violent outcome. Pregnancy complications can also produce central nervous system dysfunction. However, we have not found pregnancy complications to be consistently related to criminal outcome. This is perhaps due to a methodological confound. Pregnancy complications are more difficult to measure than delivery complications (for example, they often rely on mother's retrospective memory).

2.4. Minor Physical Anomalies and Family Instability

There is one observable indicator of fetal maldevelopment during pregnancy that does not require subjective reporting by the mother—that is the measure of minor physical anomalies (MPAs). MPAs are physical anomalies that reflect a disruption in fetal development. Consider the development of the ears—early in gestation they are seated low on the head and gradually move upwards into their normal position—if the development of the fetus is disrupted, the movement of the ears could be slowed or stopped. The result is low-set ears—a minor physical anomaly. Other minor physical anomalies include: the presence of more than one hair whorl, asymmetrical ears, attached ear lobes, curved fingers, and single palmar crease (Waldrop & Halverson, 1971). The presence of more than three minor physical anomalies suggests a disruption in fetal development. A concomitant disruption in the development of the central nervous system is assumed. Therefore the presence of minor physical anomalies can be considered an observable indicator of central nervous system dysfunction.

We examined the combined effects of minor physical anomalies and family instability on the outcome of criminal violence in a subsample of individuals from the Rigshospitalet cohort. Most of the individuals in this subsample were children of psychiatrically-ill parents. Seventy-two males were included in this analysis. Again, we have the same pattern of results that have been noted in our analysis of delivery complications. In this high

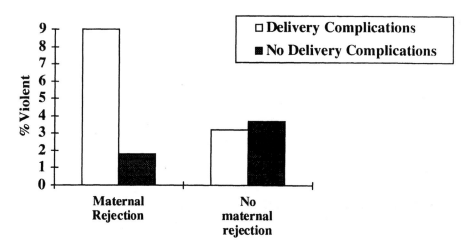

Figure 3. Delivery complications, maternal rejection and violent offending.

risk sample, males with high numbers of minor physical anomalies and an unstable family upbringing evidenced higher rates of violence than other groups in the sample (see Figure 4). In fact, rates of violence were extremely high in this subgroup, suggesting the added detrimental effect of a parent with a psychiatric illness.

Our results from the Copenhagen cohorts consistently reveal that perinatal factors interact with other risk factors to increase the likelihood of violent outcomes in adulthood. This relationship is not simply additive. The combination of central nervous system damage (caused by perinatal problems) and environmental risks is a potent combination in producing the outcome of violence. Our results are consistent with Moffitt's (1993) persistent-offender theory which posits a transactional, developmental process in which biologically vulnerable individuals find themselves ensnared in social environments which do not alleviate their vulnerabilities, but rather exacerbate them. This developmental process is reflected by a lifetime characterized by aggressive, criminal, and often violent behavior.

The consistent pattern of results that we have presented thus far support the ideas that: (1) persistent criminal offenders or violent offenders are distinct, (2) it is important to focus on cumulative risk and prevention, and (3) early life predictors may play an important role in this process. However, the results presented thus far have not been concerned with early childhood behavior—an important aspect to consider in terms of the developmental psychopathology perspective. We will next consider what the persistent or violent offender may look like in early behavioral development, and the potential role of perinatal factors in this behavioral process.

3. FOCUS ON EARLY BEHAVIORAL DEVELOPMENT

What does the persistent or violent criminal male look like in early childhood? There is reason to believe that his behavior may be disinhibited. Perinatal factors have been found to be related to behavioral disinhibition in childhood (Chandola, Robling, Peters, & Melville-Thomas, 1992; Firestone & Peters, 1983)—and a combination of disinhibition and conduct disorders predicts to adult crime (Hechtman, 1989). Disinhibition, therefore, may be the behavioral link between perinatal factors and persistent aggression.

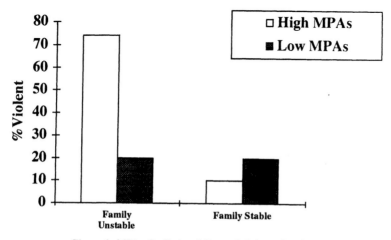

Figure 4. MPAs, family instability and violent offending.

3.1. The Role of Temperament

The hypothesis that behavioral disinhibition may be the link between perinatal factors and violence or persistent aggression is consistent with the literature on temperament and aggressive outcomes. In their longitudinal cohort in New Zealand, Moffitt and her colleagues (Henry, Caspi, Moffitt & Silva, 1996) have found that early childhood temperament characterized by a "lack of control" is associated with later convictions for violence. Antisocial personality disorder has also been linked to the temperament dimension of Novelty Seeking (Svrakic, Whitehead, Przybeck & Cloninger, 1993). The temperament dimension of inhibited to uninhibited behavior has also been correlated with later externalizing behavior problems in adolescence (Schwartz, Snidman & Kagan, 1996).

Kagan has noted that approximately 30 percent of children can be categorized as "inhibited" or "uninhibited" and that these temperament qualities are stable across a span of five years from age 21 months to age seven years (Kagan, 1989). Inhibited and uninhibited children have been found to differ in their physiological functioning (Reznick, Kagan, Snidman, Gersten, Baak, & Rosenberg, 1986). Specifically uninhibited children have been found to have lower heart rates than inhibited children. Lower heart rates have also been found to characterize individuals convicted of violence (Farrington, 1987). Taken together, these results support the hypothesis that disinhibition may characterize the behavior of later persistent or violent offenders during their childhood.

3.2. Mauritius Study

We have recently begun to examine behavioral disinhibition in the context of a longitudinal cohort in Mauritius. This is a representative sample of children born in 1969 and assessed first at the age of three. Children in this sample were born in two towns, Quatre Bornes and Vacoas, that were chosen to approximate the racial and socioeconomic distribution of the island as a whole. Multiple phases of data collection were included in my analyses, and subsamples of males who had the requisite data for these analyses were utilized.

The measures for this study included hospital records of perinatal complications at birth, occupational status of parents at birth, and measures of behavioral disinhibition at ages three and eight years of age.

3.2.1 Obstetrical Measures. Obstetrical data were available in follow-back studies of hospital records—because these data were recorded for medical rather than research purposes, information is available for only more serious or global perinatal complications. In this study, individuals were split into two groups based on their record of perinatal and birth complications—zero complications versus one or more. Complications included factors such as: abnormal bleeding during pregnancy, mother's illness during pregnancy, short gestation, use of forceps in delivery, breech presentation, and lack of oxygen to fetus during delivery.

3.2.2 Socioeconomic Status Measure. Subjects were rated as to their socioeconomic status on the basis of parent occupation. Occupations were rated on a scale of 1 (unemployed) to 8 (academic or head of moderate/large business). The higher of the mother and father's ratings was chosen to represent the child's socioeconomic level. A median split was then used to separate the sample into high and low socioeconomic groups.

3.2.3 Behavioral Disinhibition Measure. The measure of behavioral disinhibition was based on Kagan's conceptualization of shyness vs. fearlessness. This was a combined rating, developed by Angela Scarpa and colleagues (Scarpa, Raine, Venables & Mednick, 1995), based on a psychologist measure at age three and a teacher measure at age eight. At age three children were rated as inhibited or disinhibited based on psychologist ratings (in the laboratory) of their level of sociability, crying behavior, approach-avoidance, verbalizations, and social interactions with other children (alpha for scale=.66). At age eight teachers rated the children on several aspects of sociability and level of fearfulness—these ratings were combined to form a measure of inhibition-disinhibition (alpha for scale=.83). For the purposes of our analyses, children were categorized as stably disinhibited if they ranked in the top 40% of disinhibition scores at both ages three and eight.

3.2.4 Perinatal Factors, Social Risk and Behavioral Disinhibition. Consistent   with the idea that behavioral disinhibition may be a mediator between perinatal factors and aggression, we found that the disinhibited group had a higher rate of perinatal complications (31 percent versus 14 percent). We also noted that this relationship was apparent especially in the low socioeconomic group in this sample (see Figure 5). In this sample of males, perinatal complications interacted significantly with socioeconomic status in the prediction of persistent behavioral disinhibition ($\chi2$ (df=1, N=96) =6.77, $p<.01$). It should be noted that socioeconomic status did not interact significantly with maternal rejection in the prediction of violence in our Danish cohort study (Raine, et al., 1994). One explanation for these contradictory findings is that low socioeconomic status may be associated with greater social risks and difficulties in Mauritius than it is in the country of Denmark.

Perinatal factors predict to disinhibition—the other side of this developmental process is determining whether disinhibition predicts to later aggression in this sample. One recent, related follow-up study of adolescents suggests that it might (Schwartz et al., 1996). In this follow-up study of two child cohorts in adolescence (initially assessed on temperament at ages 21 months and 31 months), uninhibited children were found to have higher levels of externalizing problems than inhibited children. This result was consistent across both samples of males, with uninhibited males scoring higher than inhibited males on Achenbach behavioral ratings of aggression and delinquency.

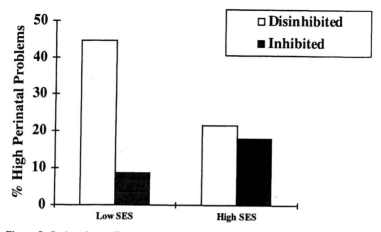

Figure 5. Perinatal complications, socio-economic status and behavioral disinhibition.

In the Mauritius cohort, aggression measures are available at age 11 in the form of parent reports of aggression and delinquency on the Achenbach Child Behavior Checklist. As presented in Table 1, uninhibited males scored higher than inhibited males on the scale of aggression. It is also important to note that it is behavioral disinhibition that is related to increased levels of aggression, rather than inhibition being related to decreased levels. When we compare stably disinhibited males to all other males in the sample, we find that stably disinhibited males have higher levels of both aggression and delinquency (see Table 1).

The results found in this sample suggest that behavioral disinhibition at ages three and eight may be more strongly related to later aggression scores than delinquency scores for males. One possible explanation for this result is the more "socialized" aspect of some of the delinquency items—such as association with delinquent peers—this socialized delinquency may be more reflective of adolescent-limited aggression. Another possible explanation is the low base rate of delinquency in the sample overall.

The results from the Mauritius cohort are consistent with the hypothesis that perinatal factors, in combination with social risks or other vulnerabilities may increase behavioral disinhibition, which in turn, increases the risk for later aggression. Our work in Denmark on the role of perinatal factors and criminal outcome suggests that these behaviorally disinhibited children may be at an increased risk especially for persistent aggression and violence later in life.

4. REMAINING QUESTIONS

We have presented results concerning the development of persistent aggression across the life span. Early life factors, such as perinatal factors appear to play a significant role in this process, when combined with other risks or vulnerabilities. Our results leave us with several remaining questions about the development of aggression and its continuity throughout the life span.

First, more information is needed concerning the mediators between perinatal problems and disinhibition/aggression—we have just completed a preliminary analysis of data from Helsinki suggesting that second trimester flu exposure is related to criminal violence. MRI data have just been collected on these individuals and it will be possible to examine what areas of the brain might be affected by second trimester flu exposure in violent individuals. Preliminary analyses by Tyrone Cannon suggest that left sulcal to brain ratios are higher for those individuals exposed to the flu in the second trimester. These results are consistent with the literature linking frontal and temporal cortex damage and violent offending.

Table 1. Behavioral disinhibition and aggression in Mauritian male children

	Stably disinhibited	Stably inhibited	F (df 1, 69)
CBCL aggression	15.40	11.36	6.85, $p<.05$
CBCL delinquency	3.11	2.33	1.71, p=ns
	Stably disinhibited	All others	F (df 1, 266)
CBCL aggression	15.40	11.77	9.16, $p<.01$
CBCL delinquency	3.11	2.28	4.14, $p<.05$

Another question that remains to be researched is whether specific social risks or protective factors may have a differential effect on this process. In our Danish study of maternal rejection (Raine et al., 1994), we found that socioeconomic status did not have the same detrimental effect as maternal rejection when it was combined with delivery complications in the prediction of criminal violence. However, in Mauritius, socioeconomic status did interact with perinatal complications to predict stably uninhibited temperament. Why would the "same" social factor have an effect in one culture and not in another? Future research must examine the specifics of social factors—what specific risks do they reflect for an individual in that particular culture?

In addition, it is important to examine how our termed "biological" and "social" risk factors may influence one another, in addition to increasing the risk for violence. It is also possible that other risk factors may have influenced both our "biological" and "social" risk factors, and may explain the increased violence as well. In our studies, delivery complications were not correlated with the social risk factors that we assessed. However, it is conceivable that other, unmeasured social or environmental risk factors may have increased the risk for both delivery complications and the social vulnerabilities that we examined. One such possible factor might be maternal alcohol abuse. A thorough analysis of these alternate explanations of results should be incorporated into future research.

Finally, we have not answered very important questions concerning the specifics of the process by which perinatal problems and social risks would work together to produce the outcome of adult violence or persistent criminal behavior. What important person-environment interactions exist in this process? In addition, how can this process be interrupted—what accounts for continuity and discontinuity of aggression over time in these at-risk individuals? A more complete understanding of these processes is necessary for the development of effective intervention and prevention strategies—the ultimate goal of this research.

REFERENCES

Brennan, P. A., Mednick, S. A., & John, R. S. (1989). Specialization in violence: Evidence of a criminal subgroup: Criminology, 27, 437–453.

Brennan, P. & Raine, A. (In Press). Recent biosocial advances in antisocial behavior. Clinical Psychology Review.

Chandola, C. A, Robling, M. R., Peters, T. J. & Melville-Thomas, G. (1992) Pre- and perinatal factors and the risk of subsequent referral for hyperactivity. Journal of Child Psychology & Psychiatry & Allied Disciplines, 33, 1077–1090.

Farrington, D. (1987). Implications of biological findings for criminological research. In S. A. Mednick, T. E. Moffitt, and S. A. Stack (Eds.) The causes of crime: New biological approaches (pp. 42–64). Cambridge: Cambridge University Press.

Farrington, D. (1991). Childhood aggression and adult violence: Early precursors and later-life outcomes. In D. J. Pepler and K. H. Rubin (Eds.) The development and treatment of childhood aggression. Hillsdale, NJ: Erlbaum.

Firestone, P. & Peters, S. (1983). Minor physical anomalies and behavior in children: A review. Journal of Autism & Developmental Disorders, 13, 411–425.

Gorenstein, E. (1990). Neuropsychology of juvenile delinquency. Forensic Reports, 3(1) 15–48.

Hechtman, L. (1989). Attention-deficit hyperactivity disorder in adolescence and adulthood: An updated follow-up. Psychiatric Annals, 19, 597–603.

Henry, B., Caspi, A., Moffitt, T. E., & Silva, P. A. (1996). Temperamental and familial predictors of violent and nonviolent criminal convictions: Ages 3 to 18. Developmental Psychology, 32, 614–623.

Kagan, J. (1989). Temperamental contributions to social behavior. American Psychologist, 44, 668–674.

Kandel, E., & Mednick, S. A. (1991). Perinatal complications predict violent offending. Criminology, 29, 519–529.

Lewis, D. O., Shanok, S. S. & Balla, D. A. (1979). Perinatal difficulties, head and face trauma, and child abuse in the medical histories of seriously delinquent children. *American Journal of Psychiatry*, *136*, 419–423.

Loeber, R., Wung, P., Keenan, K., Giroux, B., Stouthamer-Loeber, M., Van Kammen, W. B., & Maughan, B. (1993). Developmental pathways in disruptive child behavior. *Development & Psychopathology*, *5*(1–2), 103–133.

Moffitt, T. E. (1993). Adolescence-limited and life-course-persistent antisocial behavior: A developmental taxonomy. *Psychological Review*, *100*, 674–701.

Olweus, D. (1979). Stability of aggressive reaction patterns in males: A review. *Psychological Bulletin*, *86*(4), 852–875

Raine, A., Brennan, P. A., & Mednick, S. A. (1994). Birth complications combined with early maternal rejection at age 1 year predispose to violent crime at age 18 years. *Archives of General Psychiatry*, *51*, 984–988.

Reznick, J. S., Kagan, J., Snidman, N., Gersten, M., Baak, K., & Rosenberg, A. (1986). Inhibited and uninhibited children: A follow-up study. *Child Development*, *57*, 660–680.

Scarpa, A., Raine, A., Venables, P. H., & Mednick, S. A. (1995). The stability of inhibited/ uninhibited temperament from ages 3 to 11 years in Mauritian children. *Journal of Abnormal Child Psychology*, *23*, 607–618.

Schwartz, C. E., Snidman, N., & Kagan, J. (1996). Early childhood temperament as a determinant of externalizing behavior in adolescence. *Development and Psychopathology*, *8*, 527–537.

Svrakic, D. M., Whitehead, C. Przybeck, T. R. & Cloninger, C. R. (1993). Differential diagnosis of personality disorders by the seven-factor model of temperament and character. *Archives of General Psychiatry*, *50*, 991–999.

Walker, E. F., Cudeck, R., Mednick, S. A., & Schulsinger, F. (1981). Effects of parental absence and institutionalization on the development of clinical symptoms in high-risk children. *Acta Psychiatrica Scandinavica*, *63*(2), 95–109.

Waldrop, M. F. & Halverson, C. F. (1971). Minor physical anomalies and hyperactive behavior in young children. *Exceptional infant*, *2*, 343–380.

Werner, E. E. (1987). Vulnerability and resiliency in children at risk for delinquency: A longitudinal study from birth to adulthood. In J. D. Burchard, & S. N. Burchard (Eds.), *Primary prevention of psychopathology* (pp. 16–43). Newbury Park, CA: Sage.

11

UNRAVELING THE SOCIAL CONTEXT OF PHYSIQUE AND DELINQUENCY

A New, Long-Term Look at the Gluecks' Classic Study[*]

Robert J. Sampson[1] and John H. Laub[2]

[1]Department of Psychology
University of Chicago
Chicago, Illinois 60637
[2]Northeastern University
Boston, Massachusetts

In 1950, Sheldon and Eleanor Glueck published their now classic study, *Unraveling Juvenile Delinquency*. One of the most frequently cited works in the history of delinquency research, the Gluecks sought to answer a basic and enduring question—what factors differentiate boys reared in poor neighborhoods who become serious and persistent delinquents from boys raised in the same neighborhoods who do *not* become delinquent or antisocial? To address this question, the Gluecks studied in meticulous detail the lives of 500 delinquents and 500 nondelinquents who were raised in the same low-income environments of central Boston during the Great Depression era (circa 1928–1940).

One of the Gluecks' answers to the delinquency question generated controversy from its inception, a controversy that lingers to the present day. Delinquent boys, they claimed, were disproportionately mesomorphic (e.g., muscular, chunky) in body build (1950: 193–4). Adopting Sheldon's somatotype classification scheme (Sheldon et al., 1940; see also Sheldon, 1954), the Gluecks specifically argued that boys with a mesomorphic constitution were temperamentally more aggressive and consequently had a greater potential for delinquency than boys with an endomorphic (round, plump) or ectomorphic (thin, lean) build. The Lombrosian idea that a biological characteristic such as body type could explain a social phenomenon such as delinquency was, of course, anathema to sociologists. The Gluecks were thus roundly criticized by the discipline then firmly in control

---

[*] The data were derived from the Sheldon and Eleanor Glueck archives of the Harvard Law School Library, currently on long-term loan to the Henry A. Murray Research Center, Cambridge. We thank Chris Kenaszchuk for research assistance.

Biosocial Bases of Violence, edited by Raine *et al.*
Plenum Press, New York, 1997

of the intellectual reproduction of criminological knowledge (see Laub and Sampson, 1991).

Despite the critique by sociologists, especially Edwin Sutherland (see e.g., Schuessler, 1973; Sutherland, 1951), the mesomorphic explanation of delinquency has proven resilient. Empirically, a number of studies have found mesomorphic body build to be more prevalent among delinquents than nondelinquents (Gibbens, 1963; Cortes and Gatti, 1972; Hartl et al., 1982). More recently, in their highly influential *Crime and Human Nature*, Wilson and Herrnstein (1985: 89) reasserted a constitutional basis to delinquency in the form of mesomorphic body build. Controversy and sociological scoffing notwithstanding, then, mesomorphy continues to be invoked as a factor in delinquency.

The research design of the Gluecks' study provides a unique opportunity to address anew body build and its interaction with social factors in the prediction of later criminal behavior. Extending our prior work with the Glueck archives, we address four key questions. First, to what extent was the Gluecks' measurement and coding of body build reliable? Second, even with a reliable somatotype measure, can we replicate the original finding that mesomorphy is associated with adolescent delinquency? Third, does mesomorphy *predict* long-term patterns of adult violence among those boys originally defined as delinquent? Fourth, does mesomorphy *interact* with time-varying social factors (e.g., job instability) that are known to predict patterns of adult criminal offending? By addressing these four interrelated questions through a reanalysis of the Gluecks' original data, we hope to shed light on the current status of mesomorphy as a causal factor in the dynamics of crime.

1. RESEARCH DESIGN

The present paper is based on data from the Gluecks' original study of juvenile delinquency and adult crime among 1,000 Boston males born between 1924 and 1935 (Glueck and Glueck, 1950, 1968). The Gluecks' delinquent sample comprised 500 10- to 17-year old white males from Boston who, because of their persistent delinquency, had been recently committed to one of two correctional schools in Massachusetts (Glueck and Glueck, 1950: 27). The nondelinquent or "control-group" sample was made up of 500 white males age 10 to 17 chosen from the Boston public schools. Nondelinquent status was determined on the basis of official record checks and interviews with parents, teachers, local police, social workers, recreational leaders, and the boys themselves. The Gluecks' sampling procedure was designed to maximize differences in delinquency, an objective that by all accounts succeeded (Glueck and Glueck, 1950: 27–29).

A unique aspect of the *Unraveling* study was the matching design. The 500 officially defined delinquents and 500 nondelinquents were matched case-by-case on age, race/ethnicity (birthplace of both parents), measured intelligence, and neighborhood deprivation. The delinquents averaged 14 years, 8 months, and the non-delinquents 14 years, 6 months when the study began. As to ethnicity, 25% of both groups were of English background, another fourth Italian, a fifth Irish, less than a tenth old American, Slavic or French, and the remainder were Near Eastern, Spanish, Scandinavian, German, or Jewish. As measured by the Wechsler-Bellevue Test, the delinquents had an average IQ of 92 and nondelinquents 94. The matching on neighborhood ensured that both delinquents and nondelinquents grew up in disadvantaged neighborhoods of central Boston. These areas were regions of poverty, economic dependency, physical deterioration, and were usually adjacent to areas of industry and commerce (Glueck and Glueck, 1950: 29).

A wealth of information on social, psychological, and biological characteristics, family life, school performance, work experiences, and other life events was collected on the delinquents and controls in the period 1939–1948. These data were collected through an elaborate investigation process that involved interviews with the subjects themselves and their families as well as interviews with key informants such as social workers, settlement house workers, clergymen, school teachers, neighbors, and criminal justice and social welfare officials. The home-interview setting also provided an opportunity to observe home and family life (Glueck and Glueck, 1950: 41–53).

Interview data and home investigations were supplemented by field investigations that culled information from the records of both public and private agencies that had any involvement with a subject or his family. These materials verified and amplified the materials of a particular case investigation. For example, a principal source of data was the Social Service Index, a clearinghouse that contained information on all dates of contact between a family and the various social agencies (e.g., child welfare) in Boston. Similar indexes from other cities and states were utilized where necessary.

1.1. Mesomorphy and Delinquency

As noted above, a somatotype is a characterization of a person's physique in terms of three general components—endomorphy, ectomorphy, and mesomorphy (Glueck and Glueck, 1950: 193–194). The Gluecks argued that this body-type classification was especially relevant to a research design where delinquent and nondelinquent boys were matched on age and ethnicity. Using two independent methods of physique classification, the Gluecks found that delinquents were much more likely than nondelinquents to be "muscular," "masculine," or "mesomorphic"—"a predominantly bone-and-muscle, tightly knit, and energetic type, with excesses in measurements having largely to do with the torso from the waist to the shoulder, and consisting of greater lateral breadth in such regions as neck, shoulders, chest, and upper extremities, with a tapering torso" (1950: 187). The Gluecks contended that these differences in bodily morphology may "reasonably be regarded as fundamentally related to differences in natural energy-tendencies" (1950: 274).

In a later work, *Physique and Delinquency* (1956), the Gluecks argued that the special significance of physical characteristics become apparent through consideration of the "intermediary traits of temperament, personality, and character that distinguish the mesomorph from other body types" (1956: 7). The Gluecks went on to say that the mesomorphic physique type was "more highly characterized by traits particularly suitable to the commission of acts of aggression (physical strength, energy, insensitivity, the tendency to express tensions and frustrations in action), together with a relative freedom from such inhibitors to antisocial actions as feelings of inadequacy, marked submissiveness to authority, emotional instability, and the like" (1956: 226). The Gluecks also argued that there were interaction effects between mesomorphy and certain traits like adventurousness and emotional conflict (1956: 227). Regardless of the exact mechanism, then, mesomorphic physique indicated to the Gluecks a *delinquency potential*. As they argued: mesomorphic boys are "endowed with traits that equip them well for a delinquent role under the pressure of unfavorable sociocultural conditions; while endomorphs, being less energetic and less likely to act out their 'drives,' have a lower delinquency potential than mesomorphs" (1956: 270).

More recently, Wilson and Herrnstein argued that the correlation between body type and crime reflects some psychological trait, having a biological origin, that predisposes an individual to criminality (1985: 89). Based on a review of extant research, they assert that mesomorphs are more expressive, more extroverted, more likely to have domineering temperaments, and given to high levels of activity compared to non-mesomorphs.

2. MEASURES

To test in a more rigorous manner the mesomorphy hypothesis of the Gluecks and later theorists such as Wilson and Herrnstein (1985), we return to the rich data of *Unraveling Juvenile Delinquency* and its follow-up. As part of a larger, long-term project, we have reconstructed and computerized the Gluecks' data, a process that included the validation of numerous measures found in the original files. For a full description of these efforts and other procedures taken to address prior criticisms of the Gluecks' study, see Sampson and Laub (1993). For the present study, we focus mainly on measures of adolescent mesomorphy and of delinquency and violence across the life course. Like the Gluecks, we begin by looking at differences in delinquency as measured by the official criterion of the research design (delinquent group versus control group). However, we extend their analysis by examining individual-level variations in the frequency of delinquency. For example, the Gluecks searched the files of the Massachusetts Board of Probation which maintained a central file of all records from Boston courts since 1916 and from Massachusetts as a whole from 1924. These records were compared and supplemented with records from the Boys' Parole Division in Massachusetts. Out-of-state arrests, court appearances, and correctional experiences were gathered through correspondence from equivalent state depositories. From these data we constructed a measure of the number of juvenile arrests divided by the number of days each subject was not incarcerated. Similar to the criminal career notion of "lambda" (Blumstein et al., 1986), this measure is an individual crime rate "while free" (Sampson and Laub, 1993).

Of equal importance to official records was the Gluecks' collection of self-reported, parental-reported, and teacher-reported delinquency of the boy. From these "unofficial" sources we were able to develop measures of delinquency and childhood antisocial behavior that were not dependent on actions taken by the juvenile justice system. Our main measure is a summary scale of involvement in up to 26 different categories of delinquency as reported by parents, teachers, and the boy himself. We also examined offense-specific measures (e.g., truancy, fighting) and the total amount of delinquency for all crime types reported by a particular source (e.g., self-report total, parent-report total). Because the results were very similar, the present analysis is based on the sum of all delinquent behaviors that were measured consistently across reporters (range: 0–26).

Although the *Unraveling* study was not longitudinal, there is also retrospective data on three key dimensions of troublesome childhood behavior. From the parent's interview there is an indicator distinguishing those children who were overly restless and irritable from those who were not. A second measure reflects the extent to which a child engaged in violent temper tantrums and was predisposed to aggressiveness and fighting. The Gluecks collected data only on habitual tantrums—when tantrums were "the predominant mode of response" by the child to difficult situations growing up (1950: 152). The third variable is the boy's self-reported age of onset of misbehavior. We created a dichotomous variable where a 1 indexes an age of onset earlier than age eight. Those who had a later age of onset *and* those who reported no delinquency (and hence no age of onset) were assigned a zero. Suggesting construct validity, all three measures are significantly correlated. For example, of those children rated difficult in childhood, 34 percent exhibited tantrums compared to 13 percent of those with no history of difficultness. Similarly, for those with an early onset of misbehavior, 47 percent were identified as having tantrums, compared to 20 percent of those with no early onset (all p < .05).

There is evidence of the predictive validity of these "child-effects" measures as well. Fully 100 percent of those scoring high on child antisocial behavior were arrested in adolescence compared to 25 percent of those scoring low (gamma = .69). Using data on adult crime collected by the Gluecks as part of a follow-up study (see below), 60 percent of those scoring high on childhood antisocial behavior were arrested at ages 25–32, compared to less than 25 percent with no signs of early disorder. There is also a strong relationship between childhood antisocial disposition and arrests even at ages 32–45. For further evidence on the predictive validity of these measures of childhood antisocial behavior see Sampson and Laub (1994).

2.1. Follow-Up Criminal Data

In addition to asking whether variations in adolescent delinquency are related to mesomorphic body build, we examine the prediction of long-term involvement in crime. Indeed, perhaps the most interesting question is whether mesomorphy is an important factor for understanding adult criminal careers. Are mesomorphs more likely than non-mesomorphs to continue offending into adulthood? If so, is mesomorphy more predictive of adult violence than crimes against property? Does mesomorphy interact with time-varying social factors such as job instability and marital conflict to predict adult crime? There is almost no empirical research that has attempted to discern the explanatory power of body build in longitudinal perspective (cf. Hartl et al., 1988).

The major part of our analysis, therefore, focuses on the prospective follow-up of the 500 delinquent boys. As shown elsewhere (Sampson and Laub, 1993), whereas very few of the control subjects became involved in adult crime, the delinquent group subjects amassed well over 3,000 arrests after age 18. The research design of the Gluecks thus not only maximized differences in childhood *and* adult antisocial behavior, it yielded two qualitatively distinct samples that we treat separately in the analysis (see also Sampson and Laub, 1993:148). Specifically, we wish to examine the considerable variations in adult criminal careers among the delinquent group subjects.

To accomplish this goal, we draw on the Gluecks' search of criminal records of the Glueck men up to the age of 32. These data include both in-state and out-of-state arrests. In addition, we searched the Massachusetts Office of the Commissioner of Probation between January and June 1993 for 475 of the original 500 delinquents (25 had died during the Gluecks' study). This search provided updates on the criminal histories for each of the 475 subjects to an average age of 65. Operating since 1926, the Office of the Commissioner of Probation is the central repository of criminal record data for the state of Massachusetts. All criminal offenses (presentable to court) are reported to this central system (excluded items include all non-criminal motor vehicle/traffic offenses). We categorized each arrest charge as one of four offense types—violent, property, drug/alcohol, or other. The age of the subject at the time of the arrest was also coded for each of his arrests.

Of course, official criminal records are limited to offenses committed by the men that came to the attention of the criminal justice system, and hence refer only to "official" criminal histories (see Geerken, 1994). Although limited in scope, it should be noted that official data capture serious offenses fairly well (see Gove et al., 1985), and that the criminal record data from the Massachusetts Office of the Commissioner of Probation have been successfully used in prior research (see Vaillant, 1983; McCord, 1979; Sampson and Laub, 1993). Note also that we do not have any information for those subjects who moved out of state or for those men who may have committed crimes out of state. However, given low rates of inter-state mobility for this particular cohort of men plus the fact that Widom (1989: 259) found that local police departments accounted for 80 percent of the adult arrests on record in her follow-up study, we are cautiously confident of the relative accuracy of state criminal records—especially on the crucial issue of constancy in record-keeping patterns over time. Computerized national databases (e.g., FBI criminal records) are problematic in this regard (Geerken, 1994), and besides they do not cover the full period at risk of arrest for the Glueck cohorts. Despite limitations, then, these state level data provide an important foundation for a study of criminal careers over the life span.

We also rely on detailed interview-based information collected by the Gluecks as part of their follow-up of the men at ages 25 and 32. We examine indicators of excessive drinking, job stability, military service, and marital attachment (Sampson and Laub, 1993: 143–145). In addition, we control for selection bias associated with subject attrition for each wave of the follow-up (for further details see Sampson and Laub, 1993: 149–153).

3. SOMATOTYPE METHODOLOGY

The Gluecks took a 5 by 7 inch photograph of each boy in the *UJD* study for purposes of classification into a somatotype. The boys were photographed in three positions (front, side, and rear). The boy's height, weight, and age appeared on each photograph. From the inspection and measurement of the three positions in the photographs, the necessary anthropometric data to determine physique were derived. The photographs were projected on a screen and magnified 8.2 times for more accurate measurement of various body parts (e.g., chest breadth). William Sheldon and Earnest A. Hooton served as consultants to this aspect of the *UJD* project. Bryant Moulton took the photographs and Carl C. Seltzer, a former student and colleague of Hooton's, analyzed the photographs with the assistance of Ashton Tenney. The somatotype ratings were made without knowledge of the boy's background (Glueck and Glueck, 1950: 54–55; Appendix C).

3.1. Was the Gluecks' Measure Reliable?

To verify the Glueck's coding of mesomorphy, we conducted a double-blind validation test for the purposes of this paper. Fifty cases in the delinquent group were randomly selected, stratified by body type as originally coded by the Gluecks. Specifically, we selected subjects classified as endomorph (N=10), ectomorph (N=10), non-extreme mesomorph (N=15), and extreme mesomorph (N=15). Without knowledge of the Gluecks' evaluation, two raters then independently studied the original photographs for the fifty cases and coded body type.

Although the Gluecks' rating scheme was much more refined (e.g., in training and access to magnified photographs), the results of our test are nonetheless encouraging. Overall our double-blind ratings of mesomorphy agreed with the Gluecks original scoring 73 percent of the time (N= 100), and inter-rater reliability across our two judges was 70 percent (N = 50). For a "pure" measure of extreme mesomorphy where mixed types (e.g., mesomorphic endomorphs) were ignored, the agreement was even higher. Specifically, there was an 82 percent agreement with the Gluecks' measurement and 84 percent agreement among our two raters. Considering that our coders did not have recourse to magnified pictures and bodily measurements, not to mention training in anthropological measurement, these figures are high indeed. More to the point, they significantly improve upon chance (kappa > .5 for the key measure of extreme mesomorphy).

4. RESULTS

Having demonstrated the reliability of the two measures of mesomorphy, we now turn to the question of adolescent delinquency for both groups of boys (N = 1,000). To replicate the Gluecks' major finding, we first consider adolescent delinquency measured by the official criterion of the research design (1 = delinquent, 0 = control group). To add further insight on the Gluecks' analysis in *Unraveling*, we also investigate the "unofficial" delinquency scale and the three measures of early childhood antisocial behavior described above. The association of these measures with mesomorphic body build is displayed in Table 1.

The results clearly replicate the Gluecks' published findings. Sixty-seven percent of mesomorphs were officially delinquent compared with 37 percent of nondelinquents (p < .01). When we substitute the more refined measure of extreme mesomorphy, the same pattern obtains—there is a 31 percentage-point differential in delinquency. When unofficial delinquency is considered (row 2), we see a significant albeit reduced magnitude of relationship. Note that the percentage difference is on the order of 15 rather than 30. It appears that mesomorphy is more strongly related to the official criterion of delinquency than the combined self-parent-teacher reports.

The last rows of Table 1 display the results for the indicators of early conduct disorder and antisocial behavior, along with two personality measures derived from psychiatric interviews with the boys—*extroversion* and *aggression* (see Glueck and Glueck, 1950: 245). Quite clearly, both measures of mesomorphy show a weak relationship with early onset of misbehavior, violent tantrums, and difficult (restless) child behavior. The association for early onset and difficultness is in fact near zero, and, while the relationship for tantrums is statistically significant, it is quite small (less than 10 percentage point difference). Consistent with the Gluecks' hypothesis, however, boys with mesomorphic body builds were significantly more likely to be rated (by independent psychiatrists) as having "extroverted" and "aggressive" personalities. In particular, Table 1 shows that extreme mesomorphs were three times more aggressive and almost twice as extroverted in personality as non-extreme mesomorphs.

Table 1. Bivariate Association between mesomorphy and delinquency: delinquent and control groups

	Mesomorph		Extreme mesomorph	
	No (531)	Yes (448)	No (830)	Yes (149)
% Officially delinquent	37	67*	46	77*
% Unofficially delinquent[a]	28	44*	33	48*
% Early onset	7	6	7	8
% Tantrums	19	28*	22	30*
% Child difficult	46	44	44	46
% Extroverted personality	31	56*	37	72*
% Aggressive personality	6	15*	8	24*

*p < .05
[a]Percent unofficially delinquent refers to the trichotomized "high" category.

In short, our initial results do not break much new ground analytically but they are interesting nonetheless. Supporting the main story told by the Gluecks, we find that mesomorphy is associated with higher rates of official delinquency and psychiatric diagnoses of aggressive, extroverted personalities. Yet as we move farther away from officially-defined criteria and closer to individual-level variations in delinquent and anti-social *behavior*, mesomorphy is much reduced in explanatory power.

4.1. Multivariate Models

What the Gluecks did not address was the explanatory power of mesomorphy relative to social factors. Nor did they test systematically the statistical interaction of mesomorphy with social factors in a multivariate analysis (see e.g., Glueck and Glueck, 1956). We thus complete their nascent analysis of adolescent variations in mesomorphy and delinquency.

Table 2 displays substantive models of family process, school attachment, early onset of antisocial behavior, and adolescent delinquency drawn from our previous analyses (Laub and Sampson, 1988; Sampson and Laub, 1993: chapters 3–4). The first two columns display the ML logistic results for the official delinquency criterion. Columns 3 and 4 list the OLS results for the total measure of unofficially reported delinquency. By comparing these two behavioral outcomes - one official and one unofficial—we are in a position to assess the overall validity and robustness of mesomorphy as a correlate of delinquency independent of established social factors.

Table 2. OLS linear and ML logistic regression of delinquency on selected family process, school attachment, childhood antisocial behavior, and mesomorphy variables: Delinquent and control groups

	Delinquency			
	Official status ML logistic[a]		Self-parent-teacher reported OLS linear	
	b	t-ratio	B	t-ratio
A. (N = 786)				
Parent-child attachment	−.48	−3.11*	−.08	−2.93*
Erratic/harsh discipline	.42	5.45*	.15	5.37*
Maternal supervision	−1.29	−8.64*	−.31	−10.39*
School attachment	−1.11	−9.28*	−.39	−14.01*
Mesomorphy	1.07	4.78*	.05	2.08*
	ML Model X^2 = 555, 5 d.f.		OLS R^2 = .55	
B. (N = 703)				
Parent-child attachment	−.36	−2.15*	−.07	−2.72*
Erratic/harsh discipline	.38	4.38*	.11	3.79*
Maternal supervision	−1.24	−7.63*	−.30	−9.69*
School attachment	−1.00	−7.58*	−.35	−11.99*
Early onset	1.43	2.08*	.09	3.54*
Child difficult	.84	3.38*	.07	2.89*
Tantrums	1.11	3.45*	.10	3.77*
Mesomorphy	1.12	4.01*	.06	2.43*
	ML Model X^2 = 524, 8 d.f.		OLS R^2 = .58	

* $p < .05$
[a]Entries for ML Logistic "b" are the raw maximum-likelihood coefficients; "t-ratios" are coefficients divided by s.e.

Panel A begins the multivariate assessment by introducing measures of parent-child attachment, erratic/harsh family discipline, maternal supervision, school attachment, and mesomorphy. The results are consistent with our previous findings that both official and unofficially-reported (self-parent-teacher) delinquency are higher among boys with weak ties to parents, low maternal supervision, low school attachment, and exposure to erratic, harsh, and threatening family discipline. By far the largest effects are attributable to *school attachment* and *supervision*. Once these family and school factors are controlled, the direct effect of mesomorphy is quite small for between-individual variations in unofficial delinquency (beta only .05). In comparison, mesomorphy continues to have a relatively larger effect on official delinquency. These models were repeated with extreme mesomorphy but the findings were substantively the same. Namely, extreme mesomorphy had a weak direct effect on unofficial delinquency (beta = .05) compared to official delinquency status (t-ratio > 4.0).

Panel B adds to the model the measures of early misbehavior. All three indicators—tantrums, child difficultness, and early onset—have independent positive effects on adolescent delinquency regardless of their measurement source. Our family-school model remains intact, however, surviving the controls for these three dimensions of early childhood antisocial behavior/temperament. Hence one way of interpreting Panel B is that changes in adolescent delinquency (i.e., variations unexplained by early propensity to deviance) are directly related to informal processes of family and school social control in adolescence. Therefore, while "child effects" *may* be present, supporting a psychologically-based model, the "social" context of family and school has larger effects and confirms earlier patterns. Knowledge of childhood propensities is not a sufficient condition for explaining later delinquency.

Perhaps more interesting, the coefficient for mesomorphy is largely impervious to these controls as well, suggesting that early misconduct is not mediating the biological effects of body constitution on adolescent delinquency. There is some evidence that extroverted personality may play such a role, but only for unofficial delinquency. Specifically, when we introduce extroversion into the model, the (small) direct effect of mesomorphy (see columns 3–4 in Table 2) disappears entirely, and extroversion has a significant positive effect on unofficial delinquency (data not shown in tabular form). Strangely, however, mesomorphy retains its direct effect on *official* delinquency, compared to a null effect for extroversion.

We are thus inclined to discount the idea that personality variables like extroversion mediate the effect of mesomorphy on delinquency (cf. Sheldon, 1954; Wilson and Herrnstein, 1985). First, there is not much to mediate as mesomorphy has a small effect to begin with (Panel A, Table 2). Second, the instability of the mediation across measures of delinquency gives us pause, and leads us to suspect that mesomorphy's (relatively) stronger relationship with official delinquency stems in part from how juvenile justice officials reacted to large, stout, and muscular boys compared to non-mesomorphs (e.g., weak or plump boys).

Furthermore, we created a series of interaction terms to test whether the effect of mesomorphy varied by social context. It did not. Whether family process, school attachment, early misbehavior, or even the personality domain of extroversion, the interaction terms with mesomorphy were uniformly insignificant. We also reestimated each model in Table 2 for mesomorphic and non-mesomorphic boys. The parameter estimates for the set of social factors were, as a rule, not significantly different in subgroup equations defined by mesomorphy. We thus conclude that the mesomorphy relationships in Table 1–2 do not interact in any meaningful way with social context or personality.

To this point, then, our results seem to suggest that the relationship between meso-morphy and delinquency is contingent. Indeed, it is disturbing that the mesomorphy find-ing in Tables 1–2 emerges mainly for official delinquency. In our previous analyses of the adolescent data from the *Unraveling* study, the patterns for unofficial and official delin-quency have almost always been similar (see Sampson and Laub, 1993: chapters 4–5). One possibility, of course, is that of official bias. Perhaps the juvenile court was more likely to institutionalize boys with mesomorphic body types because they were perceived to be bigger, more dangerous, and more culpable. It may be no coincidence as well that mesomorphy correlates with ethnic background—37 percent of Italian boys were rated as extreme mesomorphs compared with only 18 percent of non-Italians (p < .01). Prejudice against Italians, which was known to exist in Boston at the time of the Gluecks' study (see e.g., Whyte, 1943: 273), thus may have contaminated the results despite the matching on ethnicity.

In any case, the message seems to be that while mesomorphy *is* correlated with de-linquency, it fails to consistently explain much of the between-individual variation in de-linquency once other social factors are held constant. Mesomorphy also fails to interact with known correlates of adolescent delinquency in a consistent fashion.

5. PREDICTING ADULT CRIME AND VIOLENCE

The major question that remains concerns the unfolding of delinquent careers: Is mesomorphy in adolescence a sound predictor of *adult* crime in the lives of the delinquent group men? To answer this question, Table 3 turns to the prediction of adult crime and de-viance classified by four age categories. The first category (rows 1–5) reflects crime and deviance in the transition to young adulthood (ages 17–25). Despite the considerable vari-ation in body type *within* the delinquent group, mesomorphy has no predictive power whatsoever, whether it be for violent arrests, property arrests, official charges in the mili-tary, or excessive drinking measured in the interviews.

The second set of age-specific results diverges slightly by yielding two significant relationships. Interestingly, however, the direction for both of these is exactly opposite to expectations derived from the constitutional argument of the Gluecks. Namely, meso-morphs are *less* likely than non-mesomorphs to be arrested at ages 25–32 (for any crime type), and extreme mesomorphs are *less* likely to be arrested for violence than non-ex-treme mesomorphs. Although nonsignificant, the property and drinking comparisons fol-low this same pattern as well.

The results for ages 32–45 and 46–59 continue to demonstrate the weak predictive power of individual differences in bodily constitution. In fact, none of the relationships are significant, and their direction fluctuates in a seemingly random pattern more so than the earlier ages.

The results in Table 3 are thus clear. No matter what the later age, mesomorphic body build in adolescence has no meaningful or consistent relationship with patterns of criminal offending in adulthood among the Gluecks' delinquents. The conclusion holds for every crime type, and, at the ages of 17–32, for alternative measurement strategies and even charges in the military. Therefore, while mesomorphy may distinguish official delin-quents from nondelinquents in adolescence, it does not help to explain the later incidence of crime among juvenile offenders.

Moreover, even if we pool the delinquent and control group data, mesomorphy does a poor job in predicting later outcomes. For example, in the combined sample of 1,000

Table 3. Adolescent mesomorphy and patterns of crime and deviance across the life course: delinquent group, ages 17–59

	Mesomorph		Extreme mesomorph	
	No (196)	Yes (299)	No (380)	Yes (115)
Ages 17-25				
% Arrested, total	82	80	82	79
% Arrested, violence	34	34	34	32
% Arrested, property	63	61	63	57
% Charged, military	66	56	62	55
% Excessive drinking	44	40	43	36
Ages 25-32				
% Arrested, total	69	57*	64	55
% Arrested, violence	21	16	21	9*
% Arrested, property	29	29	31	26
% Excessive drinking	40	33	37	31
Ages 32-45				
% Arrested, total	62	57	60	55
% Arrested, violence	21	19	19	24
% Arrested, property	20	22	20	23
% Arrested, drug/alcohol	36	30	33	29
Ages 46-59				
% Arrested, total	23	24	22	28
% Arrested, violence	6	6	6	5
% Arrested, property	7	7	7	5
% Arrested, drug/alcohol	13	10	12	7

*$p < .05$.

men, 42 percent of mesomorphs were arrested at age 32 compared to 35 percent of non-mesomorphs—a difference of only 7 percent. For extreme mesomorphy the corresponding difference is 6 percent. When we extend the criterion to arrest at ages 32–45, a similar pattern emerges: Forty-one percent of mesomorphs were arrested at ages 32–45 (12 percent for violence) compared to 31 percent of non-mesomorphs (8 percent for violence). When one considers the high degree of continuity in criminal offending across time for these two groups (see Sampson and Laub, 1993: chapter 6), the failure of mesomorphy to predict meaningful long-term differences in crime is particularly striking.

5.1. Interaction

Once again, it may be that we are looking at things from the wrong angle. Just because main effects fail to emerge does not mean that mesomorphy is necessarily unimportant over the long haul. Consistent with the interactional bio-psycho-social framework advanced by the Gluecks, mesomorphy may only work in concert with social factors to explain longitudinal patterns of crime. In other words, the *interaction* of mesomorphy with time-varying social factors may be where the action is during adult criminal careers. To test the longitudinal interaction hypothesis, we thus conducted a series of tests informed by prior theory and the empirical results in Tables 1–3.

A central thesis in our work has been that early tendencies to juvenile delinquency, in conjunction with weak adult social bonds, explain variations in adult crime. We provide an extended assessment of this theoretical scenario by estimating models wherein key indicators of social bonding—job stability and marital attachment —— are combined into a summary scale. For those men who were married, we summed standardized scales of job stability and marital attachment. Men who were unmarried and hence had no marital bonds were assigned their value on job stability for the measure of social bonds. This two-fold procedure retains the sample of non-married men, increases degrees of freedom, and is consistent with our theory because it weights social bonds according to the nature of institutional connections. That is, single men are characterized by their work attachment alone, whereas for men who are married we take into account strength of marital bonds in addition to their work bonds. The creation of a summary adult "social bonds" scale is supported by the significant covariation of attachment and job stability among married men (see Sampson and Laub, 1993: 169).

The model in Panel A of Table 4 estimates the effects of both prior social bonds and *change* in social bonds on the frequency of arrests while free in the community at ages 25–32. To avoid confounding mesomorphy with adolescent differences in actual delinquent behavior or official reaction bias, we control for juvenile delinquency as measured by the unofficial delinquency scale and the rate of individual offending (arrests) while free. We control for the selection hazard of attrition during follow-up as well (see Sampson and Laub, 1993: 149–153). This model provides a test of the overall effects of change in age-varying social bonds on change in adult crime.

Table 4. OLS main- and interaction-effect models predicting crime frequency at ages 25 to 32 and 32–45 by mesomorphy, juvenile delinquency, and adult social bonds (ages 17–25 and 25–32): delinquent group

	Main effect		Interaction	
Independent variables	B	t-ratio	B	t-ratio
A. Ages 25–32				
Juvenile (age <17):				
Mesomorphic	.00	.11	.00	1.00
Arrest frequency	.13	2.64*	.13	2.65*
Unofficial delinquency	.05	1.02	.05	1.02
Adult (ages 17–32)				
Exclusion risk, 17–25	.15	2.99*	.16	3.01*
Social Bonds, 17–25	−.24	−3.96*	−.32	−2.75*
▲ in Social Bonds, 25–32	−.35	−5.48*	−.27	−2.32*
Meso. x Social Bonds 17–25	—		.09	.80
Meso. x ▲ Social Bonds 25–32	—		−.07	−.66
R^2		.37		.37
N=292				
B. Ages 32–45				
Juvenile (age <17)				
Mesomorphic	−.06	−1.06	−.06	−1.08
Arrest frequency	−.01	−.19	−.01	−.17
Unofficial delinquency	.11	1.85+	.10	1.81+
Adult (ages 17–32)				
Exclusion risk, 17–25	.17	2.98*	.18	3.01*
Social Bonds, 17–25	−.14	−1.96*	−.25	−1.83+
▲ in Social Bonds, 25–32	−.23	−3.27*	−.16	−1.18
Meso. x Social Bonds 17–25	—		.12	.97
Meso. x ▲ Social Bonds 25–32	—		−.08	−.62
R^2		.19		.19
N=266				

*p < .05.   + p < .10

The results in Panel A are quite clear—social bonds in the transition to young adult-hood (ages 17–25) and changes in social bonds at ages 25–32 predict variations in crime at ages 25–32 unexplained by mesomorphy and juvenile criminal propensities. Both prior level and increases in levels of young-adult social bonding also inhibit crime frequency at ages 32–45 (see Panel B). By contrast, mesomorphy has no effect on crime at either age period. These basic results held up when we analyzed the measure of "extreme" mesomorphy, specific crime types (e.g., property and violence), and even when we controlled for prior measures of crime frequency at ages 17–25.

Columns three and four present results for the interaction of mesomorphy with time-varying social bonds. Specifically, we multiplied the dichotomous indicator of mesomorphy by each of the social-bond scales. For crime frequency at ages 25–32, nothing changes when these interactions are considered, including the percentage of variance explained by the overall model. The main effects of social bonds continue, while the insignificance of mesomorphy—whether alone or in interaction with social bonds—also continues. Moreover, we reestimated the equations in Panel A within each category of mesomorphy to test for possible interactions with other predictors such as prior misconduct. There were no statistically significant differences in coefficient estimates across the subgroup equations (data not shown), confirming the lack of interaction of mesomorphy with known predictors of adult criminal offending.

The same general pattern emerges in predicting the frequency of crime at ages 32–45, even though multicollinearity appears to have inflated the standard errors of our measures of social bonds. For example, the largest beta coefficient is for social bonds at ages 17–25, whereas its t-ratio is only significant at $p < .10$ (-1.83). Substantively, the data suggest that social bonds formed in the transition to young adulthood have a lasting impact on later crime. On the other hand, similar to the pattern for ages 25–32, the main effects of mesomorphy are insignificant and small, as are the indicators for the interaction of social bonds with mesomorphy. These results are fully replicated using ML logistic models of crime participation at ages 32–45. Subdividing the sample by body type (meso-morph, nonmesomorph) also failed to yield significant interactions with the social predictors other than social bonds.

In short, the overall pattern in Table 4 confirms earlier results that social bonds—both prior levels and concurrent change—inhibit crime and deviance throughout a major period of adult development (Sampson and Laub, 1993: 200–3). The addition of mesomorphy does nothing to change this substantive finding.

6. CONCLUSION

This paper replicated the major finding of the Gluecks that boys with a mesomor-phic body build were more likely to be officially delinquent than boys with an endomor-phic or ectomorphic body type. We were, however, unable to establish that mesomorphy made much of a difference independent of social factors. Perhaps more important, meso-morphy does not predict long-term patterns of involvement in adult violence, nor does it appear to interact with key social factors (whether in adolescence or adulthood) to predict criminal careers. Combined with its rather moderate relationship with delinquency to begin with and the differential findings in adolescence by type of delinquency measurement (official versus unofficial), mesomorphy does not seem to qualify for entry on the list of "sturdy" correlates of delinquency and adult crime.

7. REFERENCES

Blumstein, Alfred, Jacqueline Cohen, Jeffrey Roth, and Christy Visher, (Eds.). 1986. *Criminal Careers and "Career Criminals"*. Washington, D.C.: National Academy Press.

Cortes, J.B. and Gatti, F.M. .1972. *Delinquency and Crime: A Biopsychological Approach*. New York: Seminar Press.

Geerken, Michael R. 1994. "Rap Sheets in Criminological Research: Considerations and Caveats." *Journal of Quantitative Criminology* 10:3–21.

Gibbens, T.C.N. 1963. *Psychiatric Studies of Borstal Lads*. London: Oxford University Press.

Glueck, Sheldon and Eleanor Glueck. 1950. *Unraveling Juvenile Delinquency*. New York: Commonwealth Fund.

———. 1956. *Physique and Delinquency*. New York: Harper.

———. 1968. *Delinquents and Nondelinquents in Perspective*. Cambridge: Harvard University Press.

Gove, Walter R., Michael Hughes, and Michael Geerken (1985). "Are the Uniform Crime Reports a Valid Indicator of Index Crimes?" *Criminology* 23: 451–501.

Hartl, Emil, E. Monnelly, and R. Elderkin. 1982. *Physique and Delinquent Behavior: A Thirty-Year Follow-up of William H. Sheldon's 'Varieties of Delinquent Youth'*. New York: Academic.

Laub, John H., and Robert J. Sampson. 1988. "Unraveling Families and Delinquency: A Reanalysis of the Gluecks' Data." *Criminology*, 26: 355–380.

Laub, John H. and Robert J. Sampson. 1991. "The Sutherland-Glueck Debate: On the Sociology of Criminological Knowledge." *American Journal of Sociology* 96: 1402–1440.

McCord, Joan. 1979. "Some Child-Rearing Antecedents of Criminal Behavior in Adult Men." *Journal of Personality and Social Psychology* 37: 1477–1486.

Sampson, Robert J., & Laub, John. H. 1993. *Crime in the making: Pathways and turning points through life*. Cambridge: Harvard University Press.

———. 1994. "Urban Poverty and the Family Context of Delinquency: A New Look at Structure and Process in a Classic Study." *Child Development* 65:523–540.

Schuessler, Karl. 1973. *Edwin H. Sutherland on Analyzing Crime*. Chicago: University of Chicago Press.

Sheldon, William, S. Stevens, and W. Tucker. 1940. *The Varieties of Human Physique*. New York: Harper and Row.

Sheldon, William. 1954. *Atlas of Men: A Guide for Somatotyping the Adult Male at all Ages*. New York: Gramercy Publishing Co. Sutherland, Edwin H. 1951. "Critique of Sheldon's Varieties of Delinquent Youth." *American Sociological Review* 16:10–13.

Vaillant, George E. 1983. *The Natural History of Alcoholism*. Cambridge, MA: Harvard University Press.

Whyte, W.F. 1943. *Street Corner Society: The Social Structure of an Italian Slum*. Chicago: University of Chicago Press.

Widom, Cathy Spatz. 1989. "Child Abuse, Neglect, and Violent Criminal Behavior." *Criminology* 27: 251–271.

Wilson, J.Q. and Herrnstein, R.J. 1985. *Crime and Human Nature*. New York: Simon and Schuster.

12

SEROTONIN AND HUMAN VIOLENCE

Do Environmental Mediators Exist?

Markus J. P. Kruesi and Teresa Jacobsen

Institute for Juvenile Research
Department of Psychiatry
University of Illinois-Chicago
907 South Wolcott Avenue
Chicago, Illinois 60612

1. INTRODUCTION

The purpose of this chapter is to review evolving evidence for the possibility of environmental mediators of relationships between serotonin and violence in human children and adolescents. This chapter does not ask whether there are environmental mediators of violence - a variety of evidence suggests that environmental mediation of violence occurs (Tolan & Guerra 1994) and that interactions between biology and environmental events increase the risk of crime (Raine et al., 1994, 1996). Rather this paper focuses upon the possibility of environmental mediators of the relationship between serotonin and violence.

The magnitude of this task necessitates some constraints to fit within a chapter rather than an entire volume. This chapter focuses upon one measure of serotonergic function, cerebrospinal fluid (CSF) concentrations of a serotonin metabolite, 5-hydroxyindoleacetic acid (5-HIAA). CSF 5-HIAA, an indirect measure of central serotonin (Stanley et al., 1985) is thought to reflect serotonin turnover. Other serotonergic measures also have been found to have relationships to aggression (Coccaro et al., 1989, 1996; Halperin et al., 1994; Mann et al., 1992a; Moffit et al., this volume, Moss et al., 1990), but comprehensive review of all serotonergic measures exceeds the limits of this chapter. Therefore, other serotonergic parameters will only be discussed to provide perspective on the CSF 5-HIAA findings. Similarly, adult studies will be drawn upon in order to provide salient background and contrast but the focus of this chapter is on the young. Humans are also the focus of this chapter, but non-human investigations are used for reference and perspective.

When considering the relationship of neurochemistry to psychopathology, it is important to consider whether a neurochemical finding represents a trait (persistent response), a state (response only under certain conditions), or a scar (response which appears only after onset of illness or insult and then persists) (Cook & Kruesi, in press).

Biosocial Bases of Violence, edited by Raine *et al.*
Plenum Press, New York, 1997

State and trait are related to the concepts of nurture and nature. It follows that neurochemical measures may represent effects of nature or nurture. Although such questions may require study of patients affected with a psychopathological condition and in remission, whether a neurochemical finding represents a trait or state is usually determined by examining whether normal subjects vary in the measure over time. For example, plasma norepinephrine levels may double while performing mental arithmetic (Mefford et al., 1981), suggesting a state-like quality. Thus, although most people associate neurochemistry with studies of "nature," many neurochemical measures are as responsive to the environment as self-report of anxiety and mood.

In contrast, whole blood serotonin concentration is comparatively stable in humans from about age 9 onwards (see Cook & Kruesi for review). Stability over time is indicative of a trait. In addition, long-term changes in neurochemical measures representing "scar" may occur after trauma ranging from metabolic "trauma" such as congenital hypothyroidism to emotional trauma such as sexual abuse or, at least in non-human primates, adverse rearing conditions. Scar type markers are consistent with the notion of a sensitive or critical period, i.e. sensitivity to the environment at a particular phase of development. Accordingly, this chapter will examine the evidence that CSF 5-HIAA and its relationship to violence have trait, state, or scar like characteristics. Because the chapter emphasizes studies on the young, it also allows for an exploration of emerging changes in or the establishment of relationship between CSF 5-HIAA concentration and behavior over time.

1.1. CSF 5-HIAA and Its Relation to Behavior

The path from synthesis of serotonin and its subsequent metabolism to 5-HIAA is briefly reviewed below to help facilitate understanding of potential environmental influences upon serotonin violence relationships. Serotonin's synthetic pathway begins from the amino acid precursor, L-tryptophan. Tryptophan competes with other large neutral amino acids for entry into the brain and this competition can result in altered brain serotonin content (Fernstrom & Wurtman, 1972). Tryptophan is hydroxylated to 5-hydroxytryptophan by tryptophan hydroxylase. After 5-hydroxytryptophan is formed, it is rapidly decarboxylated to 5-hydroxytryptamine (serotonin). Because the pathway is limited by the hydroxylation step and the hydroxylation is limited by the availability of L-tryptophan (substrate), an increase in tryptophan concentration in the nerve cell may result in an increase in neuronal serotonin synthesis. Serotonin is principally metabolized to 5-hydroxyindoleacetic acid, although the primary method of inactivation at the synapse is through uptake back into the presynaptic neuron.

CSF 5-HIAA refers to the concentration of the metabolite in samples of cerebrospinal fluid. It is still unclear what the biologic meaning of CSF 5-HIAA is (Potter & Manji 1993). Low CSF 5-HIAA concentrations do not specify the site of abnormality in 5-HT pathways or indicate whether the problem is primary or secondary (Pranzatelli 1994). As described later, relatively low CSF 5-HIAA does have substantial consistency as a biologic marker for certain forms of aggressive behavior. Samples are usually obtained from humans via lumbar puncture and from non-human primates by sampling from the cisterna magna.

Recognition of the relationship of CSF 5-HIAA to aggressive behavior came first in the landmark observation by Asberg et al. (1976, 1976a) of a bimodal distribution of CSF 5-HIAA concentrations among depressed patients. The patients with lower concentrations were significantly more likely to have attempted suicide. Subsequent studies across continents and differing investigator groups have confirmed this. A recent meta-analysis of 27 research reports on the cerebrospinal fluid levels of neurotransmitter metabolites involving 1202 psychiatric patients found strong evidence for attempted suicides, especially those using violent methods, having lower levels of CSF 5-HIAA as compared to psychiatric controls (Lester 1995). Furthermore, those who made subsequent suicidal actions were found to have lower levels of CSF 5-HIAA.

Significant inverse correlations between CSF 5-HIAA and lifetime history of aggression, particularly aggression towards others was initially established by Brown et al. (1979,1982). Later investigations have clarified that it is impulsive/affective aggression rather than more predatory aggression that is linked with low CSF 5-HIAA (Linnoila et al., 1983; Virkkunen et al., 1989,1995; Brown & Linnoila 1990; Golden et al., 1991; Mehlman et al., 1994). Not all arsonists or homicide offenders have low 5-HIAA, but those with affect laden impulsive acts are significantly more likely to have low concentrations (Linnoila et al., 1983).

The consistency of findings relating abnormalities of serotonergic measures in adults to impulsive aggressive behavior (Brown et al., 1986; Coccaro et al., 1992, 1996; Linnoila & Virkkunen 1992; Virkkunen et al., 1989, 1995, 1996) has encouraged the ongoing search for serotonergic relationships with aggression in children and adolescents (Kruesi & Lelio 1996). CSF concentrations of monoamine metabolites in pediatric samples have only recently come under systematic study. Early CSF investigations of children within the disruptive spectrum focused upon hyperactivity/minimal brain dysfunction (Shetty & Chase 1976; Cohen et al., 1977; Shaywitz et al., 1977). The lack of positive findings in these initial studies may have been due to methodologic differences and/or lack of systematic measurement of aggression. Reports of low 5-HIAA associating with conduct disorder and cruelty to animals by children raised the possibility of links between serotonin and childhood aggression (Kruesi et al., 1985, 1989). A subsequent study of 29 children and adolescents with disruptive behavior disorders (16 of whom had conduct disorder) which sought to include children with severe physical aggression found both cross sectional (Kruesi et al., 1990) and longitudinal (Kruesi et al., 1992) relationships between lower CSF 5-HIAA concentrations and greater physical aggression. By contrast, a recent study of 29 significantly less aggressive children with attention-deficit hyperactivity disorder (11 of whom had conduct disorder) found positive correlations between 5-HIAA concentration and lifetime history of aggression (Castellanos et al., 1994). Impulsivity, although found to relate to CSF monoamine concentrations in adults (Asberg et al., 1987; Coccaro et al., 1989), did not correlate with CSF monoamine levels in either of the child studies (Kruesi et al., 1990; Castellanos et al., 1994).

The relationship between low CSF 5-HIAA and other directed aggression found in some children and adolescents (Kruesi et al., 1985, 1992), is consistent with studies of adults, in that the aggression that is related to low 5-HIAA was more severe, affective, and physical. Less aggressive children with attention- deficit hyperactivity disorder had an opposite direction to their correlation with 5-HIAA, with more aggression associated with higher 5-HIAA concentrations (Castellanos et al., 1994). Moreover, the relationship of low 5-HIAA and pathologic aggression in humans involves lifetime history and it may be that the Castellanos sample will display the relationship over time. In support of this, relationships in the Kruesi et al. original sample (1990), intensified over time (1992).

In summary, across human and non-human primate studies there is a fair amount of consensus that low CSF 5-HIAA is associated with poorly modulated, socially unproductive, impulsive and/or affect laden aggression (Mehlman et al., 1994; Stein et al., 1993; Kruesi et al., 1992; Lester 1995).

2. EMPIRICAL FINDINGS

Empirical support for external environmental events mediating or influencing relationships between CSF 5-HIAA and aggression are discussed below. First, evidence regarding CSF 5-HIAA as a trait is considered. Next, CSF 5-HIAA's state-like relationship with aggression is reviewed. Then, support for the possibility of CSF 5-HIAA having scar marker status is reviewed.

2.1. Trait

Trait type relationships of biologic markers are marked by persistence over time. The following types of investigations provide support for the trait hypothesis of CSF 5-HIAA aggression relationships: investigations of CSF 5-HIAA heritability and stability over time plus investigations of the prognostic value of CSF 5-HIAA for future aggression.

CSF 5-HIAA as a trait marker has considerable support from non-human primate studies of heritability. In contrast to findings from humans, the CSF 5-HIAA concentration of rhesus monkeys exhibits strong genetic influence. Higley and his colleagues (1992) examined the genetic contributions to CSF 5-HIAA concentrations in 55 young rhesus monkeys studied at age 6 months. Genetic contributions were examined by correlating the 5-HIAA concentrations to half-siblings (sired by same father) and to siblings fostered to an unrelated, lactating mother. The authors found large genetic contributions to 5-HIAA, with maternal and paternal genetic influences accounting for over a third of the variance in 5-HIAA. A separate study of 54 monkeys with known pedigrees found significant differences among sire families in CSF levels of 5-hydroxyindoleacetic acid (Clarke et al., 1995).

The genetic contribution to human CSF 5-HIAA concentrations is less clear. The only study of heritability in humans found a larger contribution from "cultural heritability" than from genetic contribution (Oxenstierna et al., 1986). Analysis of the study of mono and di-zygotic twins, siblings and unrelated men was complicated and potentially confounded by higher 5-HIAA concentrations in the twins. Whilst this study did not conclusively rule out a significant genetic contribution to human CSF 5-HIAA concentrations, it did raise the possibility of a significant environmental contribution.

That concentrations of 5-HIAA have trait-like quality is supported by studies showing that individual differences in concentrations of 5-HIAA show considerable stability over time (Higley et al., 1992a). For instance, in rhesus monkeys 5-HIAA levels have been shown to correlate strongly from infancy (age six months) through the second and third years of life (Higley et al., 1992). Human studies also show considerable within individual stability for CSF 5-HIAA concentrations. Intraclass correlations for CSF 5-HIAA concentrations obtained from pediatric patients having multiple lumbar punctures ranged from .82 to .97 with time intervals ranging from 69 to 980 days (Riddle et al., 1986). This also suggests trait-like stability for CSF 5-HIAA concentrations. Nonetheless, within subject variation was observed in that study: the mean coefficient of variation for CSF 5-HIAA concentration within patients having 3 lumbar punctures was 23%.

Studies of the prognostic stability of CSF 5-HIAA behavior relationships are consistent with trait-like properties. Multiple adult follow-up studies (Brown et al., 1979; Traskman et al., 1981; Roy et al., 1989; Virkkunen et al., 1989, 1996; Koponen et al., 1994; Nordstrom et al., 1994; Faustman et al., 1993), two pediatric follow-up reports (Kruesi et al., 1989, 1992), and a four year follow up of non-human primates (Higley et al., 1996), consistently find that low CSF 5-HIAA is associated with a poorer prognosis than higher concentrations. Most individuals in the human follow-up studies received treatment(s) for their psychiatric disorders. However, despite treatment, the CSF 5-HIAA aggression relationship remained unimproved and actually worsened over time. This immutability of behavior-biologic marker association is consistent with a trait (or scar) rather than state relationship.

2.2. State

State dependence can refer to both internal as well as external conditions which act as change events. Some environmental influences are exteroceptive: season, weather, and atmospheric conditions are examples. Another frequent possibility involves ingested agents such as diet and pharmaceuticals.

2.2.1. Season, Weather and Atmospheric Influences. Studies have found associations between season, weather, and atmospheric variables and CSF 5-HIAA as well as a variety of other serotonergic measurements. Measures other than CSF 5-HIAA are included here because of the relative consistency with which seasonal influences are seen and because some measures like plasma tryptophan reflect events having potential influence on CSF 5-HIAA. Investigations into these relationships have been prompted by well known seasonal fluctuations in suicide and affective disorder (Durkheim 1897/1952, Rosenthal et al., 1983). Seasonal variations have been repeatedly demonstrated in platelet 5-HT levels, (Arendt et al., 1977; Mann et al., 1992; Wirz-Justice et al., 1977), platelet 5-HT uptake (Arora et al., 1984; Egrise et al., 1986; Swade & Coppen, 1980) and in cerebrospinal fluid 5-HIAA (Brewerton et al., 1988; Czernansky et al., 1988) in studies of adults. One study also demonstrated a seasonal variation in whole blood serotonin, with levels being highest in the winter and lowest in the summer (Mann et al., 1992). Not all studies have found seasonal variation in serotonergic parameters. No significant variation in CSF 5-HIAA was seen in a study of 135 adult alcoholics (Roy et al., 1991) nor in a study of 114 healthy individuals (Blennow et al., 1993).

Two studies have examined the seasonal variation of the serotonergic measures in pediatric patients. Platelet 5-HT showed seasonal variation in a pediatric sample (Brewerton et al., 1993). However, the seasonal variation seen in most adult studies of CSF 5-HIAA was not found in CSF from children and adolescents (Swedo et al., 1989). Neither study followed a within subjects design. Thus, the apparent seasonal variation, or lack thereof, might instead be due to differences between the individuals tested during the different seasons.

A more recent study of adults used a within subjects design to investigate seasonal variation in plasma L-tryptophan availability in healthy volunteers and relationships to violent suicide occurrence (Maes et al., 1995). Twenty-six healthy adult volunteers (13 men and 13 women) had monthly blood samplings for assays of L-tryptophan and competing amino acids (valine, leucine, isoleucine, tyrosine, and phenylalanine) during 1 calendar year. Significant annual rhythms were detected in L-tryptophan, and in the L-tryptophan/competing amino acid ratio. Both were significantly lower in the spring than in the other seasons. An important portion of the variance in L-tryptophan availability (ie, 12% to 14%) could be explained by the composite effects of present and past atmospheric factors; higher ambient temperature and relative humidity in the face of lower air pressure are the most important predictors of low L-tryptophan availability. In Maes et al's study, the bimodal seasonal pattern in the availability of plasma L-tryptophan matches seasonal patterns in the prevalence of violent suicide in populations from which the study volunteers were drawn as well as depression in other studies. Moreover, in another study from Belgium, non-violent suicides did not show a seasonal pattern (Linkowski et al., 1992).

In contrast, studies of aggression directed at others do not appear to track as neatly with seasonal serotonergic changes as do studies of self-directed aggression (suicide) does. A study of over 27,000 reports about women abused by their live-in male partners in the United States found statistically significant annual rhythms in the frequencies of abuse, with maxima in the summer (Michael & Zumpe 1986). The rhythms were closely related to annual changes in ambient temperature in these locations, and the time of the maxima was similar to those previously reported for rapes (Michael & Zumpe 1983a). Homicides, however, did not show a seasonal pattern in one study (Michael & Zumpe 1983). However, another investigation of archival data did find links between violent crime and hotter months (Anderson 1987). The peak months for aggressive incidents occur in summer. However, studies of CSF 5-HIAA do not find summer to be the nadir for concentrations (Brewerton et al., 1988). This does not however rule out the possibility that atmospheric pressure, temperature, or fluctuations thereof relate to a serotonergic mediation of some forms of violence in the summer.

Atmospheric pressure changes more frequently than season. Both the Maes et al. study (1995) of plasma tryptophan cited above and some CSF studies suggest that CSF 5-HIAA may be influenced by atmospheric pressure. Significant inverse correlations between atmospheric pressure and CSF concentrations of 5-HIAA have been reported (Garelis et al., 1974; Nordin et al., 1992). However, a more recent study failed to replicate the CSF 5-HIAA atmospheric pressure relationship (Faustman et al., 1993). Once again, the use of an across individuals rather than a within subjects design limits clarity.

2.2.2. Pharmacologic/Dietary Alteration of Serotonin Aggression Relationships. Another set of inquiries into the malleability of serotonin aggression relationships comes from pharmacologic treatment or dietary manipulation. For example, a possible biochemical rationale for treating impulsive aggression with newer antidepressants which have prominent serotonergic effects comes from adult and pediatric studies finding associations between aggressive and/or disruptive behavior and altered serotonergic function (Brown et al., 1986; Kruesi et al., 1985, 1989, 1990, 1992; Stoff et al., 1992; Coccaro et al., 1992, 1996; Linnoila & Virkkunen 1992).

2.2.2.1. Pharmaceuticals Altering Serotonergic Activity. Trazodone alone or in combination with the amino acid precursor of serotonin, tryptophan, has serotonergic effects. Open trials in adults (O'Neill et al., 1986; Pinner & Rich 1988) and in pediatric age groups found similarly encouraging results in reducing aggression. In an open trial of three children (ages 7–9) with diagnoses of disruptive disorder, all three showed a significant decrease of aggression using 75 mg of trazodone (Ghaziuddin & Alessi 1992). In another trazodone study, the item "often argues with adults" improved by 67% and "often loses temper" improved 60% in 22 hospitalized children (ages five-12) described as behaviorally-disturbed and treatment-resistant (Zubieta & Alessi 1992). The most frequent improvement in the conduct-disordered group was in "cruelty to people" and "frequency of fights" (63% and 57% of cases respectively). Improvement of outward aggression was found most frequently in 70% of patients (14 of 20) as evaluated by the Iowa Conners.

Fluoxetine was the first selective serotonin re-uptake inhibitor (SSRI) marketed in the U.S. Thus, it is not surprising that the bulk of information on SSRI's involves fluoxetine. Markowitz (1992) studied 21 severely to profoundly mentally retarded individuals, ranging in age from 17 to 56 years, who demonstrated self-injurious, emotionally labile, and aggressive behaviors and were treated in an open study with fluoxetine. Nineteen subjects (86%) were reported to be responders to fluoxetine administration (13-marked, four-moderate, and two-mild) by global ratings of behavior. In contrast, Rosenberg, Johnson, and Sahl (1992) described a 13 year-old boy, diagnosed with attention-deficit hyperactivity disorder (ADHD), conduct disorder, and moderate mental retardation, who was treated with fluoxetine for symptoms related to his ADHD, continuing aggression (initiating physical fights, etc.), irritability, and dysphoria. The case report showed a worsening in the patient's symptoms, believed to be a result of induction of manic symptoms by fluoxetine and the medication was eventually discontinued.

Buspirone is an antianxiety agent that is neither pharmacologically nor chemically related to barbiturates or benzodiazepines. Its mechanism of action is unknown, but has been associated with an increased affinity to the serotonin (5-HT1A) receptors of the brain, while not affecting GABA binding. Buspirone has been reported to decrease aggression in open trials (Ratey et al., 1989). Also, buspirone was reported to be effective in reducing both frequency and duration of aggressive outbursts (baseline: 24 outbursts /13.7 min. average vs. buspirone: 11 outbursts / 4.8 min. average) in a 39-year-old, moderately retarded treatment resistant (the patient had been treated with haloperidol, phenobarbital, and thioridazine) female in a partially-blind, placebo-controlled study (Gedye 1991). An additional study by Realmuto and colleagues (1989), reported the effects of buspirone in autistic children, utilizing an open, four-week design, comparing buspirone to either methylphenidate or fenfluramine. The targeted symptoms included aggression in their population of four autistic children (three males, one female) ranging in age from nine to ten and a half years. Two of the children showed at least mild improvement in aggression when receiving buspirone. In contrast, methylphenidate and fenfluramine had not been beneficial to the patients studied. These two studies do suggest possible benefit in buspirone use for aggressive symptoms in some persons, but further study with adequate controls and blinding is needed.

Fenfluramine, a sympathomimetic amine with antiserotonergic properties has had different effects upon aggression in retarded versus normal intelligence patients. One study (Donnelly et al., 1989) reported no clinical benefits on behavioral parameters including aggression or hyperactivity in non-retarded children with ADHD. However, two studies have found decreases in aggression and/or conduct problems in controlled trials with mentally retarded subjects (Seilkowitz et al., 1990; Aman et al., 1993).

Beta blockers may also reduce symptoms of aggression (Kuperman & Stewart 1987; Greendyke, Kanter, Schuster et al., 1986) in some populations of children, adolescents, and adults. Although the effects on noradrenergic functions are prominent, beta-blocking drugs have been found to alter serotonin mediated behavior but not serotonin receptor number (Eison et al., 1988). Problematic aggression has generally been studied with beta-blockers in populations of organically brain-damaged individuals (Greendyke & Kanter 1986; Yudofsky, Williams, & Gorman 1981; Ratey, Morrill, & Oxenkrug 1983). An open-design study using propranolol was reported by Kuperman and Stewart (1987), who were interested in the treatment of physically aggressive behavior in 16 subjects, ranging in age from four to 24 years (12 were aged four to 17 years) and all demonstrated physically aggressive behavior as their main problem. Ten of the 16 subjects (62.5%) were rated moderately or much improved according to the ratings by parents, teachers, and clinicians following at least a three month trial of propranolol. Two of the nonresponders had shown an initial, marked decrease in their aggressive behaviors, but became noncompliant when discharged from the unit. This study suggests a role for propranolol in the treatment of aggression. Despite evidence available for the utility of propranolol, pediatric studies utilizing atenolol, nadolol, or pindolol have not been conducted in a psychiatric population under double-blind conditions. Matthews-Ferrari and Karroum (1992) reported on using metoprolol in an 11-year-old male for control of severe aggressive outbursts that had been previously treated unsuccessfully with methylphenidate and carbamazepine. Outbursts of aggression diminished to zero and improvement was also noted in peer relationships and self-esteem. Another group reported a significant decline in the frequency of aggressive outbursts and rating scale scores in 16 chronic, psychiatric inpatient adults (most were diagnosed with schizophrenia) treated with nadolol in double-blind conditions, when compared to a control group (Ratey et al., 1992).

A very clever study by Raleigh, et al. (1983, 1991) of the effects of pharmacologic manipulation of serotonin upon dominance in non-human primates warrants attention because of the provocative implication that relationships between serotonin and behavior are modifiable with a combination of social and pharmacologic maneuvers. Moreover, relationships between social competence, which may be thought of as a reflection of the dominance hierarchy, and CSF 5-HIAA have been found in human children (Kruesi et al., 1990). A monkey's stature in a dominance hierarchy is important as it may play a role in whether the monkey is the exhibitor or recipient of aggression. In the Raleigh et al. study, thirty-six adult male vervet monkeys were housed in 12 social groups, containing 3 adult males, at least 3 adult females and immature offspring. The groups were all established at least 6 months prior to study. After the dominant males were removed from each social group, for the next 4 weeks, one of the remaining subordinate males in each group received tryptophan and fluoxetine to increase serotonin, or they received fenfluramine or cyproheptadine to decrease serotonin. Increasing serotonergic activity via tryptophan and fluoxetine facilitated the acquisition of dominance in males. Once the treated subjects became dominant they *also remained so* during the entire experimental period. On the other hand, treatment with fenfluramine or cyproheptadine drastically decreased likelihood of becoming dominant. It is important to note that the behavioral changes exhibited by the tryptophan and fluoxetine treated subjects paralleled those shown by animals who acquired dominance in drug-free conditions. However, the pharmacologic effects were apparently mediated by the males relationship to females in group. This led Raleigh et al. (1991) to conclude that serotonergic systems may promote the acquisition of male dominance in unstable social conditions.

In another study of vervets, Raleigh et al. (1984) found whole blood serotonin concentrations were two times higher in dominant than in subordinate members of established vervet groups. When a dominant monkey was removed from social position, whole blood serotonin decreases within a week to level of the subordinate group members. Moreover, dominant social status was found to facilitate the behavioral effects of serotonergic agonists in adult vervet monkeys (Raleigh et al., 1985).

Although these findings suggest that the serotonergic system may promote dominance in vervets, it may also be the case that being dominant or subordinate affects the functioning of the serotonergic system as well. Thus, the direction of causality is unclear at present, but there are suggestions of a reciprocal interaction.

Recent study of crayfish indicates that the modulatory effect of serotonin on dominant behavior depends on the social experience of the animal (Yeh et al., 1996). Serotonin injected into the circulatory system causes crayfish to assume a dominant posture. When a pair of previously isolated crayfish of similar size are placed together in a small aquarium, the pair will agonistically interact to determine which crayfish is dominant. After the agonistic interaction, the subordinate moves by retreating or tailflipping to avoid contact with the dominant animal. The effect of serotonin on the neural circuit for tailflip behavior was found to depend on the animal's social experience. Serotonin persistently enhanced the response to sensory stimuli of the lateral giant tailflip command neuron in socially isolated crayfish, reversibly inhibited it in subordinate animals, and reversibly enhanced it in socially dominant crayfish. Serotonin receptor agonists had opposing effects: m-CPP, a 5HT-1 agonist, had no effect on the lateral giant neuron's responses in social isolates but it reduced excitatory postsynaptic evoked potentials in both dominant and subordinant crayfish. In contrast, a 5HT-2 agonist produced similar electrophysiologic changes in isolates, dominants, and subordinants. Additional experiments (Yeh et al., 1996) demonstrated that the modulatory effect of serotonin can depend not only on the crayfish's current social status, but also on its *prior* status.

Provocative as the Yeh et al. findings are in arguing for social experience as a modulator of serotonergic neural plasticity relevant to aggressive behavior, some cautions are necessary. The serotonergic receptors in the stomatogastric nervous system of crustaceans differ from vertebrate types (Zhang & Harris-Warrick 1994).

2.2.2.2 Diet. Diet is often considered a methodologic variable because the dietary amino acid, tryptophan, is the precursor for serotonin. However, dietary impact upon serotonergic systems and their behavioral relations are complicated. Tryptophan loading and depletion provide examples, but other less obvious dietary influences deserve consideration.

Oral tryptophan loading with up to 100 mg/kg of tryptophan increased plasma tryptophan levels, yet whole blood 5-HT levels were unaffected (Cook et al., 1992). Because most blood 5-HT is within platelets, this indicates that acute loading of the amino acid precursor to 5-HT does not alter intracellular (platelet) 5-HT concentrations. However, oral tryptophan supplementation has been seen to decrease aggression in some individuals under some conditions (O'Neil et al., 1986; Volavka et al., 1990; Cleare & Bond, 1994).

Acute tryptophan depletion has also been used for investigation of serotonergic function. Tryptophan depletion worsens mood in affectively ill subjects (Henninger et al., 1992), as well as in normals (Young et al., 1985) but did not impact aggression in 14 adults with intermittent explosive disorder (Salomon et al., 1994). In the Salomon et al. study, following a low tryptophan diet the adults were challenged with a tryptophan depleting amino acid mixture or a control drink. No significant increase in Buss Durkee Hostility Inventory was seen. However, findings may have been missed because of the lack of provocation or interaction with others that usually occur in the ecological contexts surrounding explosive aggressive events. In contrast, a study of 24 men with high levels of trait aggression on the Buss Durkee Hostility Inventory found that a tryptophan depleting amino acid mixture led to significantly more angry, aggressive, annoyed, quarrelsome, and hostile responses (Cleare & Bond, 1994).

Dietary tryptophan alteration has also been shown to impact behavior of non-human primates. Chamberlain et al. (1987) found that by increasing or decreasing tryptophan in the diet of vervet monkeys, they could produce either a decrease or increase in levels of aggression. In another study, when the dominant male vervet was removed from a social group which included 3 males and at least 3 adult females, the administration of tryptophan (or fluoxetine) to one of the previous subordinate males caused that animal to ascend to dominance (Raleigh et al., 1991), whilst reduction in serotonergic function via cyproheptadine or fenfluramine treatment led to subordinate status.

An additional line of evidence from diet studies supports an environmental mediation of serotonin violence relationships. An excess of unnatural death (e.g. suicide, violence or accident) has been seen in human trials aimed at lowering cholesterol to decrease the mortality associated with coronary artery disease (Muldoon et al., 1990; Engleberg 1992; Neaton et al., 1992). Associations between low plasma cholesterol and aggressive behavior have been reported in monkeys, adult and pediatric human samples (Virkkunen 1979, 1983; Virkkunen & Penttinen 1984; Kaplan et al., 1991; S.A. Mednick personal communication). However, others have failed to find this association (Stewart & Stewart, 1981; Rao et al., 1991). In a recent study, juvenile cynomolgus monkeys (eight female and nine male) were studied over an 8 month period while they were fed a diet high in fat and either high or low in cholesterol (Kaplan et al., 1994). Animals that consumed a low-cholesterol diet were more aggressive, less affiliative, and had lower cerebrospinal fluid concentrations of 5-hydroxyindoleacetic acid than did their high-cholesterol counterparts (p < .05 for each). The association among dietary cholesterol, serotonergic activity, and social behavior suggests that alteration of dietary lipids can influence brain neurochemistry and behavior. A human study of CSF 5-HIAA blood cholesterol correlation did not find strong support for this hypothesis (Ringo et al., 1994), but the study did not involve dietary manipulation.

2.3. Scar

The evidence in support of a possible scar-like relationship between CSF 5-HIAA and violence is as follows: By adolescence, if not before, low concentrations of CSF 5-HIAA, which for purposes of this discussion are defined as about 80 picomolar or less in human samples, appear to be both relatively stable over time and accompanied by a poor prognosis. This may reflect a trait, but questions exist. Some developmental findings are also indicative of a scar-like relationship. The stability of individuals CSF 5-HIAA concentration exists within older time frames (Higley 1992) as opposed to from birth onwards. Concentrations are known to decrease with development from infancy to adolescence in both humans and non-humans (Kruesi et al., 1988; Riddle et al., 1986; Seifert et al., 1980; Higley et al 1992; Langlais et al., 1985), but whether the interindividual rank order stability exists from birth to adulthood is not clear and whether there are CSF 5-HIAA behavior relationships in infants that parallel those in older individuals is also not clear. Human aggression shows significant stability as a trait from about age 3 onwards (Olweus 1979; Huesmann et al., 1984), but whether this behavior is already set in infancy is an open question. CSF 5-HIAA concentrations appear to stabilize somewhere between infancy and adolescence. Therefore, scar relationships are not ruled out because there may be a period during which the relationship is not, as yet established.

Recent developmental studies show striking parallels between the emergence and subsequent development of aggressive behavior, on the one hand, and of changes in concentrations of 5-HIAA on the other. Briefly, aggressive behavior in the rhesus monkey typically emerges at low levels during late infancy. It subsequently increases at a steady rate during the juvenile period, peaking in early adolescence (Bernstein et al., 1983). Concentrations of 5-HIAA, on the other hand, are highest in infancy, and decline thereafter (Higley et al., 1992; Langlais et al., 1985). Interestingly, stable correlations between CSF 5-HIAA and other metabolites (HVA, NE) are first attained during the time period when the infant rhesus *changes* from spending most of its time in contact with its mother to spending increasing amounts of time playing with peers (3–5 months of age). The timing of this change suggests a possible link to the developing mother-infant attachment bond. As the infant begins to leave a protective base, the serotonergic system undergoes important change as does the infant's behavioral repertoire, allowing increased possibility of self-protection. In humans, early attachment has been shown to relate to longer term attentional, behavioral (Jacobsen & Hofmann, in press), and cognitive trajectory (Jacobsen et al., 1994).

Post-natal environmental manipulations in non-human primates suggest alteration of serotonin behavior relationships can occur. For example, female rhesus monkeys who were peer reared were found to have *higher* concentrations of CSF 5-HIAA than mother reared females (Higley et al., 1992). However, males did not, in that study, evidence the same effect. More compelling are the effects described by both Kraemer (this volume, Kraemer & Clarke in press) and Higley et al. (1992) of peer rearing upon CSF 5-HIAA behavior relationships. Kraemer notes that 24 mother reared monkeys have significant correlations between CSF 5-HIAA, social behavior and aggression (Kraemer & Clark, in press; Kraemer this volume) whilst the 23 peer reared monkeys do not have the same significant correlations.

The impact of rearing on serotonin behavior relationships is likely to benefit from additional scrutiny. Further analyses of these data sets may prove even more informative, because the authors might control for the CSF 5-HIAA of the biologic parents of the differentially reared monkeys. This could shed light upon the degree to which rearing impacts serotonin systems. This would be helpful particularly because there are no CSF 5-HIAA samples prior to the implementation of the rearing change. Future studies which document serotonergic parameters pre- and post-peer rearing will be critical for addressing how rearing may impact serotonin systems.

An additional potential confound in understanding the environmental elements that mediate serotonin behavior relationships also deserves examination. The peer reared animals were "bottle fed" meaning raised on commercial formula (Similac), whereas the mother raised monkeys continued to be breast fed. The potential relevance is that breast feeding has been documented to improve intelligence compared to formula feeding (Lucas et al., 1992, 1994). The decrement in intelligence associated with formula feeding is of the same magnitude as that shown to exist between conduct disordered and control samples (Moffitt 1993). Does the difference in breast milk versus formula alter the relationships between serotonin and behavior? Again, studies that address these questions may have important implications for treatment intervention.

3. SYNTHESIS

There is some evidence in support of all three types of CSF 5-HIAA behavior relationships: trait, state and scar. The relationship need not be exclusively of one type - all three types may well be present. For example, while a non-human primate's CSF 5-HIAA is dependent upon his biologic parents and has a trait like relationship, maternal deprivation may result in scar like changes and feeding that animal a diet with altered cholesterol content can alter serotonergic parameters in a state like fashion. Studies to date have not yet systematically altered all three types of variables and then assessed their contributions. As noted earlier, studies with already collected data sets which used peer rearing may well have data available on the biologic parents 5-HIAA and could control for such influence.

Because available evidence suggests that the prognosis for low 5-HIAA individuals is poor (Brown et al., 1979; Traskman et al., 1981; Roy et al., 1989; Virkkunen et al., 1989, 1995, 1996; Koponen et al., 1994; Nordstrom et al., 1994; Faustman et al., 1993; Kruesi 1989; Kruesi et al., 1992; Mehlman et al., 1995), the critical questions to be solved are: (a) what, descriptively, is the relation between 5-HIAA levels and behavior in early development, and when is this relation first established? (b) how can the relation be altered for low 5-HIAA individuals? (c) is there a critical or sensitive period in which this alteration has to occur? (d) how can the adverse relationship be prevented from developing in the first place?

The evidence for trait and scar type relationships suggests that substantial alteration in the neural processes represented by CSF 5-HIAA concentrations would need to occur early in life or prenatally in order to ameliorate risk. Evidence for the stability of human aggression from age 3 or so onward (Olweus 1979) is consistent with this idea as is evidence of stable 5-HIAA concentrations from late childhood onward (Riddle et al., 1986). How much alteration is necessary to make significant changes is an open question. Nonetheless, the crustacean (Yeh et al., 1996) and non-human primate studies (Kraemer & Clarke, in press; Raleigh et al., 1991; Yodinyuad et al., 1985) raise the intriguing possibility of socioenvironmental alteration of what might otherwise be viewed as trait risk. Studies of state dependent relationships suggest that there may also be comparatively subtle alterations that can alter risk for better or worse. For example, plasma tryptophan which is associated with fluctuations in suicide has peak-trough differences in yearly variation of about 17% of the mean (Maes et al., 1995).

A case for environmental mediation of both state and scar type relationships between serotonin and aggression is evident, although there is inconsistency in the evidence. The most compelling evidence for environmental mediation of scar type associations comes from the peer reared monkey model. The pharmacologic/dietary manipulation of serotonergic function and impact of seasonal/barometric phenomena support the environmental mediation of state type serotonin aggression relationships.

The evidence suggests that environmental mediation of serotonin behavior relationships is a real possibility. Understanding these relationships may benefit future efforts.

REFERENCES

Aman, M.G., Kern, R.A., McGee, D.E., & Arnold, L.E. (1993). Fenfluramine and methylphenidate in children with mental retardation and ADHD: Clinical and side effects. *Journal of American Academy of Child Adolescent Psychiatry, 32*(4), 851–859.

Anderson, C.A. (1987). Temperature and aggression: effects on quarterly, yearly, and city rates of violent and nonviolent crime. *Journal of Pers Sociology and Psychology, 52*, 1161–73.

Arendt, J., Wirz-Justice, A., & Bradtke, J, (1977). Annual rhythm of serum melatonin in man. *Neuroscience Letters, 7*: 327–330.

Arora, R.C., Kregel, L., & Meltzer, H. (1984). Transsynaptic control of gene expression. *Annual Review of Neuroscience, 16*:17–29.

Asberg, M., Traskman, L., & Thoren, P. (1976). 5-HIAA in the cerebrospinal fluid: a biochemical suicide predictor? *Archives of General Psychiatry, 33*:1193–7.

Asberg, M., Thoren, L., & Traskman, P. (1976a). Serotonin depression: a biochemical subgroup within affective disorders. *Science, 191*:478–80

Asberg, M., Schalling, D., Traskman-Bendz, L., & Wagner, A. (1987). Psychobiology of suicide, impulsivity and related phenomena. In *Psychopharmacology: The Third Generation of Progress.* H.Y. Meltzer (Ed.). Raven Press, New York.

Bernstein, I., Williams, L., & Ramsay, M. (1983). The expression of aggression in old world monkeys. *International Journal of Primatology, 4*:113–124.

Blennow, K., Wallin, A., Gottfries, C.G., Karlsson, I., Mansson, J.E., Skoog, I., Wikkelson, C., & Svennerholm, L. (1993) Cerebrospinal fluid monoamine metabolites in 114 healthy individuals 18–88 years of age. *European Neuropsychopharmacology, 3*(1):55–61.

Brewerton, T.D., Berrettini, W.H., Nurnberger, J.I., & Linnoila, M. (1988). Analysis of seasonal fluctuations of CSF monoamine metabolites and neuropeptides in normal controls: Findings with 5-HIAA and HVA. *Psychiatry Research, 23*:257–265.

Brewerton, T.D., Flament, M.F., Rapoport, J.L., & Murphy, D.L. (1993). Seasonal effects on platelet 5-HT content in patients with OCD and controls. *Archives of General Psychiatry, 50*(5):409.

Brown, G.L., Goodwin, F.K., Ballenger, J.C., Goyer, P.F., & Major, L.F. (1979). Aggression in humans correlates with cerebrospinal fluid amine metabolites. *Psychiatry Research, 1*:131–139.

Brown, G.L., Ebert, M.H., Goyer, P.F., Jimerson, D.C., Klein, W.J., Bunney, W.E., & Goodwin, F.K. (1982). Aggression, suicide, and serotonin: Relationships to CSF amine metabolites. *American Journal of Psychiatry, 139*:741–746.

Brown, G.L., Kline, W., Goyer, P., Minichello, M., Kruesi, M., & Goodwin, F.K. (1986). Relationship of childhood characteristics to cerebrospinal fluid 5-hydroxyindoleacetic acid in aggressive adults. In C Shagass, RC Josiassen, WH Bridger, KJ Weiss, D Stoff, GM Simpson (Eds.) *Biological Psychiatry.* (chap 7, pp.177–179). New York: Elsevier Science Publishing.

Brown, G.L., & Linnoila, M.I. (1990). CSF serotonin metabolite (5- HIAA) studies in depression, impulsivity, and violence. *Journal of Clinical Psychiatry, 51*:31–41.

Castellanos, F.X., Elia, J., Kruesi, M.J., Gulotta, C.S., Mefford, I.N., Potter, W.Z., Ritchie, G.F., & Rapoport, J.L. (1994) Cerebrospinal fluid monoamine metabolites in boys with attention-deficit hyperactivity disorder. *Psychiatry Research, 52*(3):305–316.

Chamberlain, B., Ervin, F.R., Pihl, R.O., & Young, S.N. (1987). The effect of raising or lowering tryptophan levels aggression in vervet monkeys. *Pharmacology Biochemical Behavior, 28*:503–510.

Clarke, A.S., Kammerer, C.M., George, K.P., Kupfer, D.J., McKinney, W.T., Spence, M.A., & Kraemer, G.W. (1995) Evidence for heritability of biogenic amine levels in the cerebrospinal fluid of rhesus monkeys. *Biological Psychiatry, 38*(9):572–577. Cleare, A.J., & Bond, A.J. (1994, December). Effects of alterations in plasma tryptophan levels on aggressive feelings [Letter to the editor]. *Archives of General Psychiatry, 51*:1004–1005.

Coccaro, E.F., Siever, L.J., Klar, H.M., Maurer, G., Cochrane, K., Cooper, T.B., Mohs, R.C., & Davis, K.L. (1989). Serotonergic studies in patients with affective and personalty disorders: Correlates with suicidal and impulsive aggressive behavior. *Archives General Psychiatry, 46*(7):587–99.

Coccaro, E.F., Kavoussi, R.J., & Lesser, J.C. (1992). Self- and other-directed human aggression: The role of the central serotonergic system. *International Clinical Psychopharmacology, 6*, Suppl. (6)70–83.

Coccaro, E.F., Kavoussi, R.J., Sheline, Y.I., Lish, J.D., & Czernasky, J.G. (1996). Impulsive aggression in personality disorder correlates with tritiated paroxetine binding in the platelet. *Archives of General Psychiatry, 53*:531–536.

Cohen, D.J., Caparulo, B.K., Shaywitz, B.A., & Bower, M.B.J. (1977). Dopamine and serotonin metabolism in neuropsychiatrically disturbed children: CSF homovanillic acid and 5-hydroxyindoleacetic acid. *Archives of General Psychiatry, 34*:545–550.

Cook, E.H., Anderson, G.M., Heninger, G.R., Fletcher, K.E., et al. (1992). Tryptophan loading in hyperserotonemic and normoserotonemic adults. *Biological Psychiatry, 31*:525–8.

Cook, E., & Kruesi, M.J.P. (in press). Neurochemical measures. In D. Shaffer, J. Richters (Eds.) *Assessment and Diagnosis in Child Psychopathology.* Second Edition.

Czernansky, J.G., Faull, K.F., & Pfefferbaum, A. (1988). Seasonal changes in CSF monoamine metabolites in psychiatric patients: What is the source? *Psychiatry Research, 25*:361- 363.

Donnelly, M., Rapoport, J.L., Potter, W.Z., Oliver, J., Keysor, C.S., & Murphy, D.L. (1989). Fenfluramine and dextroamphetamine treatment of childhood hyperactivity. *Archives of General Psychiatry, 46*:205–212.

Durkheim, E. (1952). *Suicide: A study in sociology* (J.A. Spaulding & C. Simpson, Trans.). London: Routledge and Kegan Paul. (Original work published in 1897)

Egrise, D., Rubenstein, M., Schoutens, A., Cantraine, F., & Mendelewicz, J. (1986). Seasonal variation of platelet serotonin uptake and ^{3}H-imipramine binding in normal and depressed subjects. *Biological Psychiatry, 21*:283–293.

Eison, A.S., Yocca, F.D., & Gianutsos, G. (1988). Noradrenergic denervation alters serotonin$_2$-mediates behavior but not serotonin$_2$ receptor number in rats: modulatory role of beta adrenergic receptors. *Journal of Pharmacologic Experimental Therapy, 246*:571–577.

Engleberg, H. (1992). Low serum cholesterol and suicide. *Lancet, 339*:727–729.

Faustman, W.O., Ringo, D.L., & Faull, K.F. (1993). An association between low level of 5-HIAA and HVA in cerebrospinal fluid and early mortality in a diagnostically mixed psychiatric sample. *British Journal of Psychiatry, 163*:519–21.

Fernstrom, J.D., & Wurtman, R.J. (1972). Brain serotonin content: Physiological regulation by plasma neutral amino acids. *Science, 178*:414–416.

Garelis, E., Young, S.N., Lal, S., Sourkes, T.L. (1974). Monoamine metabolites in lumbar CSF: The question of their origin in relation to clinical studies. *Brain Research, 79*:1–8.

Ghaziuddin, N., & Alessi. N. (1992). An open clinical trial of trazodone in aggressive children. *Journal of Child and Adolescent Psychopharmacology, 2* 291–298.

Gedye, A. (1991). Buspirone alone or with serotonergic diet reduced aggression in a developmentally disabled adult. *Biological Psychiatry, 30*:88–91.

Golden, R.N., Gilmore, J.H., Corrigan, M.H., Ekstrom, R.D., Knight, B.T., & Garbutt, J.C. (1991). Serotonin, suicide and aggression: clinical studies. *Journal of Clinical Psychiatry, 52*, suppl:61–69.

Greendyke, R.M., & Kanter, D. (1986). Therapeutic effects of pindolol on behavioral disease: A double blind study. *Journal of Clinical Psychiatry, 47*(8): 423–426.

Greendyke, R.M., Kanter, D.R., Schuster, D.B., Verstraete, S., & Wooton, J.A. (1986). Propranolol treatments of assaultive patients with organic brain disease. *Journal of Nervous and Mental Disease*, 174(5):290–294.

Halperin, J.M., Vanshdeep, S., Siever, L.J., Schwartz, S.T., Matier, K., Wornell, G., & Newcorn, J.H. (1994). Serotonergic function in aggressive and nonaggressive boys with attention deficit hyperactivity disorder. *American Journal of Psychiatry, 151*:243–248.

Heninger, G.R., Delgado, P.L., Charney, D.S., Price, L.H., & Aghajanian, G.K. (1992). Tryptophan deficient diet and amino acid drink deplete plasma tryptophan and induce a relapse of depression in susceptible patients. *Journal of Chemical Neuroanatomy, 5*:347–8

Higley, J.D., Mehlman, P.T., Taub, D.M., Higley, S.B., Suomi, S.J., Vickers, J.H., & Linnoila, M. (1992): Cerebrospinal fluid monoamine and adrenal correlates of aggression in free- ranging rhesus monkeys. *Archives of General Psychiatry, 49*: 436–441.

Higley, J.D., Suomi, S.J., & Linnoila M (1992a): A longitudinal assessment of CSF monoamine metabolite and plasma cortisol concentrations in young rhesus monkeys. *Biological Psychiatry, 32*: 127–145.

Higley, J.D., Mehlman, P.T., Higley, S.B., Fernald, B., Vickers, J., Lindell, S.G., Taub, D.M., Suomi, S.J., & Linnoila, M. (1996). Excessive mortality in young free-ranging male nonhuman primates with low cerebrospinal fluid 5- hydroxyindoleacetic acid concentrations. *Archives of General Psychiatry, 53*:537- 543.

Huesmann, L.R., Eron, L.D., Lefkowitz, M.M., & Walder, L.O. (1984). Stability of aggression overtime and generations. *Developmental Psychology, 20*:1120–1134.

Jacobsen, T., Edelstein, W., & Hofmann, V. (1994). A longitudinal study of the relation between representations of attachment in childhood and cognitive functioning in childhood and adolescence. *Developmental Psychology, 30*(1):112–14.

Jacobsen, T., Hofmann, V. (in press). Children's attachment representations: Longitudinal relations to school behavior and academic competency in middle childhood and adolescence. *Developmental Psychology.*

Kaplan, J.R., Manuck, S.B., Shively, C. (1991). The effects of fat and cholesterol on social behavior in monkeys. *Psychosomatic Medicine, 53*:634–642.

Kaplan, J.R., Shively, C.A., Fontenot, M.B., Morgan, T.M., Howell, S.M., Manuck, S.B.,Muldoon, M.F., & Mann, J.J. (1994). Demonstration of an association among dietary cholesterol, central serotonergic activity, and social behavior in monkeys. *Psychosomatic Medicine, 56*(6):479–84.

Koponen, H.J., Lepola, U., Leinonen, E. (1994). A long-term follow- up study of cerebrospinal fluid 5-hydroxyindoleacetic acid in delirium. *European Archives of Psychiatry and Clinical Neuroscience, 244*(3):131–134.

Kraemer, G.W., & Clarke, A.S. (in press). Social attachment, brain function and aggression. *Annals of New York Academy of Science.*

Kruesi, M.J.P., Linnoila, M., Rapoport, J.L., Brown, G.C., & Petersen, R. (1985). Carbohydrate craving, conduct disorder and low CSF 5-HIAA. *Psychiatry Research, 16*:83–86.

Kruesi, M.J.P., Swedo, S.E., Hamburger, S.D., Potter, W.Z., & Rapoport, J.L. (1988). Concentration gradient of CSF monoamine metabolites in children and adolescents. *Biological Psychiatry, 24*:507–514.

Kruesi, M.J.P. (1989). Cruelty to animals and CSF 5-HIAA. Psychiatry Research, 28, 115–116.

Kruesi, M.J.P., Rapoport, J.L., Hamburger, S., Hibbs, E., Potter, W.Z., Lenane, M., & Brown, G.L. (1990). Cerebrospinal fluid monoamine metabolites, aggression, and impulsivity in disruptive behavior disorders of children and adolescents. *Archives of General Psychiatry, 47*, 419–426.

Kruesi, M.J.P., Hibbs, E.D., Zahn, T.P., Keysor, C.S., Hamburger, S.D., Bartko, J.J., & Rapoport, J.L. (1992). A 2-year prospective follow-up study of children and adolescents with disruptive behavior disorders. Prediction by cerebrospinal fluid 5-hydroxyindoleacetic acid, homovanillic acid, and autonomic measures. *Archives of General Psychiatry, 49*, 429- 435.

Kruesi, M.J.P., & Lelio, D.F. (1996). Disorders of conduct and behavior. In J. Weiner (Ed.), *Diagnosis and Psychopharmacology of Childhood and Adolescent Disorders* (pp. 401–447) Second Edition. New York: John Wiley & Sons, Inc.

Kuperman, S., & Stewart, M.A. (1987). Use of propranolol to decrease aggressive outbursts in younger patients. *Psychosomatics, 28*(6), 315–319.

Langlais, P.J., Walsh, F.X., Bird, E.D. & Levy, H.L. (1985). Cerebrospinal fluid neurotransmitter metabolites in neurologically normal infants and children. *Pediatrics, 75*:580–86.

Lester, D. (1995). The concentration of neurotransmitter metabolites in the cerebrospinal fluid of suicidal individuals: a meta-analysis. *Pharmacopsychiatry, 28*(2):45–50.

Linkowski, P., Martin, F., & De Maertelaer, V. (1992). Effect of some climatic factors on violent and non-violent suicides in Belgium. *Journal of Affective Disorders, 25*(3):161–6

Linnoila, M., Virkkunen, M., Scheinin, M., Nuutila, A., Rimon, R., & Goodwin, F.K. (1983). Low cerebrospinal fluid 5- hydroxyindoleacetic acid concentration differentiates impulsive from nonimpulsive violent behavior. *Life Sciences, 33*:2609–2614.

Linnoila, V.M., & Virkkunen, M. (1992). Aggression, suicidality, and serotonin. *Journal of Clinical Psychiatry, 53* (Suppl.), 36–51.

Lucas, A., Morley, R., Cole, T.J., Lister, G., & Leeson-Payne, C. (1992). Breast milk and subsequent intelligence quotient in children born preterm. *Lancet, 339*:261–4.

Lucas, A. Morley, R., Cole, T.J., & Gore, S.M. (1994). A randomised multicentre study of human milk versus formula and later development in preterm infants. *Arch Dis Child Fetal Neonatal Ed, 70*:141–6.

Maes, M., Scharpe, S., Verkerk, R., D'Hondt, P., Peeters, D., Cosyns, P., Thompson, P., De Meyer, F., Wauters, A., & Neels, H. (1995). Seasonal Variation in Plasma L-Tryptophan Availability in Healthy Volunteers. *Archives of General Psychiatry, 52*:937–946.

Mann, J., McBride, P., Anderson, G., & Mieczkowski, T. (1992). Platelet and whole blood serotonin content in depressed inpatients: Correlations with acute and life-time psychopathology. *Biological Psychiatry, 32,* 243–257.

Mann, J.J., McBride, P.A., Brown, R.P., Linnoila, M., Leon, A.C., DeMeo, M., Mieczkowski, T., Myers, J.E., & Stanley, M. (1992a). Relationship between central and peripheral serotonin indexes in depressed and suicidal psychiatric inpatients. *Archives of General Psychiatry, 49*(6):442–6

Markowitz, P. (1992). Effect of fluoxetine on self-injurious behavior in the developmentally disabled: A preliminary study. *Journal of Clinical Psychopharmacology, 12*(1), 27–31.

Matthews-Ferrari, K., & Karroum, N. (1992). Metoprolol for aggression. In: Letters to the editor. *Journal of the American Academy of Child and Adolescent Psychiatry, 31*(5), 994.

Mefford, I.N., Ward, M.M., Miles, L., Taylor, B., Chesney, M.A., Keegan, D.L., & Barchas, J.D. (1981). Determination of plasma catecholamines and free 3,4-dihydroxyphenolacetic acid in continuously collected human plasma by high performance liquid chromatography with electrochemical detection. *Life Science, 28,* 477–483.

Mehlman, P.T., Higley, J.D., Faucher, I., Lilly, A.A., Taub, D.M., Vickers, J., Suomi, SJ., Linnoila, M. (1994). Low CSF 5-HIAA concentrations and severe aggression and impaired impulse control in nonhuman primates. *American Journal of Psychiatry, 151*:1485–91.

Mehlman, P.T., Higley J.D., Faucher, I., Lilly, A.A., Taub, D.M., Vickers, J., Suomi, S.J., Linnoila. M. (1995). Correlation of CSF 5-HIAA concentration with sociality and the timing of emigration in free-ranging primates. *American Journal of Psychiatry, 152*:907–13.

Michael, R.P., & Zumpe, D. (1983). Annual rhythms in human violence and sexual aggression in the United States and the role of temperature. *Social Biology, 30*(3):263–78.

Michael, R.P., & Zumpe, D. (1983a). Sexual violence in the United States and the role of season. *American Journal of Psychiatry, 140*:883–6.

Michael, R.P., & Zumpe, D. (1986). An annual rhythm in the battering of women. *American Journal of Psychiatry, 143*(5):637–40.

Moffitt, T.E. (1993). The neuropsychology of conduct disorder. *Development and Psychopathology, 5*:135–151.

Moffitt, T., Caspi, A., Fawcett, P., Brammer, G.L., Raleigh, M., Yuwiler, A., & Silva, P. (in press). Whole blood serotonin and family background relate to male violence. In A. Raine, D. Farrington, P. Brennan, & S.A. Mednick (Eds.), *Biosocial bases of violence.* New York: Plenum Publishing.

Moss, H.B., Yao, J.K., & Panzak, G.L. (1990). Serotonergic responsivity and behavioral dimensions in antisocial personality disorder with substance abuse. *Biological Psychiatry, 28*:325–338.

Muldoon, M.F., Manuck, S.B., Matthews, K.A. (1990). Lowering cholesterol concentrations and normality: A quantitative review of primary prevention trials. *British Medical Journal, 301*:309–314.

Neaton J.D., Blackburn, H., Jacobs, D., et al. (1992). Serum cholesterol level and mortality findings for men screened in the multiple risk factor intervention trial. *Archives of Internal Medicine, 152*:1490–1500.

Nordin, C., Swedin, A., Zachau, A. (1992). CSF 5-HIAA and Atmospheric Pressure. *Biological Psychiatry, 31*:644–645.

Nordstrom, P., Samuelsson, M., Asberg, M., Traskman-Bendz, L., Nordin, C., & Bertilsson, L. (1994). CSF 5-HIAA predicts suicide risk after attempted suicide. *Suicide and Life Threatening Behavior, 24*(1):1–9.

Olweus, D. (1979). Stability of aggressive reaction patterns in males: a review. *Psychological Bulletin, 86*:852–75.

O'Neill, M., Page, N., & Adkins, W.N. (1986). Tryptophan-trazodone treatment of aggressive behavior. *Lancet, 19*(11), 859–860.

Oxenstierna, G., Edman, G., Iselius, L., Oreland, L., Ross, S.B., & Sedvall, G. (1986). Concentrations of monoamine metabolites in the cerebrospinal fluid of twins and unrelated individuals: a genetic study. *Journal of Psychiatry Research, 20*:19–29.

Pinner, E., & Rich, C. (1988). Effects of trazodone on aggressive behavior in seven patients with organic mental disorders. *American Journal of Psychiatry, 145,* 1295–1296.

Potter, W., & Manji, H. (1993). Are monoamine metabolites in cerebrospinal fluid worth measuring? *Archives of General Psychiatry, 50,* 653–656.

Pranzatelli, M.R. (1994). Serotonin and human myoclonus. *Archives of Neurology, 51*:605–617.

Raine, A, Brennan, P., & Mednick, S.A. (1994). Birth complications combined with early maternal rejection at age 1 year predispose to violent crime at age 18 years. *Archives of General Psychiatry, 51*:984–988.

Raine, A., Brennan, P, Mednick, B., & Mednick, S. (1996). High rates of violence, crime, academic problems, and behavioral problems in males with both early neuromotor deficits and unstable family environments. *Archives of General Psychiatry, 53*:544–549.

Raleigh, M.J., Brammer, G.L., & McGuire, M.T. (1983). Male dominance, serotonergic systems, and the behavioral and physiological effects of drugs in vervet monkeys (cercopithecus aethiops sabaeus) in *Ethopharmacology: Primate models of neuropsychiatric disorders.* K. Miczek (Ed.). Alan R. Liss, Inc. New York.

Raleigh, M.F., McGuire, M.T., Brammer, G.L., & Yuwiler, A. (1984). Social and environmental influences on blood serotonin concentrations in monkeys. *Archives of General Psychiatry, 41*:405–410.

Raleigh, M.J., Brammer, G.L., McGuire, M.T., & Yuwiler, A. (1985). Dominant social status facilitates the behavioral effects of serotonergic agonists. *Brain Research, 348*:274–83.

Raleigh, M.J., McGuire, M.T., Brammer, G.L., Pollack, D.B. & Yuwiler, A. (1991). Serotonergic mechanisms promote dominance acquisition in adult male vervet monkey. *Brain Research, 559,* 181–190.

Rao, U., Carson, G.A., Rappaport, M.D. (1991). Serum cholesterol and aggressive behavior in psychiatrically hospitalized children. *Acta Psychiatr Scand, 83*:77–78.

Ratey, J.J., Morill, R., & Oxenkrug, G. (1983). Use of propranolol for provoked and unprovoked episodes of rage. *American Journal of Psychiatry, 140*(10), 1356–1357.

Ratey, J.J., Sovner, R., Mikkelson, E., & Chmielinski, H.E. (1989). Buspirone therapy for maladaptive behavior and anxiety in developmentally disabled persons. *Journal of Clinical Psychiatry, 50,* 382–384.

Ratey, J.J., Sorgi, P., O'Driscoll, G.A., Sands, S., Daehler, M. L., Fletcher, J.R., Kadish, W., Spruiell, G., Polakoff, S., Lindem, K.J., Bemporad, J.R., Richardson, L., & Rosenfeld, B. (1992). Nadolol to treat aggression and psychiatric symptomatology in chronic psychiatric inpatients: A double-blind, placebo-controlled study. *Journal of Clinical Psychiatry, 53*(2), 41–46.

Realmuto, G.M., August, G.J., & Garfinkel, B.D. (1989). Clinical effect of buspirone in autistic children. *Journal of Clinical Psychopharmacology, 9,* 122–125.

Riddle, M.A., Anderson, G.M., McIntosh, S., Harcherik, D.F., Shaywitz, B.A., & Cohen, D.J. (1986). Cerebrospinal fluid monoamine precursor and metabolite levels in children treated for leukemia: age and sex effects and individual variability. *Biological Psychiatry, 21*:69–83.

Ringo, D.L., Lindley, S.E., Faull, K.F., & Faustman, W.O. (1994). Cholesterol and serotonin: seeking a possible link between blood cholesterol and CSF 5-HIAA. *Biological Psychiatry, 35*:957–9.

Rosenberg, D.R., Johnson, K., & Sahl, R. (1992). Evolving mania in an adolescent treated with low-dose fluoxetine. *Journal of Child and Adolescent Psychopharmacology, 2,* 299–306.

Rosenthal, N.E., Sack, D.A., Wehr, T.A. (1983). In T.A. Wehr & F.K. Goodwin (Eds.), *Circadian Rhythms and Affective Disorders* (pp. 185–201). Boxwood Press, Pacific Grove, CA

Roy, A., DeJong, J., & Linnoila, M. (1989). Cerebrospinal fluid monoamine metabolites and suicidal behavior in depressed patients. *Archives of General Psychiatry, 46*:609–612.

Roy, A., Adinoff, B., DeJong, J., & Linnoila, M. (1991). Cerebrospinal fluid variables among alcoholics lack seasonal variation. *Acta Psychiatr Scand, 84,* 579–582.

Roy, A. (1993). Serotonin, suicide and schizophrenia (letter). *Canadian Journal of Psychiatry, 38*(5): 369.

Salomon, R.M., Mazure, C.M., Delgado, P.L., Mendia, P., Charney, D.S. (1994). Serotonin function in aggression: the effect of acute plasma tryptophan depletion in aggressive patients. *Biological Psychiatry, 35*:570–2.

Seifert, W.E., Foxx, J.F., Butler, I.J. (1980). Age effect on dopamine and serotonin metabolite levels in cerebrospinal fluid. *Annals of Neurology, 8*:38–42.

Seilkowitz, M., Sunman, J., Pendergast, A., et al. (1990). Fenfluramine in prader-willi syndrome: A double-blind, placebo controlled trial. *Archives of Disease of Childhood, 65,* 112- 114.

Shaywitz, B.A., Cohen, D.J., & Bowers, M.B., Jr. (1977). CSF monoamine metabolites in children with minimal brain dysfunction. Evidence for alteration of brain dopamine. A preliminary report. *Journal of Pediatrics, 90,* 76–71.

Shetty, T., & Chase, T.N. (1976). Central monoamines and hyperkinase of childhood. *Neurology, 26,* 1000–1002.

Smith, S.E., Pihl, R.O., Young, S.W., & Ervin, F.R. (1985). *Psychopharmacology, 87*:173–177.

Stanley, M., Traskman-Bendz, L., & Dorovini-Zis, K. (1985). Correlations between aminergic metabolites simultaneously obtained from human CSF and brain. *Life Science, 37*, 1279- 1286.

Stein, D.J., Hollander, E., & Liebowitz, M.R. (1993). Neurobiology of impulsivity and the impulse control disorders. *Journal of Neuropsychiatry and Clinical Neurosciences, 5*:9–17.

Stewart, A.M., Stewart, S.G. (1981). Serum cholesterol in antisocial personality: failure to replicate earlier findings. *Neuropsychobiology, 7*:9–11..

Stoff, D.M., Pasatiempo, A.P., Yeung, J., Cooper, T.B., Bridger, W.H., & Rabinovich, H. (1992). Neuroendocrine responses to challenge with dl-fenfluramine and aggression in disruptive behavior disorders of children and adolescents. *Psychiatry Research, 43*(3), 263–276.

Swade, C., & Coppen, A. (1980). Seasonal variations in biochemical factors related to depressive illness. *Journal of Affective Disorders, 2*, 249–255.

Swedo, S.E., Kruesi, M.J.P., Leonard, H.L., Hamburger, S.D., Cheslow, D.L., Stipetic, M., & Potter, W.Z. (1989). Lack of seasonal variation in pediatric lumbar cerebrospinal fluid neurotransmitter metabolite concentrations. Acta Psychiatr Scand, 80:644–649.

Tolan, P.H., & Guerra, N.G. (1994). *What works in reducing adolescent violence: An empirical review of the field.* Monograph prepared for the Center for the Study and Prevention of Youth Violence. Boulder, CO: University of Colorado.

Traskman L., Asberg, M., Bertilsson, L., & Sjostrand, L. (1981). Monoamine metabolites and suicidal behavior. *Archives of General Psychiatry, 38*: 631–636.

Virkkunen, M. (1979). Serum cholesterol in antisocial personality. *Neuropsychobiology, 5*:27–30.

Virkkunen, M. (1983). Serum cholesterol levels in homicidal offenders: Aloe cholesterol level is connected with a habitually violent tendency under the influence of alcohol. *Neuropsychobiology, 10*:65–69.

Virkkunen, M., & Penttinen, H. (1984). Serum cholesterol in aggressive conduct disorder: A preliminary study. *Biological Psychiatry, 19*:435–439.

Virkkunen, M., Eggert, M., Rawlings, R., & Linnoila, M. (1989). A prospective follow-up study of alcoholic violent offenders and fire setters. *Archives of General Psychiatry, 53*(6):523–529.

Virkkunen, M., Goldman, D., Nielsen, D.A., & Linnoila, M. (1995). Low brain serotonin turnover rate (low CSF 5-HIAA) and impulsive violence. *Journal of Psychiatry and Neuroscience, 20*(4):271–5.

Virkkunen, M., Eggert, M., Rawlings, R., & Linnoila, M. (1996). A prospective follow-up study of alcoholic violent offenders and fire setters. *Archives of General Psychiatry, 53*:523–529.

Volavka, J., Crowner, M., Brizer, D., Convit, A., et al. (1990). Tryptophan treatment of aggressive psychiatric inpatients. *Biological Psychiatry, 28*:728–32.

Wirz-Justice, A., Lichsteiner, M., & Feer, H. (1977). Diurnal and seasonal variations in human platelet serotonin in man. *Journal of Neural Transmission, 41*:7–15.

Yeh, S., Fricke, R.A., & Edwards, D.H. (1996). The effect of social experience on serotonergic modulation of the escape circuit of crayfish. *Science, 271*:366–69.

Yodyinyuad, U., DeLaRiva, D.H., Herbert, J., & Keverne, E.B. (1985). Relationship between dominance hierarchy, cerebrospinal fluid levels of amine transmitter metabolites (5-hydroxyindoleacetic acid and homovanillic acid) and plasma cortisol in monkeys. *Neuroscience, 16*:851–8.

Young, S.N., Smith, S.E., Pihl, R., & Ervin, F.R. (1985). Tryptophan depletion causes a rapid lowering of mood in normal males. *Psychopharmacology, 87*:173–177.

Yudofsky, S., Williams, D., & Gorman, J. (1981). Propranolol in the treatment of rage and violent behavior in patients with chronic brain syndromes. *American Journal of Psychiatry, 138*:218–220.

Zhang, B., & Harris-Warrick, R.M. (1994). Multiple receptors mediate the modulatory effects of serotonergic neurons in a small neural network. *Journal Exp Biology, 190*:55–77.

Zubieta, J., & Alessi, N. (1992). Acute and chronic administration of trazodone in the treatment of disruptive behavior disorders in children. *Journal of Clinical Psychopharmacology, 12*:346–351.

13

SOCIAL ATTACHMENT, BRAIN FUNCTION, AGGRESSION AND VIOLENCE

Gary W. Kraemer

Department of Kinesiology and
Harlow Primate Laboratory
22 N. Charter St.
University of Wisconsin
Madison, Wisconsin 53715

ABSTRACT

In primates intraspecies aggression can be viewed as a purposeful social behavior. Many species of nonhuman primates establish relatively stable social groups. Intragroup aggressive encounters generally occur when the intragroup dominance hierarchy is not stable, and such encounters more often than not have the effect of restoring stability. Hence, aggression is usually both socially regulated and regulatory as far as intragroup and interindividual interactions are concerned. "Violence" can be defined as unregulated aggression, and it is lack of regulation that distinguishes violence as being disruptive and anti-social. One way to address and understand the causes of violence is to investigate the mechanisms by which social behavior per se usually comes to be regulated. "Attachment" has been traditionally viewed as being the process by which the infant bonds to a caregiver and thereafter develops and maintains affiliative intraspecies relationships. An implication of this view is that social behavior is ultimately both caused and regulated by guided internal motivations to develop affiliative social relationships. Investigations of the psychobiology of attachment mechanisms indicate, however, that the neurobiological development of the primate infant is considerably more plastic and less internally guided than previously thought. While past theories suggested that species specific neurobiological mechanisms develop autonomously and enable the infant to engage in regulated social interactions, the more current view is that neurobiological systems that regulate behavior develop their usual mature stature as a result of experiencing affiliative interactions. Disruption of usual affiliative attachments produces what amounts to neurobiological dysregulation. This translates to failure to regulate usual social behaviors, and one result of this may be violence.

Biosocial Bases of Violence, edited by Raine *et al.*
Plenum Press, New York, 1997

1. INTRODUCTION

It seems to be likely that views of animal social behavior and aggression developed in the first several decades of this century have strongly influenced Western European and American ideas about the causes of violence among humans. One objective of this essay is to explain how these views may hinder as much as help our contemporary understanding of aggression and violence. A second objective is to propose that the causes of violence among humans can be better understood by considering what is known about the psycho-biology of social affiliation (attachment) in non-human primates. Maternal deprivation produces increases in the incidence of violence among rhesus monkeys and a "dysregulation" among several brain neurochemical systems (Kraemer, & Clarke, 1996). The emotional and cognitive domains of brain function that are affected by maternal deprivation, or lack of early socialization, are those that enable individuals to cope with social stressors and regulate their behavior in the social environment. Hence, violence among social primates may be related to objectively definable and verifiable abnormalities in brain function, the etiology of which is primarily social environmental rather than genetic. Discussing the implications of adopting a neurobiologically based view of the causes of human violence and its prevention and treatment is the final objective of this paper.

1.1. Lack of Structure versus Lack of Love

In his influential book, *On Aggression*, Konrad Lorenz (later to win the Nobel Prize for his work) proposed that species specific aggressive behavior and motivations to express such behavior are "built-in" to the nervous system. If aggression is prevented then the drive to aggress continues to build unless there is some societally acceptable way to release, channel, or sublimate the drive (Lorenz, 1966; Storr, 1970). "Violence" occurs when pent-up motivations to aggress are not diverted to socially acceptable venues and cannot be contained. Hence, violence can be defined as aggression occurring out of context and perhaps with inordinate intensity. Lorenz's explanations of the causes of aggression and violence were based on ethological studies of aggressive behavior in a variety of animal species, but he and others have suggested that the basic causes of aggression in animals and humans are analogous if not homologous in many respects (Storr, 1970). From this viewpoint, aggression among humans in some form is inevitable, but may be expressed in socially acceptable forms. The primary explanation for the existence of violence, then, is that environmental or social regulation of aggression has failed in some way.

This logic is followed even in psychological theories that are not explicitly based on Lorenzian constructs. For example, theories proposed by Eysenck (Eysenck, 1977), Mednick (Mednick, 1977), and Buikhuisen (Buikhuisen, 1988) differ by suggesting respectively that variation in sensitivity to punishment, susceptibility to avoidance conditioning, or inhibition learning ability, may explain individual differences in the effectiveness of socialization in controlling criminal behavior and violence. What they share is the implicit premise that humans are naturally aggressive, and that civilization of the individual by sanctioning of aversive consequences for aggression is the way in which society usually counteracts this.

Given this premise, theorizing about why some individuals become violent criminals often focuses on how the society might fail to socialize the individual, how the individual might fail to respond and conform to social sanctions, or some combination of the two. In the former case, where societal failure seems to be primary, we can identify social risk factors such as child abuse and enculturation to violence that would affect most individuals (Johnson, 1996; Staub, 1996). On the other hand, if we focus on individual constitutional factors then it seems that individuals might vary in their intrinsic aggressiveness, and may be more or less conditionable, responsive to punishment, or cognitively able to inhibit their aggressive motivations. It is reasonable to suppose that this is true because their brain perceptual, cognitive, emotional, and neuroendocrine mechanisms differ, and understanding the genetic and environmental causes of individual neurobiological differences becomes an issue. Ultimately, we need to understand how the biology of the individual interfaces or fails to interface with social behavioral regulatory mechanisms. This would provide us with a "biosocial" view of the causes of aggression and violence.

In subsequent sections of this paper, aspects of the literature on the effects of rearing animals, primarily rodents and nonhuman primates, in different rearing environments will be considered. What the evidence suggests is that among social mammals exhibition of adult social behaviors, such as aggression, in their appropriate contexts depends in large measure on prior experiences in a domain of social interaction that is conceptually distinguishable from aggression, that is affiliation. Affiliation is the most primary process by which the biology and behavior of the individual is integrated into a larger social domain and the rules for conduct in that domain. The contention to be presented herein is that brain mechanisms that usually regulate aggression come into existence as a result of affiliative transactions of the individual with its caregiver, with peers, and then with the larger society (Kraemer, 1992b; Kraemer, & Clarke, 1996). If we begin to ponder issues at this "biosocial" level of analysis, however, then it seems that the basis for the pervasive assumptions about the biological substrates of the motivation to aggress needs to be revisited.

1.2. Biological Motivation?

Ethological views of the causes and nature of human aggression are based on observations of animals in their natural environment, with some human introspection thrown in for good measure. The drive to aggress has been likened to the drive to engage in sexual behavior. In his book, *Human Aggression*, Storr asks us to consider, introspectively, whether it is reasonable to propose that our *urge* to engage in sexual behavior is learned, or that it could ever be unlearned. Presumably our answer should be "No", and if so, Storr's point is that we have an analogous basic urge to aggress. One implication of this rhetorical questioning is that the motivation to exhibit aggressive behavior in response to environmental and/or social stimuli, and to seek such stimuli, is intrinsic to normal brain function. Violence can occur if internally motivated attempts to express usual and appropriate aggressive behavior are frustrated.

An often cited example is that of territorial male cichlid fish, which if provided an environment bereft of other males, will eventually turn on and kill females and even his own offspring. Lorenz suggested that male cichlids "need" hostile male neighbors upon which they can vent their aggression, otherwise they become violent (Lorenz, 1966). This behavior seems to occur in the absence of any learning that could explain it, and thus to be innate, also meaning genetically determined. This further implies that the brain mechanisms that mediate both the intraspecies territorial aggression and the destruction of females and offspring are genetically determined. The neurobiological organization of the fish that fights with males is not different from the fish that kills females and offspring. Hence it seems that cichlid fish brain mechanisms do not need prior social experience to command the appropriate aggressive behavior, nor in all likelihood would any sort of prior social experience inhibit the violent behavior. All the self-energized brain mechanisms of the fish have to do in order to function is to mature for a sufficient period in a physiologically competent fish body.

In this particular example we are constrained to disregard prior experience of the fish or differences underlying neurobiological organization as possible determinants of whether aggression or violence is exhibited. However, that a mammal is alive in the wild means it *necessarily* had certain prior social experiences and has modified its exhibition and inhibition of behaviors in relation to those experiences, otherwise it would not be alive. We will never observe a juvenile, adolescent, or adult rodent or primate in which prior experience can be disregarded as a determinant of ongoing behavior. Among the necessary experiences every mammal has are those social interactions that sustain developing offspring until they reach the physiological and behavioral stature necessary to act independently of a mother or caregiver. Among primates it also appears that early social experience organizes neurobiological mechanisms to the extent of quantitatively and qualitatively modifying aspects of brain function. This would be the usual or normal case. It is also true that among non-human primates, exhibition of purposeless, antisocial, or counter-survival aggression, i.e., aggression out of context (violence), is extremely rare. A critical question, then, is how the concept of an "appropriate context" for aggression might be embodied in neural mechanisms regulating behavior.

1.3. Violence as Aggression Out of Context—Biological Considerations

Aggressive behavior can be globally defined as a set of actions/motor patterns that are capable of damaging or degrading the integrity other beings or objects. Many species bite, slash, and claw. Others gore, trample, or pummel. The motor pattern used to inflict damage will be referred to as "aptic" behavior in many species. "Aptic" means "prepared" or "pre-set." What is being avoided here is characterizing aggression as a "fixed-action pattern." Referring to a behavior as a fixed action pattern implies that the neurobiological substrates of the behavior and the causes of activation of those substrates are mechanistic, immutable, and genetically determined. The behavior pattern by which species-typical aggression is expressed may be largely determined by genetic endowment. The idea that the context in which the behavior is expressed is also genetically encoded deserves further scrutiny, though.

Most aggression amongst animals occurs in contexts that are explainable and understandable by humans (Collias, 1944; Eibl-Eibesfeldt, 1961). Moyer developed an aggression taxonomy (Moyer, 1968) based on the idea that different kinds of aggression are elicited by different stimulus contexts. He proposed that the more global concept of aggression among animals could be operationally subtyped into: predatory, intermale, fear-induced, instrumental, territorial, maternal, and irritable aggression. One implication of such a view is that the sensory context in which aggression is usually exhibited is encoded in some categorical way in the nervous system, as are the neurochemical and neuroendocrine responses that attend engagement in aggression. Indeed, in many species the situations, stimuli, and the topography of aggressive behavior seems to be relatively stereotyped. For example, introduction of an "intruder" rat into a colony of laboratory rats reliably elicits attack by the dominant male of the colony, and the topography of attack and defensive behaviors exhibited by the laboratory rats is similar to that observed in feral rats (Blanchard, & Blanchard, 1977). It seems reasonable to suppose that there is a neural basis for this type of aggression, the formation of which is guided by the rat genome regardless of whether the rat is a member a laboratory or feral colony.

What is often overlooked, however, is that prior social experience common to the species is critical for stimulus contexts to maintain a species specific topography of elicited aggressive adult behavior. Stated another way, tailoring of what seems to be stereotyped exhibition of aggression to its appropriate social and environmental context depends in large measure on prior social experience. Hence, isolating rodents or primates from social interactions with conspecifics for sufficient periods of time produces a proclivity to exhibit aggression that does not conform to the usual stimulus context(s) or behavioral topography (Harlow, Harlow, & Suomi, 1971; Kraemer, & Clarke, 1990; Miczek, 1979; Valzelli, & Garattini, 1972). In primates, exhibition of competent sexual behavior also depends on prior social experience (Harlow, et al., 1971). If rhesus monkeys are entirely deprived of maternal care and experience with peers female solicitation and copulatory behavior is totally abolished. Male sexual behavior is fragmented, that is, behaviors such as penile erection and pelvic thrusting occur in unusual contexts and are directed at improbable and even inanimate objects. This point is raised here because of Storr's rhetorical question about the sexual urge or drive cited before, and its postulated comparability with an urge or drive to aggress. What we see in laboratory social deprivation studies is that aptic behavior patterns may emerge relatively intact, while the structuring of aptic behavior patterns into a cohesive "whole" that amounts to competent aggression or sexual behavior in context is absent. We do not have to suppose that the aptic behavior patterns are "motivated", that there is some drive that forces their exhibition, only that they occur with some probability in an active and genetically pre-set nervous system. What this suggests, then, is that we need to take a closer look at how social experience affects the brain mechanisms that regulate aggression so that most often it is understandable.

1.4. Primate Social Structure and Aggression

In this section reference will be made primarily to studies of primate behavior completed some decades ago. One reason to do this, of course, is because many definitive studies that establish characteristics of primate societies were conducted some time ago. Another reason is to illustrate that the portrayal of primate social structure provided by many of these studies bears little relationship to popularized or "commonsensical" views of primate social structures. The latter often liken nonhuman primate social dominance systems to social political systems resembling human monarchies. That is, a structure in which there is one aggressive dominant and privileged leader, and a host of less privileged and subordinate followers. This does not mean that examples of such social structures cannot be found among nonhuman primate societies, rather that such structures are far from representative. That this is not widely appreciated cannot be explained by suggesting that there has not been sufficient time for research findings in primatology to make their way into the mainstream of cultural views. Rather, it may be that what science tells us about nonhuman primate social structures in general runs counter to basic views about the nature of human aggression and dominance. Some of these long held ideas include the proposition that peck-order is a ubiquitous social ordering principle (Schjelderup-Ebbe, 1931) and that dominance is a "social instinct" (and is thereby heritable) (McDougall, 1908). The idea that cultural "filtering" may interfere with our scientific understanding of aggression and violence will be addressed more extensively in a later section.

1.5. Primate Politics

The rules for intragroup primate competition and aggression are not easily categorized using taxonomies focusing primarily on environmental conditions and resources.

Competition for environmental resources (e.g., sustenance, access to mates) plays a role in aggression to be sure, but humans and baboons engage in social aggression in the midst of plenty (Sapolsky, 1990). Early on in development, individuals engage in play, and this has little to do with acquisition of environmental resources. Nevertheless, play and intra-specific competition usually seem to go hand in hand. When aggressive play is concluded among immature monkeys, one actor is likely to be advantaged or disadvantaged in terms of social stature or "dominance" (Harlow, 1969). In some species when aggression among adults is concluded, one combatant may be wounded more than the other and physiologically disadvantaged as well (Sapolsky, 1990). The latter outcome seems to be more the exception than the rule, however.

Dominant stature has classically been viewed as providing the dominant individual with greater access to resources (territorial, nutritional, procreative) (Chance, 1956). One teleological goal of achieving dominance might be to be in a position to command an advantage in the future even though the specific rewards may not be present or evident at the time that competition for dominance occurs. In this view, attaining dominance is but a means to a more materialistic end. Implicit in this idea is that individuals competing for dominance would realize that they will have more access to resources as a result of their social dominance, and this is their goal. Access to physiologically sustaining resources, in turn, would improve one's chances in further challenges. The ultimate reward, of course, would be an increase in reproductive fitness. If dominant individuals gain access to preferred mates then one might suppose that dominant lineages would be positively selected. The physiological foundations of dominant stature would be genetically transmitted and thus inherited to some degree. The combined tactical and strategic cognitive states required to seek and preserve dominance on this order ask a lot of most non-human primates, however.

Non-human primates may defend a territory, but dominant individuals do not appear to intentionally sequester or mobilize resources gained in one encounter to improve their chances in subsequent challenges. Beyond this, studies of both feral and captive groups of a number of primate species indicate that "dominant" individuals do not necessarily gain greater access to either food or mates as has often been proposed (Gartlan, 1968). More specifically, it is not necessarily true that the dominant individual has greater access to mates per se, or to preferred mates (Kummer, 1957). A rigid and controlling dominance hierarchy is most likely to be observed in captive groups, and by comparison to free ranging groups, aggression levels per se are also likely to be higher in captive groups (Gartlan, 1968). Nonetheless, a dominant individual is not necessarily the most aggressive individual (Hall, & DeVore, 1965). Indeed, in many cases the technically dominant individuals appear to serve at the pleasure of the subordinates who may be more influential in defining the overall social structure (De Waal, 1977; Rowell, 1966). Finally, the ideas that there is one dominant individual and all the rest are subordinate, that there is a linear "pecking-order" among subordinates, or that a dominant individual in one social context is dominant in all social contexts do not generalize well to non-human primate social structures *per se*. For example, a "senate" (several individuals) may collectively exercise dominance (DeVore, 1965), and individuals may be dominant in terms of regulating intragroup aggression but subordinate otherwise (Simonds, 1965). Finally, it seems that many aspects of "being dominant" (or subordinate) are primarily learned or bestowed upon the individual, rather than inherited genetically (Gartlan, 1968).

Collectively, studies of primate social groups cast doubt on whether there could be a drive to become dominant based on any sort of internal "needs" related to sexuality, aggression, or nutritional resources (Gartlan, 1968; Rowell, 1966). If a captive group is well provisioned, competing for resources or territory (other than personal space) is not an issue. Yet there is aggression and social hierarchies of some sort are often formed. What this suggests is that what we see and can study in captive primate groups is the social dominance system displayed relatively free of many of the environmental factors that have been proposed to justify the existence of such systems. In human terms, it seems likely that existence of or lack of respect, safety, and affiliation (attachment, love) are the most proximal social stature perceptions of the primate nervous system. In both feral and captive environments, vocal, facial, and whole body social signaling of stature and intention is well practiced among primates. Gartlan (Gartlan, 1968) suggests that dominance and subordination are most profitably viewed as a "roles" in which both constitutional and social experience factors contribute in some proportion to make one role or another more suitable for some individuals than others. Despite what may seem commonsensical, it seems that dominant individuals may not have as much choice about assuming their roles as is commonly thought, and there are costs of being dominant just as there are rewards.

1.6. Social Goals and Aggression

What is being suggested, then, is that primates will aggress (or submit) to achieve a certain profitable perception of the social environment, and their place in it. It is reasonable to suggest that in some circumstances a male monkey may engage in aggression in order to perceive deferential behavior of other males, and solicitous behavior of females. Acting in order to perceive one's self and others in a certain way is an intrinsically social yet intrasubjective goal. Being aggressive may not be the most common or even direct means of attaining such goals, nor do the various aspects of desirable consequences necessarily go together. Deferential behavior of subordinate males does not guarantee solicitous behavior of females. Indeed, intragroup aggressive encounters seem to occur when the dominance hierarchy is not stable, that is, when individual perceptions of hierarchical relationships might be uncertain. This, rather than squabbling over resources or favors, seems to be the critical precipitating factor.

Two examples of many permutations will be offered. First, it may be that the technically dominant individual cannot or does not exhibit the appropriate role and stature of dominance. In this case the subsequent challenges by subordinates are not inhibited. The dominant individual is not protected by social stature and is likely to be repeatedly challenged. Second, it may be that the subordinates are not protected from harassment and threat by subordination behavior. If subordination behavior does not inhibit intrusions by the dominant monkey, the subordinates have nothing to lose by forming coalitions and challenging the dominant individual. In each of these case what we see is a case of dominant individuals skirting or failing to act according to protocol, and a resulting counter-action by subordinates or coalitions of them.

Such encounters more often than not have the effect of restoring stability, because the "new" dominant individual may follow the protocol in a way that his or her predecessor did not. One has to learn from experience what the probable consequences of an aggressive encounter might be at a social level. Engaging in aggression is only one of many behaviors that may increase one's stature. Winning, and remaining less wounded, is likely to increase one's stature and ability to thwart the next challenge. Winning a battle often reduces the probability that one will be attacked again. On the other hand, losing a tooth and nail battle is likely to degrade one's social stature and physiological capacity to withstand another bout. Deference of subordinates affords the dominant monkey increased social freedom without challenge. Acknowledgment of subordination affords reprieve from harrassment as long as the dominant individual follows the "rules." Hence, aggression, when it is exhibited, is usually both socially regulated and regulatory as far as intragroup and interindividual interactions are concerned. There is nothing unusual about the behavior of humans or non-humans engaging in aggression in this context. What about unusual behavior though — that is, violence?

2. AGGRESSION AND VIOLENCE — NEUROBIOLOGICAL FOUNDATIONS

One way to present a new idea is to contrast it with an old idea. In many domains of research on animal behavior we find that a "behavioral system" has been described at the level of observation (and perhaps experimentation). Examples would be research on the causes of aggression or affiliation as far as social and environmental stimuli seem to be required for these behaviors to be exhibited. Explaining why any individual acts or responds affiliatively or aggressively to stimuli depends on some underlying theoretical construct of brain function even if this is not explicitly stated. One underlying construct prominent in both the literature on aggression and affiliation is that exhibition of particular behaviors to specific classes of stimuli is instinctive, and that the cause of behavioral action per se is attributable to in-born "drives" or "motivations." The views of Konrad Lorenz, presented before and to be expanded upon here, provide an illustrative example encompassing both ethological research and the translation of this research into societal and cultural perspectives on human aggression and violence. Understanding how these views are embedded in culture and society is important because it enables us to see why contemporary neurobiological theories ought not to be based on the same premises, and why reasoning based on these theories presents a view of the causes of violence that are different from and even at odds with might be commonly accepted.

2.1. Lorenzian Aggression and Violence

According to Lorenz (Lorenz, 1966) engaging in aggression diminishes the drive to aggress over variable interludes. But then, the drive drives, and the need arises again. If the individual is prevented from aggressing the drive becomes stronger. As the urge to aggress builds the dependency on specific environmental contexts to release aggression diminishes until aggression can occur in almost any context and be directed at objects that would not ordinarily be attacked. This ethological view combined with more general aspects of psychoanalytic theory passes as common sense in some venues of European and American society. One societal example and some English language examples will be cited to illustrate the point. In schools in the United States, juvenile and adolescent boys not only can but are encouraged to engage in physical contact sports such as American football. One justification for this is that the combination of sexual and aggressive drives is poorly sublimated in adolescent males (Kleiber, & Kelly, 1980). Socially acceptable aggressive sports provide one means of dissipating the energy or drive that might otherwise surface in less socially acceptable behaviors. Whilst playing football, a sexually and aggressively frustrated young man might also be "blowing-off steam." This colloquialism is replete with the metaphor of energy induced build up of pressure and its release or venting in an aggressive act. Indeed, verbal tirades are also referred to as "venting." These kinds of explanations of the causes of aggression or violence are embedded in some societal practices, language, and what we refer to as common sense or knowledge.

It is often difficult to reason beyond or perhaps under this deep structure to see that other conceptualizations are not only possible, but perhaps more useful. Societies and cultures appear to tolerate contradictions in explanations of the underlying causes of behavior with some aplomb, at least for awhile. By the logic presented before, pent-up sexual and aggressive motivations seem to be one key cause of violence. Lifting sanctions on copulation and providing increased opportunity for acceptable forms of combat should reduce violence. In the United States it is a fair statement that young men (and women) are not sexually frustrated these days as they may have been some 30–40 years in the past. They have increased access to participation in contact sports. Nevertheless, the incedence of violence especially among juveniles 13–22 years of age in the United States is increasing dramatically.

A core premise that seems to provide the biggest blockade to an understanding of the neurobiological causes and purpose of behavior is related to the idea that "psychic energy", or "motivation", or "drive" causes behavior (Kraemer, 1995). Energy, motivation, and drive, however, are no more or less than "place holders" denoting "unknown causes of behavior." If we start with or try to add another premise, which is that "neuronal mechanisms in brain cause behavior", then it becomes obvious that we are not looking for or likely to find psychic energy in the brain. As Hebb (Hebb, 1949) pointed out many years ago, we are not going to find any motivations or drives in the brain either. Once the neuronal causes of behavior are understood the need for these place holders also evaporates. In advance of understanding the neuronal basis of brain function, however, what can be done is to acknowledge that theories based on unseen and unverifiable causes of behavior are incompatible with our understanding of behavior in objective terms.

For example, and as alluded to in the introduction, one implication of a Lorenzian view of violence, being just the result of unguided release of a drive to aggress, is that violence is the product of a normally formed nervous system. That is, at some level the underlying "rules" for nervous system function should be the same whether the behavior expressed is aggression or violence. To provide an analogy, we might examine clocks that have a spring or a battery to provide energy (or a "drive"). The rate of function of the clocks caused by the energy is regulated internally. One clock might be set to a societal standard called "correct time" by a human on a regular basis. This clock and the chimes it emits to mark the time are socially regulated. Another and virtually identical clock may be left on its own and not set by a human. This clock is not socially regulated. Its exhibited behavior will be different than the other clock, as will the purposes to which its behavior might be societally useful. The rules by which the clocks function and the characteristics of their inner workings do not differ because one was externally regulated, and the other was not. The chimes (behavior) of one clock occur in context and serve a purpose, however, whereas the behavior of the other is out of context and potentially disruptive.

There are two implications of this view if we consider clock chiming behavior as an analog of aggression. First, the capacity for violence (emitting chimes out of context) is a natural consequence of "building in" the capacity to exhibit aggression (chimes in context) in the first place. Second, prevention or termination of violent behavior follows rules. Since the inner workings of the violent clock are no different than the aggressive clock, the rules for setting (regulating) the clocks should be the same. That is, the violent individual should respond to the stimuli and contexts that regulate aggression when these are imposed, i.e., exhibition of violent behavior should respond to social sanctions. Hence, one explanation for the continued or even escalating level of violent behavior in some societies is that "social sanctions" have not been effectively or uniformly applied. The "cure" for violence is a more structured society and/or a more rigorous penal system.

A contention to be made herein is that the idea that social sanctions are the primary way in which society controls violence is wrong. Its wrongness is not likely to be proved empirically, however, because the idealized society that would prove the point by banishing all violence through sanctions is just that, an ideal. In the real world there is always some chink in the structure and some laxity in the imposition of consequences that can be cited as the cause of what violence exists. The idea that social sanctions control violence cannot be disproven, and that has to be acknowledged. It seems however, that the idea itself may be built on faulty assumptions about the nature of violence, and that is why the idea should be set aside.

2.2. Biosocial Distinctions between Aggression and Violence

Valzelli (Valzelli, 1981) suggested that violence is an expression of aggressive behavior patterns which have ceased to be regulated by normal situational and social stimuli. This view differs substantially from the Lorenzian definition of violence and the clock analogy presented before. By this definition, a violent cichlid fish would be one that kills females and offspring even when their are other males to fight with. A violent clock would be one that emits chimes unpredictably even after it has been set to societal time, or a clock that cannot be set to societal time. Valzelli's definition implies that violence is not comprehensible using rules that account for aggression at cultural, social, and interpersonal levels of analysis. Violence is perpetrated using aggressive behavior patterns to be sure, but violence and aggression are different, and their causes and regulators are likely to be different as well. If this is true then any plans we might make to limit or control violence cannot be based only on an understanding of how aggression is limited or controlled. Ideally, we would like to have a theory that encompasses the difference between aggression and violence, and one that would tell us how the rules differ. When we consider "causes" or "regulators" of behavior per se, however, it seems that violence may differ from aggression at a level yet to be considered — brain function and neurobiology.

Returning to the non-human primate social "rules" for aggression and dominance presented before, it seems evident that individual perceptions of and reactions to social structure are critical in following the "rules." A considerable body of evidence suggests that the mechanisms underlying social cognition and the regulation of emotion develop in particular social affiliative contexts and not others. The socially deprived monkey is likely to perceive the social world quite differently than the socially reared monkey and also to be more likely to engage in violent behavior. The major question is whether violent monkeys differ neurobiologically from socialized monkeys. That is, whether violence associated with social deprivation in monkeys is associated with brain changes that would prevent responses to social sanctions that might regulate such behavior. The implication of this would be that once violent behavior has been induced, it cannot be stopped or controlled by social sanctions that would usually control or regulate aggressive behavior. Once the neurobiology of the individual has changed the usual rules for regulation of behavior do not apply, and other methods must be devised. In order to substantiate the argument that social attachment is critical for development of neurobiological systems that regulate aggression, a review of psychobiological attachment theory and research is warranted.

2.3. Prior and Contemporary Theories of Attachment

Bowlby proposed that a genetically encoded drive to survive motivates infants to attach to an object that will reliably sustain it nutritionally and provide for its safety (a secure base)(Bowlby, 1988). By implication, infants are motivated to identify their mother, to make contact with her a goal, and to exhibit attachment behaviors (visually tracking and maintaining proximity to or contact with the caregiver) (Bowlby, 1988). As long as an infant develops an affiliative relationship with its caregiver, usually its mother, the infant will be protected and nourished and pre-programmed behavioral and biological development can proceed on schedule.

Harlow's studies indicate, however, that rhesus infants will reliably direct attachment behavior toward inanimate objects (Harlow, et al., 1971). Among these they choose objects with soft and furry surface characteristics, rather than those capable of sustaining the infant by feeding it. Of course rhesus infants are always born to a real mother that is furry and she usually feeds the infant; they are rarely given a choice. The point, though, is that the maternally deprived infant will direct its species typical attachment behavior towards an object that is in no way shape or form a mother, or even a rhesus monkey. Indeed, rhesus infants prefer a terri cloth surrogate as their attachment object even when humans or a less comfortable wire-covered surrogate feeds them. This counts as a counter-survival choice because the infant develops an attachment to an object that cannot sustain it in lieu of one that can. Rhesus infants also attach to abusive or neglectful rhesus mothers, and thereafter exhibit exaggerated attachment behavior towards these individuals even when they have other choices. In sum, the rhesus neonate's nervous system is not tuned to perceive and respond to a rhesus monkey mother (Kraemer, 1992b). Instead, it seems that the neonate's nervous system is pre-set to exhibit attachment behaviors towards a select and probably species specific set of stimuli. Many of these, such as contact comfort, have no direct survival value, but are usually expressed by a rhesus monkey mother. These stimuli and their effects on the nervous system can be viewed as causal factors in the infant's further development and exhibition of cohesive patterns of aptic behavior (Kraemer, 1992b).

Psychobiological Attachment Theory (PAT) does not depend on internal "drives" or "instincts" to explain the cause of behavior (Kraemer, 1992b). Instead it seems that brains are composed of neurons that are spontaneously active and that form sensory-motor neural nets as a result of both experience expectant and dependent processes (Greenough, Black, & Wallace, 1987). Some brain mechanisms are set by the genetic program to expect certain experiences at particular stages of biological development in order to mature. Others depend on certain experiences to guide and mold their formation. Some of the exhibited aptic behaviors enable the developing nervous system to entrain to an external regulator of experience and development. In many species this is a mother. This external regulator then plays a major role in how neural mechanisms set to express other aptic behaviors form and organize. If all goes well, species specific aptic behaviors are exhibited in what we would define as their "usual" or "normal" context. These behaviors usually help to sustain further maturation and incorporation of both sustenance and experience. By this reasoning, we have to be interested in mother-infant "attachment" if we want to understand how the individual usually comes to express and regulate aggression. From this perspective we can begin to build a catalog of what perceiving and responding to maternal stimuli usually affords the infant.

2.4. What Attachment Affords

The infant's nervous system requires maternal stimuli and interactions in order to develop normally (Kraemer, 1992b). The monkey mother is usually a potent regulator of the infants behavior and physiology through exhibition of stimuli that the infant is set to respond to in a particular way. It would be a mistake to think of this as a "reflex" in the way we typically think of spinal cord mechanisms. Instead, the idea is that the infant's behavior usually becomes entrained to the mother's, and vice versa, in a transactional relationship occurring over time. The effects of maternal stimuli are enduring, and it is well established that monkeys reared without monkey mothers differ physiologically, behaviorally, and cognitively from maternally reared monkeys.

Of particular importance here is the fact that maternal privation disrupts basic aspects of brain neurochemical and neuroendocrine function (Clarke, 1993; Kraemer, Ebert, Schmidt, & McKinney, 1989). Kraemer (Kraemer, 1992b) provided a summary of many of the domains in which total disruption of mother-infant attachment and socialization has been demonstrated to affect the functional biological development of the rhesus infant. Such effects include altered: temperature regulation, immune system function, food consumption and body weight regulation; increased exhibition of motor stereotypies; failure to be able to use facial expression as a discriminative stimulus in learning tasks, failure to inhibit approach responses to aversive stimuli, failure to be able to detect or respond appropriately to minor but significant changes in the environment; and disrupted sexual and maternal behavior. In sum, privation of maternal care can affect almost every aspect of what it means to be a social monkey (Kraemer, 1992a). If the duration of complete privation extends over the first 6–9 months of postnatal life, comparable to a human age of 6–7 years, the majority of effects persist into adulthood.

Maternally deprived monkeys are often aggressive towards themselves (self-injurious) (Kraemer, & Clarke, 1990), and do not regulate intraspecific aggression in the way mother-reared monkeys usually do (Harlow, et al., 1971; Kraemer, & Clarke, 1996). What is developmentally usual is that a rhesus monkey is born, cared for by its biological mother, plays with peers, learns the social rules, and becomes an adequate member of its society. Interactions with the mother are critical in the offspring's learning of the "when" and "how" , and "how vigorously" to defend and aggress in monkeys. A mature member of rhesus society is usually prepared to engage in aggression or defense when those behaviors are called upon by social or environmental circumstances. When can we begin to refer to aggression as being unusual and perhaps conforming to a definition of violence? The following characteristics are suggested as being characteristic of unusual aggression in rhesus monkeys and perhaps humans:

1. It is not predictable. It occurs out of the usual social context or in the absence of antecedent social signals. In monkeys, usual social aggression and defense is preceded and accompanied by facial (threat face, ear retraction, flushing), somatic (piloerection), and vocal (barking) displays of "threat." Unusual aggression may not be preceded or accompanied by these signals.

2. It is out of proportion and does not terminate appropriately. Unusual aggression may escalate to severe or lethal wounding and persist even after the recipient acknowledges the threat or responds to the attack with submissive gestures.

3. It does not contribute to retention or attainment of social status (dominance). The object of aggression is either not threatening (an infant for example), or, is threatening and for all practical purposes is invincible (a clearly dominant male for example).

In sum, monkeys observed in their free-ranging environment and mother-reared monkeys observed in the laboratory exhibit aggression that is understandable to most human observers. This behavior is consonant with a social context, it is regulated in effort, severity, persistence, and it is directed toward antagonists. Maternally deprived monkeys exhibit aggression that is not predictable, out of proportion in severity and duration, and directed towards improbable objects. Stimuli that would normally inhibit aggression fail to have this effect (Anderson, & Mason, 1978; Harlow, 1969; Harlow, et al., 1971; Mason, 1985). The form of such aggression is consistent with referring to it as "violence" by Valzelli's definition presented before.

2.5. Neurobiology of Aggression and Regulation of Aggressive Behavior

A clock analogy was used before to portray the "ethological control systems" view of the way in which social stimuli might regulate brain mechanisms. A clock mechanism can be set, reset, or left unset without producing changes in the mechanism itself or the rules by which it functions. This does not seem to be true of rhesus monkeys. Instead, it appears that the central nervous system is genetically primed to express certain basic regulatory mechanisms early in development and to incorporate additional regulatory mechanisms as a result of experience. The regulatory mechanisms acquired through experience have been more traditionally thought of as being in the cognitive domain. That is, having to do with information processing in relation to memory. More recently it has been found that experience also affects physiological regulatory systems in brain stem and limbic system as well.

Genetic and environmental influences on brain function play out in ontogeny, and there are basic structural and physiological substrates of brain function that usually form according to a genetic program (Nowakowski, 1987). Substrates in spinal cord, brain stem and limbic system run the body and express aptic behaviors. If this level of function is well-formed, upper-level brain structures can form in relation to experience, and less according to the dictates of a genetic program. Indeed, one idea is that cortical levels of brain function, especially those responsible for social interaction, are adaptive because their microstructure is not dictated by genetic mechanisms (Eisenberg, 1995). So the upper-most levels form mostly in relation to experience, while lower levels form mostly in relation to genetic endowment. Finally, it is supposed that the upper levels usually gain inhibitory control over the lower levels, most of the time (Jackson, 1958).

Research in rodents and other mammalian species demonstrates that there are anatomical structures primarily in limbic system (amygdala, septum, and hypothalamus) that are involved in and necessary for the expression of aggressive behavior (Ferris, & De Vries, in press; Valenstein, Cox, & Kakolewski, 1969). Beyond this there are cell populations in these structures with specific receptor characteristics, both neurochemical and neuroendocrine, that could serve to integrate function across hormonal and neuronal processing systems (Ferris, & De Vries, in press; Miczek, Weerts, Haney, & Tidley, 1994).

It has been known for a long time that there are ways to promote aggression in animals when it would not normally occur by manipulation of sub-components of brain function. Electrical stimulation of the brain can precipitate aggression and many features of sympathetic activation usually associated with aggression (Delgado, 1980). There is a further neurochemical specificity in activation of brain neurochemical systems, particularly among the biogenic amine neurotransmitter systems. That is, neuronal systems using norepinephrine (NE), dopamine (DA), or serotonin (5HT) as a neurotransmitter substance. These neurotransmitter are thought to play a role in brain mechanisms regulating reward, information gating and impulse control, and response to and adaptation to stress (Kraemer, 1992b).

2.6. Aggression Neurochemistry

NE and DA systems have been shown to play a role in the initiation and expression of aggression under certain circumstances of relevance to the present discussion. "Rage-like" aggression produced by brain electrical stimulation is thought to be due to increased release of NE (Reis, 1971). Amphetamine is a drug that promotes increase of both NE and DA from neuronal terminals in brain. In rodents and primates that have experienced the usual conditions of being reared by a mother and then living in a social group, low doses of amphetamine may produce modest increases in aggression while larger doses have no effect or inhibit aggression (Kraemer, & Clarke, 1990; Miczek, 1979). The latter effect probably occurs because motor-stereotypies are also produced by amphetamine, and this repetitive and compulsive-like behavior is incompatible with aggression. In both rodents and primates that have been socially deprived, however, amphetamine and other drugs that activate brain NE and DA systems or their receptors produce dramatic increases in aggression frequency, duration, and intensity (Kraemer, Ebert, Lake, & McKinney, 1983; Welch, & Welch, 1971). Social deprivation appears to reduce the activity of brain NE transmitter systems (Kraemer, et al., 1989) and this may lead to "up-regulation" or hyper-sensitivity of NE receptor systems (Hegstrand, & Eichelman, 1983). In general, social deprivation dramatically increases the probability of expression of violence when the deprived individual is exposed to drugs or social stimuli that would not normally trigger aggression. This effect appears to be due in part to changes in brain NE and/or DA systems. In monkeys, the effects of early social deprivation are extremely persistent. Adult rhesus monkeys that were deprived in infancy and early childhood remain violence-prone or unusually sensitive to drugs like amphetamine even if they have lived socially for long periods of time as juveniles and adolescents (Kraemer, Ebert, Lake, & McKinney, 1984).

The 5HT system appears to play a major role in the in inhibition of aggression. The 5HT system may regulate sensory-motor gating in such a way that non-aggressive behavioral responses to stressors are more readily accessible if this system is active, and this would be one of its functions (Miczek, Weerts, Vivian, & Barros, 1995; Soubrie, 1986). Pharmacological treatments that increase the amount of 5HT available to act on 5HT receptors, or directly activate primarily $5HT_1$ receptors generally reduce a number of kinds of aggression in rodents (Miczek, et al., 1994; Olivier, Mos, van Oorschot, & Hen, 1995). It has also been suggested that the brain arginine-vasopressin (AVP) directly activates brain mechanisms that mediate the expression of aggression, and the brain 5HT system acts to inhibit the activity of this system and thus the expression of aggression. The activity of the AVP system, in turn, has been shown to be affected or "set" in many respects by prior experiences with social aggression in hamsters (Ferris, & De Vries, in press).

2.7. Neurochemistry and Human Behavior

The idea that some human psychiatric disorders might be attributable to malfunction in one neurochemical system has been prominent in biological psychiatry theorizing in the past. For example, too much brain DA activity or too little NE activity have been cited as probable causes of schizophrenia or depression respectively (Schildkraut, & Kety, 1967; Snyder, 1973). It has been suggested that criminally violent and/or suicidal humans may have low brain 5HT activity, and perhaps cannot inhibit aggressive outbursts for this reason. This hypothesis is primarily based on finding a negative correlation between 5-hydroxyindoleacetic acid (5HIAA) in cerebrospinal fluid (CSF) and measures of violent behavior (Brown, Goodwin, Ballenger, Goyer, & Major, 1979; Coccaro, 1992; Linnoila,

& Virkkunen, 1992). 5HIAA is the major metabolite of 5HT. Thus, if one assumes that the concentration of metabolite in CSF is proportional to the amount of 5HT released, then relatively low CSF 5HIAA concentration can be taken as an indicator of low activity of the brain 5HT system. Violent by comparison to non-violent control subjects tend to have lower mean levels 5HIAA in cerebrospinal fluid (Linnoila, & Virkkunen, 1992). For a variety of reasons, however, the notion that there may be a reliable one-to-one match-up between specific human behavioral disorders and increased or decreased levels of activity in only one neurotransmitter system or another seems to be unlikely, or perhaps only part of the story (Kraemer, & Clarke, 1996; Kraemer, & McKinney, 1988).

Another idea is that behavior disorders might be attributable to disorganization of function among several brain systems (Kraemer, 1982; Siever, 1987). Biogenic amine systems regulate how brain cognitive and emotional systems change in relation to challenges. Changes in how these systems interact and perform their regulatory roles could also be an underlying cause of behavior disorders. The big question is what affects the development and interactions of regulatory systems. Current research in psychobiology indicates that the usual organization of interrelationships among the more basic physiological regulatory processes is more attributable to environmental rather than genetic factors. That is, the systems that regulate various aspects of body function and aptic behavior develop according to a plan, but their interactions need to be organized as well. One hypothesis is that the external social environment affects the development of internal interactions among neurobiological regulatory systems (Kraemer, 1992a; Kraemer, 1992b; Kraemer, 1995).

3. BRAIN MECHANISMS AFFECTED BY MATERNAL PRIVATION IN MONKEYS

Some neurotransmitter / neuromodulator systems are differentially affected by past social experience and could play a role in regulating the subsequent expression aggressive behavior over the course of social development. As noted before, monkeys reared without monkey mothers exhibit unusual patterns of aggression. "Peer-reared" monkeys are separated from their biological mothers shortly after birth and reared in a primate nursery for 30 days. During this time and thereafter they are housed with 2–3 like-aged cage-mates. Peer-reared monkeys are not deprived of social stimulation and interaction per se, and they develop strong attachments to their cage-mates. Being attached to peers, they are engaging in social behavior with animate warm furry members of their own species. What is missing then is the regulatory characteristics of an adult female rhesus monkey.

If we consider the social behavioral development of peer-reared monkeys by comparison to mother-reared monkeys there are six differences of significance to the present discussion that have been cited (Harlow, 1969; Harlow, et al., 1971; Higley, Linnoila, & Suomi, 1994; Higley, Suomi, & Linnoila, 1991; Kraemer, & Bachevalier, in press; Kraemer, & Clarke, 1996; Kraemer, & McKinney, 1979; Kraemer, & McKinney, 1988; Suomi, Harlow, & Domek, 1970; Suomi, & Ripp, 1983).

> 1. As infants and juveniles, peer-reared monkeys engage in an extraordinary amount of clinging to one another, and in self-directed finger and toe sucking (self-directed behavior). This is especially evident if the peer group is exposed to an external "threat" such as a human observer. Mother-reared juvenile monkeys, in contrast, cling or huddle together primarily during rest or sleep periods, may threaten and move about individually if threatened, and rarely are to be

found with a finger or toe in their mouths. In general, experienced observers characterize peer-reared monkeys as being considerably more timid, fearful, and emotionally labile than mother-reared monkeys.

2. Intragroup interactions among peer-reared monkeys are chaotic. Individuals are either inordinately separated from group activity, or intensely engaged, and there are rapid fluctuations between the two. Experienced human observers characterize groups of peer-reared monkeys as being in a chronically higher "tension" or "stress" state than groups of mother-reared monkeys.

3. Peer reared juvenile monkeys are more likely to have a more severe emotional response to separation from their cage-mates than mother-reared monkeys. In general, it seems that peer-reared monkeys are less able to self-regulate behavioral and physiological responses to social stressors.

4. Peer-reared monkeys are inordinately reclusive and defenseless at some times, and inordinately aggressive at others. Although aggression frequency in groups of peer-reared monkeys is generally lower than in groups of mother-reared monkeys, when aggressive bouts do occur among peer-reared monkeys they are likely to be more prolonged and vigorous, and more likely to produce wounding and severe injury than among mother-reared monkeys.

5. Peer-reared monkeys differ from mother-reared monkeys in their performance of basic cognitive tasks such as delayed non-matching to sample (DNMS). Peer-reared monkeys actually achieve a higher percent correct performance criterion on this task than mother-reared monkeys. This has been interpreted as being consistent with another characteristic, that is, peer-reared monkeys are inordinately distracted by novelty and seem to be unable to inhibit "impulsive" behaviors. This would provide an advantage on some cognitive tasks, but hinder performance on others.

6. As adults, peer-reared rhesus females are considerably more likely to reject or neglect their infants.

3.1. Peer-Rearing—Psychobiology

A number of psychobiological studies in mother-reared and mother-less rhesus monkeys have focused on the development of the brain biogenic amine neurotransmitter systems and the hypothalamic-pituitary-adrenal axis (Clarke, 1993; Kraemer, in press). In non-human primates the dependent measures of activity level in these systems are usually derived from measures of the neurotransmitter, or hormone, or their metabolites, in cerebrospinal fluid or blood. Considering these measures, it appears that peer-reared monkeys are biologically different from mother-reared monkeys, but the magnitude and even direction of difference in dependent measures may vary over development and in relation to prior or ongoing environmental circumstances (Kraemer, in press). For example, juvenile peer-reared monkeys housed in stable social groups generally maintain lower baseline levels of CSF NE (and/or NE metabolite) and the DA metabolite HVA than group housed juvenile mother-reared monkeys (Higley, et al., 1992; Kraemer, & Clarke, 1996). When challenged however, the behavioral and neurochemical response of peer-reared monkeys to social stressors or drugs that affect NE and DA neurotransmitter systems is exaggerated by comparison to mother-reared monkeys (Kraemer, & McKinney, 1979).

As noted before, the idea here is that neurotransmitter-receptor systems auto-regulate, but in monkeys the "level" of auto-regulation may be set early in development in relation to social rearing experiences (Kraemer, 1986). On the other hand, juvenile peer-reared monkeys housed in stable social groups exhibit a *blunted* neuroendocrine hypothalamic-pituitary-adrenal response to psychosocial stressors such as separation from cage-mates (Clarke, 1993). Thus, the exaggerated behavioral response to stressors in peer-reared monkeys, perhaps mediated by brain NE and DA neurochemical systems, is not paralleled by comparably enhanced or even normal neuroendocrine responses. Indeed, behavioral, neurochemical, and neuroendocrine responses to stress appear to be dissociated in peer-reared monkeys (Kraemer, in press; Kraemer, & Clarke, 1996).

For example, Kraemer et al. (Kraemer, et al., 1989) and Kraemer and Clarke (Kraemer, & Clarke, 1996) determined whether measures of neurotransmitter system activity were correlated with one another or with behavior in monkeys reared in different conditions. In the latter study comparing juvenile monkeys that were reared by their mothers or peer-reared, 9 out of a possible 21 intercorrelations among behavioral and neurochemical variables were significantly different from zero in mother-reared monkeys. Inactivity, prosocial behavior, and aggression were negatively correlated with NE levels, and prosocial behavior and aggression were negatively correlated with 5HIAA levels as well. Among neurochemical measures, positive correlations were observed between NE and 5HIAA, and homovanillic acid (HVA, the major metabolite of dopamine) and 5HIAA. Among the behaviors, prosocial behavior was positively correlated with both inactivity and frequency of aggression in mother-reared monkeys. Overall, the degree to which mother-reared monkeys evenly divided their behavior into domains of sitting quietly (inactive), or engaging in social behavior or aggressive bouts (maximizing the measures of each), was inversely correlated with measures of activity levels in the NE and 5HT systems. The latter two measures were, in turn, significantly correlated with one another.

By contrast, only 2 of 21 correlations were significantly different from zero in peer-reared monkeys. The only significant neurochemical relationship was between HVA and 5HIAA. The only behaviors that were intercorrelated in peer-reared monkeys were activity and inactivity. These seemingly "inverse" measures were not negatively correlated in mother-reared monkeys.

Overall, the picture one gets is that the manner in which peer-reared monkeys divide their time is not related to measures of neurochemical activity as it is in mother-reared monkeys. Similarly, that a peer-reared monkey engages in more or less prosocial behavior does not translate into also spending more or less time sitting quietly or engaging in aggression, as is the case with mother-reared monkeys. Perhaps the best way to summarize, then, is that the social behavior of peer-reared monkeys strikes the human observer as being chaotic and not patterned and regulated in a fashion comparable to mother-reared monkeys. Measures of biogenic amine system activity may reflect actions of underlying behavior regulatory systems, and the activity in these systems does not correspond to behavior in peer-reared monkeys in the same way it does in mother-reared monkeys.

4. AFFILIATION REGULATES AGGRESSION AND VIOLENCE

In rhesus monkeys, expression of aptic aggressive behavior patterns occurs whether the monkey has a mother or not. The social control of these behaviors depends on interactions with a mother as does the development of internal neurobiological regulatory systems. There is an overwhelming amount of evidence that brain biogenic amine systems usually play a role in regulation of behavior in a variety of mammalian species. The usual behavioral regulatory characteristics of these systems are not entirely determined by genetic endowment, nor "hard-wired" into mechanisms that mediate ongoing sensory-motor processing and behavior. Instead, it seems that development of functional competence of these systems, in rhesus monkeys at least, depends on early interactions with a maternal caregiver. What we see in peer-reared monkeys is they do not follow the rules set by mother-reared monkeys at both behavioral and biological levels of analysis. Despite the continuing desire in some quarters to try to pair one cause and type of disorder with one neurochemical system, the neurobiological effects of social environmental factors repeatedly shown to adversely affect social development in rhesus monkeys are not restricted to one neurobiological system. Once changes in multiple systems are observed, usually significant interrelationships among measures of the activity of different systems are absent as well, and the usual relationship of activity in any particular system to behavior also degrades. These conclusions taken together with the psychobiological studies of rodents reviewed before suggest two major areas in which our basic premises underlying theories about the causes of aggression and violence in animals and humans the ought to be modified.

1. *The nature of "motivation:"* Many theories about the causes of aggression, or affiliation, employ a construct of an innate "drive" to survive which may be further subdivided into a "drive to aggress" or a "drive to affiliate" as an explanation of the cause of infant behavior. "Drives" denote unknown causes of the initiation and regulation of behavior. At this point, however, the internal causes of behavior are more profitably conceived of as reflecting the actions of neural mechanisms with measurable characteristics and levels of function.

2. *Determinants of the constitutional nature of the infant:* If one assumes that significant physiological differences among individuals are always attributable to genetic endowment or perinatal physiological insults or challenges, then biological variation attributable to environmental experiential factors may go unrecognized. Persistent alteration of physiological responsiveness to stressors due to having some experiences and not others counts as an interface domain between social cognitive regulation of behavior and more basic regulation of response proclivities which may be genetic in origin.

Implicit in the biosocial theories of aggression and violence cited before is that there are fundamental biological mechanisms underlying the proclivity to act aggressively, ability to be aversively conditioned, and ability to learn and retain behavioral inhibitions. At one level the neurobiological structures that mediate these fundamental characteristics of the nervous system have to be genetically determined. What the psychobiological perspective suggests however is that multiple basic systems are genetically preset to interact, but the way in which this organization takes place may be largely determined by social environmental factors.

Alteration of basic physiologic responses to stimuli could be one way in which early maternal caregiving (attachment) persistently affects how the juvenile, adolescent, and adult primates respond to social stimuli and stressors. One way in which we might understand the relationship between aggression and violence, then, is that the violent individual has not incorporated the mechanisms that would allow them to follow the societal rules for aggression. From this viewpoint exhibition of violent behavior is attributable to an underlying neurobiological disorganization which prevents perception or response to usual social regulators. Basically, what is being suggested is that the usual social sanctions that are often cited as the major controlling factor for violence cannot work until the nervous system is first socialized, that is organized through affiliation.

5. CLINICAL IMPLICATIONS

Understanding the mechanisms of affiliation and the effects of failure of these mechanisms appears to be particularly important when the otherwise competent infant is neglected or abused; or when the neurobiologically compromised infant fails to establish an affiliative relationship with a caregiver.

If the competent infant does not attach to a caregiver that provides adequate regulation as part of a transactional relationship, then some of the effects of this will surface in what is traditionally considered to be the individual's basic cognitive processing ability and physiological constitution. The way in which the individual orients to and responds physiologically to social stressors may be altered. The kind of behavioral differences that seem to be most important would be those usually assigned to the domain of "temperamental" variation, that is to approach or withdraw, to be inhibited or uninhibited, to be shy or reckless, extraverted or neurotic (Kraemer, 1992a). This is not to say that the *dimensions* of temperament are environmentally determined, rather that the expression of temperamental qualities may be altered in children who are neglected or abused, or institutionalized for one reason or another.

On the other hand, another implication of psychobiological attachment theory is that "failure to attach" to a competent caregiver is more likely due to failure of the infant to perceive, process, and respond to stimuli in the usual way, rather than aberrations in "motivation." Many infants may not be able to perceive or tune to social regulators that may be available because of genetic factors, birth trauma, or other perinatal insults such as prenatal stress. Indeed, many of the behaviors and neurobiological characteristics associated with peer-rearing are observed in prenatally stressed mother-reared rhesus monkeys (Clarke, Wittwer, Abbott, & Schneider, 1994; Schneider, & Clarke, 1993; Schneider, & Coe, 1993). Moffitt (Moffitt, 1993) has demonstrated that neuropsychological deficits manifesting themselves in terms of motor maturation delays, difficult temperament, and cognitive deficits are often precursors of later antisocial behavior in humans. A critical point is that the compromised child may inordinately challenge even the most competent parent(s), and probability of exposure of the developing fetus and infant to insults increases in socially stressful and economically impoverished environments. If the parent has inherited or been exposed to environmental risk factors that compromise their competence, then it is easy to see how the situation becomes compounded by factors cited in the preceding paragraph.

6. SUMMARY AND CONCLUSION

The idea that animals and humans are driven by internal motivations to aggress is incompatible with explanations of the causes of aggression based on neurobiology. Though unfortunate in many respects, the idea that behavior is caused by organized neural net actions analogous to computations is more useful than believing that it is caused by the ebb and flow of hydraulic-like energy manifested in motivations and drives. It seems that the nervous system is preset to exhibit aggressive behavior towards classes of stimuli, often noxious, and to more general classes of stimuli when learning or cognitive problem solving do not generate alternatives. The organization of neural mechanisms that express assertive social behaviors and cognitive coping strategies most often occurs in the context of affiliative social interactions first with a caregiver and then in a larger social environment. Much of this organization and learning, such as conforming to the role of being dominant or subordinate for example, does not involve punitive aggression, aversive conditioning, or imposition of social sanctions. The later seem to come into play only when the more primary social learning or social cognition mechanisms do not work or require tuning. It may be that aggression and violence are rare in primate societies, not because there are penalties for such behavior, but because the individual usually builds up a repertoire of behavior that is both positively reinforced and incompatible with aggression.

If social attachment usually affords what is necessary to organize basic sensory motor and physiologic mechanisms, then therapy for the unusually aggressive individual should focus on providing a prosocial external regulatory system that is both environmental and interpersonal. It is important to acknowledge that the individual may not be open to or able to process the usual social stimuli, and that aggression or violence may be triggered by improbable stimuli and in unusual social contexts. Finding out what therapeutic approaches can be used to entrain the "disorganized" nervous system thus becomes the focus of new research and development of clinical paradigms based on psychobiological attachment theory.

It is unlikely that "pharmacological fixes" focusing on one or two aspects of neurochemical function or physiology are going to restore what is basically an organizational problem. One speculation is that the systems that regulate aptic behavior can always be entrained to social regulators once we have a clearer understanding of the dimensions of neural and system plasticity in adolescence and adulthood. Whatever the reverse entraining social stimuli might be, they will probably bear little relationship to how many societies treat juvenile delinquents and criminally violent offenders these days. Societal practice suggests that these individuals are still being viewed as being "motivated" to be violent, and that violent behavior might be limited by the rule that behavior that is punished is less likely to occur. The monkeys are telling us that these rules do not apply if the cause of violence is related to disruption of early attachments and socialization. Indeed from the perspective presented herein, isolating an aggressive individual from an environment in which prosocial behavior is a major contextual feature, or housing them with other violent individuals would be contraindicated.

REFERENCES

Anderson, C. O., & Mason, W. A. (1978). Competitive social strategies in groups of deprived and experienced rhesus monkeys. Developmental Psychobiology, 11, 289–299.
Blanchard, R. J., & Blanchard, D. C. (1977). Aggressive behavior in the rat. Physiology and Behavior, 1, 197–224.
Bowlby, J. (1988). A Secure Base . New York: Basic Books.
Brown, G. L., Goodwin, F. K., Ballenger, J. C., Goyer, P. F., & Major, L. F. (1979). Aggression in humans correlates with cerebrospinal fluid amine metabolites. Psychiatry Research, 1, 131–139.
Buikhuisen, W. (1988). Chronic juvenile delinquency: A theory. In W. Buickhuisen, & S. A. Mednick (Ed.), Explaining Criminal Behavior (pp. 27–50). Leiden, The Netherlands: E. J. Brill.
Chance, M. R. A. (1956). Social structure of a colony of Macaca mullata. British Journal of Animal Behaviour, 4, 1–13.
Clarke, A. S. (1993). Social rearing effects on HPA axis activity over early development and in response to stress in young rhesus monkeys. Developmental Psychobiology, 26, 433–447.
Clarke, A. S., Wittwer, D. J., Abbott, D. H., & Schneider, M. L. (1994). Long-term effects of prenatal stress on HPA axis activity in juvenile rhesus monkeys. Developmental Psychobiology, 27, 257–270.
Coccaro, E. F. (1992). Impulsive aggression and central serotonergic system function in humans: An example of a dimensional brain-behavior relationship. International Clinical Psychopharmacology, 7(1), 3–12.
Collias, N. E. (1944). Aggressive behavior among vertebrate animals. Physiological Zoology, 17(1), 83–123.
De Waal, F. B. M. (1977). The organization of agonistic relations within two groups of Java-monkeys (Maccaca fascicularis). Zeitshcrift fur Tierpsychologie, 44, 225–282.

Delgado, J. M. R. (1980). Neuronal constelations in aggressive behavior. In I. Valzelli, & I. Morgese (Ed.), *Aggression and Violence: A psychobiological and clinical approach* (pp. 82–97). St. Vincent: Edizione.

DeVore, I. (1965). Male dominance and mating behavior in baboons. In F. Beach (Ed.), *Sex and Behavior* New York: John Wiley.

Eibl-Eibesfeldt, I. (1961). The fighting behavior of animals. *Scientific American, 205,* 470–482.

Eisenberg, L. (1995). The social construction of the human brain. *American Journal of Psychiatry, 152*(11), 1563–1575.

Eysenck, H. J. (1977). *Crime and Personality* (3rd ed.). Albans, England: Paladin.

Ferris, C. F., & De Vries, G. J. (in press). Ethological models for examining the neurobiology of aggressive and affiliative disorders. In

Gartlan, J. S. (1968). Structure and function of primate society. *Folia Primatologica, 8,* 89–120.

Greenough, W. T., Black, J. E., & Wallace, C. S. (1987). Experience and brain development. *Child Development, 58,* 539–559.

Hall, K. R. L., & DeVore, I. (1965). Baboon social behavior. In I. DeVore (Ed.), *Primate Behavior: Field Studies of Monkeys and Apes* (pp. 53–110). New York: Holt, Rinehart, and Winston.

Harlow, H. F. (1969). The age-mate or peer affectional system. In E. B. Foss (Ed.), *Advances in the Study of behavior* (pp. 333–383). New York: Academic Press.

Harlow, H. F., Harlow, M. K., & Suomi, S. J. (1971). From thought to therapy: Lessons from a primate laboratory. *American Scientist, 59,* 538–549.

Hebb, D. O. (1949). *The Organization of Behavior* . New York: John Wiley and Sons.

Hegstrand, L., & Eichelman, B. (1983). Increased shock induced fighting and supersensitive beta-adrenergic receptors. *Pharmacology, Biochemistry, and Behavior, 19,* 313–320.

Higley, J. D., Linnoila, M., & Suomi, S. J. (1994). Ethological contributions: Experiential and genetic contributions to the expression and inhibition of aggression in primates. In M. Hersen, R. T. Ammerman, & L. Sission (Ed.), *Handbook of aggressive and destructive behavior in psychiatric patients* (pp. 17–32). New York: Plenum Press.

Higley, J. D., Mehlman, P. T., Taub, D. M., Higley, S. B., Suomi, S. J., Linnoila, M., & Vickers, J. H. (1992). Cerebrospinal fluid monoamine and adrenal correlates of aggression in free ranging rhesus monkeys. *Archives of General Psychiatry, 49,* 436–441.

Higley, J. D., Suomi, S. J., & Linnoila, M. (1991). CSF monoamine metabolite concentrations vary according to age, rearing, and sex, and are influenced by the stressor of social separation in rhesus monkeys. *Psychopharmacology, 103,* 551–556.

Jackson, J. H. (1958). *Selected Writings (edited by J. Taylor)* . New York: Basic Books.

Johnson, H. C. (1996). Violence and biology: A review of the literature. *Journal of Contemporary Human Services, 77*(1), 3–18.

Kleiber, D. A., & Kelly, J. R. (1980). Leisure, socialization, and the life cycle. In J. E. Isothila (Ed.), *Social Psychological Perspectives on Leisure and Recreation* New York: Charles C. Thomas.

Kraemer, G., & Bachevalier, J. (in press). Cognitive changes associated with persisting behavioral effects of early psychosocial stress in rhesus monkeys: The view from psychobiology. In B. Dohrenwend (Ed.), *Adversity,Stress, and Psychopathology* Washington: American Psychiatric Press.

Kraemer, G. W. (1982). Neurochemical correlates of stress and depression: Depletion or disorganization? *The Behavioral and Brain Sciences, 5,* 110.

Kraemer, G. W. (1986). Causes of changes in brain noradrenaline systems and later effects on responses to social stressors in rhesus monkeys: The Cascade Hypothesis. In *Antidepressants and Receptor Function (CIBA Foundation Symposium 123)* (pp. 216–233). Chichester: Wiley.

Kraemer, G. W. (1992a). Psychobiological Attachment Theory (PAT) and psychopathology. *Behavioral and Brain Sciences, 15*(3), 525–534.

Kraemer, G. W. (1992b). A psychobiological theory of attachment. *Behavioral and Brain Sciences, 15*(3), 493–511.

Kraemer, G. W. (1995). The significance of social attachment in primate infants: The caregiver-infant relationship and volition. In C. R. Pryce, R. D. Martin, & D. Skuse (Ed.), *Motherhood in Human and Nonhuman Primates: Biological and Social Determinants* (pp. 152–161). Basel: Karger.

Kraemer, G. W. (in press). The psychobiology of early attachment in nonhuman primates: Clinical Implications. *Annals of the New York Academy of Sciences,* ,

Kraemer, G. W., & Clarke, A. S. (1990). The behavioral neurobiology of self-injurious behavior in rhesus monkeys. *Progress in Neuro-Psychopharmacology and Biological Psychiatry, 14 (suppl),* 141–168.

Kraemer, G. W., & Clarke, A. S. (1996). Social attachment, brain function, and Aggression. *Annals of the New York Academy of Sciences, 794,* 121–135.

Kraemer, G. W., Ebert, M. H., Lake, C. R., & McKinney, W. T. (1983). Amphetamine challenge: Effects in previously isolated monkeys and implications for animal models of schizophrenia. In K. Miczek (Ed.), *Ethopharmacolgy: Primate Models of Neuropsychiatric Disorders* (pp. 199–218). New York: Alan R. Liss.

Kraemer, G. W., Ebert, M. H., Lake, C. R., & McKinney, W. T. (1984). Hypersensitivity to d-amphetamine several years after early social deprivation in rhesus monkeys. *Psychopharmacology, 82*, 266–271.

Kraemer, G. W., Ebert, M. H., Schmidt, D. E., & McKinney, W. T. (1989). A longitudinal study of the effects of different rearing environments on cerebrospinal fluid norepinephrine and biogenic amine metabolites in rhesus monkeys. *Neuropsychopharmacology, 2*, 175–189.

Kraemer, G. W., & McKinney, W. T. (1979). Interactions of pharmacological agents which alter biogenic amine metabolism and depression: An analysis of contributing factors within a primate model of depression. *Journal of Affective Disorders, 1*, 33–54.

Kraemer, G. W., & McKinney, W. T. (1988). Animal models in psychiatry: Contributions of research on synaptic mechanisms. In A. K. Sen, & T. Lee (Ed.), *Receptors and Ligands in Psychiatry and Neurology* (pp. 459–483). Cambridge: Cambridge University Press.

Kummer, H. (1957). Soziales verhalten einer matelpavian-gruppe. *Zeitschrift fur Psychologie, 33*, 1–91.

Linnoila, V. M., & Virkkunen, M. (1992). Aggression, suicidality, and serotonin. *Journal of Clinical Psychiatry, 53*(Supplement), 46–51.

Lorenz, K. (1966). *On Aggression* . London: Methuen.

Mason, W. A. (1985). Experiential influences on the development of expressive behaviors in rhesus monkeys. In G. Zivin (Ed.), *The Development of Expressive Behavior: Biology Environment Interactions* (pp. 117–152). New York: Academic Press.

McDougall, W. (1908). *Social Psychology: An introduction* . London: Methuen & Company.

Mednick, S. A. (1977). A biosocial theory of the learning of law abiding behavior. In S. A. Mednick, & K. O. Christiansen (Ed.), *Biosocial Bases of Criminal Behavior* (pp. 1–7). New York: Gardner.

Miczek, K. A. (1979). A new test for aggression in rats without aversive stimulation: Differential effects of d-amphetamine and cocaine. *Psychopharmacology, 60*, 253–259.

Miczek, K. A., Weerts, E., Haney, M., & Tidley, J. (1994). Neurobiological mechanisms controlling aggression: Preclinical developments for pharmacotherapeutic interventions. *Neuroscience and Biobehavioral Reviews, 18*, 97–110.

Miczek, K. A., Weerts, E. M., Vivian, J. A., & Barros, H. M. (1995). Aggression, anxiety and vocalizations in animals: GABAa and 5-HT anxiolytics. *Psychopharmacology, 121*, 38–56.

Moffitt, T. E. (1993). Adolescence-limited and life-course persistent antisocial behavior: a developmental taxonomy. *Psychology Review, 100*, 674–701.

Moyer, K. E. (1968). Kinds of aggression and their physiological basis. *Communications in Behavioral Biology, 2*(65–87),

Nowakowski, R. S. (1987). Basic concepts of CNS development. *Child Development, 58*, 568–595.

Olivier, B., Mos, J., van Oorschot, R., & Hen, R. (1995). Serotonin receptors and animal models of aggressive behavior. *Pharmacopsychiatry, 28(supplement)*, 80–90.

Reis, D. (1971). Brain monoamines in aggression and sleep. *Clinical Neurosurgery, 18*, 471–502.

Rowell, T. E. (1966). Hierarchy in the organization of a captive baboon group. *Animal Behaviour, 14*, 430–443.

Sapolsky, R. M. (1990). Stress in the wild. *Scientific American, 262*, 116–123.

Schildkraut, J. J., & Kety, S. S. (1967). Biogenic amines and emotion. *Science, 156*, 21–30.

Schjelderup-Ebbe, T. (1931). Die despote im sozialen leben der vogel. *Volker-psychologie und Sozialogie, 10*(2), 77–140.

Schneider, M. L., & Clarke, A. S. (1993). Prenatal stress has long-term effects on behavioral responses to stress in juvenile rhesus monkeys. *Developmental Psychobiology, 26*(5), 293–304.

Schneider, M. L., & Coe, C. L. (1993). Repeated social stress during pregnancy impairs neuromotor development of the infant primate. *Developmental and Behavioral Pediatrics, 14*(2), 81–87.

Siever, L. J. (1987). Role of noradrenergic mechanisms in the etiology of the affective disorders. In H. Y. Meltzer (Ed.), *Psychopharmacology: The Third Generation of Progress* (pp. 493–504). New York: Raven Press.

Simonds, P. E. (1965). The bonnet macaque in South India. In I. DeVore (Ed.), *Primate Behavior: Field Studies of Monkeys and Apes* (pp. 175–196). New York: Holt, Rinehart, and Winston.

Snyder, S. H. (1973). Amphetamine psychosis: A "model" schizophrenia mediated by catecholamines. *American Journal of Psychiatry, 130*, 61–67.

Soubrie, P. (1986). Reconciling the role of central serotonin neurons in human and animal behavior. *The Behavioral and Brain Sciences, 9*, 319–364.

Staub, E. (1996). Cultural-societal roots of violence: The examples of genocidal violence and of contemporary youth violence in the United States. *American Psychologist, 51*(2), 117–132.

Storr, A. (1970). *Human Aggression* . New York: Bantam.

Suomi, S. J., Harlow, H. F., & Domek, C. J. (1970). Effect of repetitive infant-infant separation of young monkeys. *Journal of Abnormal Psychology*, *76*, 161–172.

Suomi, S. J., & Ripp, C. (1983). A history of motherless monkey mothering at the University of Wisconsin Primate Laboratory. In M. Reite, & N. Caine (Ed.), *Child Abuse: The non-human primate data* (pp. 49–78). New York: Alan R. Liss.

Valenstein, E. S., Cox, V. C., & Kakolewski, J. W. (1969). The hypothalamus and motivated behavior. In J. C. Tapp (Ed.), *Reinforcement and Behavior* (pp. 242–287). New York: Academic Press.

Valzelli, I. (1981). Aggression and Violence: A biological assay on the distinction. In I. Valzelli, & I. Morgese (Ed.), *Aggression and Violence: A Psychobiological and Clinical Approach* (pp. 134–156). Milano: Edizioni-Saint Vincent.

Valzelli, L., & Garattini, S. (1972). Biochemical and behavioral changes induced by isolation in rats. *Neuropharmacology*, *11*, 17–22.

Welch, A. S., & Welch, B. L. (1971). Isolation, reactivity and aggression: Evidence for an involvement of brain catecholamines and serotonin. In J. Eleftheriou, & J. P. Scott (Ed.), *The physiology of Aggression and Defeat* (pp. 91–142). New York: Plenum Press.

14

WHOLE BLOOD SEROTONIN AND FAMILY BACKGROUND RELATE TO MALE VIOLENCE

Terrie Moffitt,[1] Avshalom Caspi,[1] Paul Fawcett,[2] Gary L. Brammer,[3] Michael Raleigh,[4] Arthur Yuwiler,[3] and Phil Silva[5]

[1]University of Wisconsin
Madison, Wisconsin 53706
[2]University of Otago School of Pharmacy
Otego, New Zealand
[3]West Los Angeles VA Medical Center
Los Angeles, California
[4]UCLA School of Medicine
Los Angeles, California
[5]University of Otago Medical School
Otego, New Zealand

ABSTRACT

Clinical and animal studies suggest that brain serotonergic systems may regulate aggressive behavior. However, the serotonin/violence relation has not been assessed at the epidemiological level. For study of an epidemiological sample we examined blood serotonin; certain physiological and behavioral data suggested that it might serve as an analogue marker for brain serotonergic function. Whole blood serotonin was measured in a representative birth cohort of 781 21-year-old women (48%) and men (52%). Violence was measured using cumulative court conviction records and participant's self-reports. Potential intervening factors addressed were: gender, age, diurnal variation, diet, psychiatric medications, illicit drug history, season of phlebotomy, plasma tryptophan, platelet count, body mass, suicide attempts, psychiatric diagnoses, alcohol and tobacco dependence, socio-economic status, IQ, and overall criminal offending. Whole blood serotonin related to violence among men but not women. Violent men's mean serotonin level was .56 *SD* above the mean of nonviolent men. The finding was specific to violence, as opposed to general crime, and it was robust across two different methods of measuring violence. Together, the intervening variables accounted for 25% of the relation between serotonin and violence. Developmental context interacted significantly with serotonin; serotonin was linked to violence primarily among men who grew up in families with little cohesion and much conflict. To our knowledge, this is the first demonstration that altered

Biosocial Bases of Violence, edited by Raine *et al.*
Plenum Press, New York, 1997

blood serotonin concentration is related to violence in the general population, and that the relation may depend on family origins.

WHOLE BLOOD SEROTONIN RELATES TO VIOLENCE IN AN EPIDEMIOLOGICAL STUDY

The serotonin system has been postulated to constrain or inhibit behavior (Depue & Spoont, 1986; Soubrie, 1986). Sub-optimal serotonergic function is thought to disinhibit aggression against the self and others, perhaps by sharpening sensitivity to stimuli that elicit aggression and blunting sensitivity to cues that signal punishment (Plutchik & Van Praag, 1989; Spoont, 1992). This study was prompted by reports that brain serotonergic dysfunction compromises the regulation of human aggressive behavior (Coccaro, 1989; Eichelman, 1993; Virkkunen & Linnoila, 1993). Specifically, we tested the hypothesis that young adults in the general population who have committed violent crimes have higher concentrations of whole blood serotonin than do their nonviolent age peers. Technological developments in the measurement and interpretation of serotonin in whole blood allowed us to conduct the first test of the serotonin/violence hypothesis in an epidemiological sample. The hypothesis was tested while controlling for gender, age, diurnal variation, diet, psychiatric medications, illicit drug history, season, precursor plasma tryptophan concentration, blood platelet count, body mass, suicide attempts, psychiatric diagnoses, alcohol and tobacco dependence, socio-economic status, IQ, and frequency of overall nonviolent criminal offending. In addition, we conducted an exploratory analysis to determine whether or not childhood family context might influence any relation between serotonin and adult violent behavior.

The serotonin/aggression hypothesis has been studied in vivo among human clinical subjects using three primary indices of serotonergic dysfunction: the primary serotonin metabolite in cerebro-spinal fluid (CSF 5-HIAA), endocrine responses to acute drug challenges, and serotonin in blood (platelet 5-HT, whole blood 5-HT, or platelet 5-HT uptake). Although many experiments have shown that manipulations of the serotonergic system can influence aggression in rodents (Cases et al., 1995; Eichelman, 1993; Pucilowski & Kostowski, 1986; Spoont, 1992) and nonhuman primates (Brammer et al., 1991; Chamberlain, Ervin, Pihl & Young, 1987), reviews of the few such experimental studies with human patients describe results as inconclusive (Coccaro, 1989; Virkkunen & Linniola, 1993). In non-experimental studies, CSF 5-HIAA has been reported to discriminate between aggressive patients and matched controls (Coccaro, 1989; Virkkunen & Linnoila, 1993; Virkkunen et al., 1994a & 1994b). Within clinical samples of aggressive patients low CSF 5-HIAA has been empirically linked with impulsive varieties of antisocial behavior and has been shown to differentiate alcoholic, personality-disordered and suicidal subgroups (Brown, Goodwin & Bunney, 1982; Schalling, 1993). Moreover, two longitudinal follow-up studies of patient samples have demonstrated that variation in CSF 5-HIAA can predict aggression over the course of two to three years in children (Kruesi et al., 1992) and adults (Virkkunen, DeJong, Bartko, Goodwin & Linnoila, 1989). In research that assesses the responsivity of the serotonin system by challenging it with drugs such as fenfluramine, blunted plasma prolactin responses to the fenfluramine challenge have been linked to patients' suicide, alcohol history, and aggressive personality (Coccaro et al., 1989), to antisocial personality disorder (Moss, Yao & Panzak, 1990; O'Keane et al., 1992), and to problem behavior in patients' family members (Coccaro, Silverman, Klar, Horvath, & Siever, 1994). Blood studies have been far fewer and findings less consistent

(Rogeness, Hernandez, Macedo & Mitchell, 1982), but blood 5-HT measures have been reported to relate to aggression among depressed patients (Mann, McBride, Anderson & Mieczkowski, 1992), conduct-disordered boys in detention (Pliszka, Graham, Rogeness, Renner, Sherman & Broussard, 1988), and aggressive men (Brown et al., 1989). These few blood findings encouraged us to examine whole blood serotonin in a human representative sample.

To date, the samples used in tests of the serotonin/violence hypothesis have been constrained to inpatients as a result of the exacting and intrusive procedures and ethical considerations inherent in studying CSF 5-HIAA or conducting drug challenges. As a result, prior studies have had to rely on relatively small samples, which are further divided into multiple comparison groups (an average of 28 subjects and 4.6 comparison groups per report), perhaps compromising statistical power and heightening the risk of deductive error by conducting many statistical tests on few subjects. Not all studies of the serotonin/violence relation report consistent findings, and reviews implicate inconsistencies in samples (Berman, Kavoussi & Coccaro, in press; Coccaro, 1989). Analyses of variation within a patient sample can be distorted by the restricted range of scores, sometimes producing inaccurate estimates of the sensitivity of a biological marker (Berk, 1983; Rasmusson, Riddle, Leckman, Anderson & Cohen, 1990). Clinical studies that use matched non-patient controls risk deductive error by failing to represent the healthy population as faithfully as they represent the population of patients (Mednick, 1978). Further, clinical samples are known to be vulnerable to selection bias that may compromise the generalizability of findings (Cohen & Cohen, 1984; Ransohoff & Feinstein, 1978). A test of the serotonin/violence hypothesis in an epidemiological sample provides a timely complement to the clinical literature by addressing these methodological concerns. Such a study can also provide new information about how abnormal the serotonin status of violent individuals is when compared against the critical comparison standard of norms for the general population (Rasmusson et al., 1990).

This epidemiological study was made feasible by the relative ease of sampling whole blood which, unlike CSF, can be readily drawn from non-patients. Serotonin in whole blood can be assayed quickly, inexpensively, and accurately (Brammer, unpublished; Yuwiler, Plotkin, Geller & Ritvo, 1970). The assays are highly reliable; repeated measures show small intra-individual and large stable inter-individual variation (Yuwiler, Brammer, Morley, Raleigh, Flannery & Geller, 1981). For interpretation, *high* concentrations of serotonin in blood are taken to correspond to low level of serotonin release in the brain and low level of 5-HIAA in CSF (Anderson, Freedman & Cohen, 1987; Hanna, Yuwiler & Cantwell, 1991); thus the hypothesis of this study was that violent adults would have higher concentrations of whole blood serotonin than nonviolent adults. Serotonin in blood is largely produced in the gut, wholly contained within platelets, and does not cross the blood-brain barrier, so it is at best an indirect index of serotonergic function in the brain (Rasmusson et al. 1990). The rationale for examining blood serotonin as a possible index of brain serotonin function is based on similarities in metabolic control of serotonin synthesis, similarities in the serotonin re-uptake systems of platelets and neurones, and the presence of 5HT2 receptors on platelets. Thus both brain and blood indices are affected by treatments that load serotonin precursors, disrupt serotonin uptake and storage, and alter serotonin degradation (Cook, Fletcher, Wainwright, Marks, Yan & Leventhal, 1994; Geller, Ritvo, Freeman et al., 1982; Given & Longenecker, 1985; Kremer, Goekoop & VanKempen, 1990; Mann, McBride, Anderson, & Mieczkowski, 1992; Pletscher, 1978; Raleigh, McGuire, Brammer & Yuwiler, 1984; Stahl, 1985; Von Hahn, Honegger & Pletscher, 1980).

In an epidemiological study, measures of violence differ from the measures used in clinical studies. Clinical studies have variously defined aggression as high scores on personality checklists tapping irritable or hostile attitudes, as observed aggressive actions on the inpatient ward, or sometimes on the basis of the violent infraction that precipitated the current hospitalization or incarceration. Personality checklists and inpatient ward behaviors bear imperfect or unknown relations to potential for criminal violence in the streets or at home. Designations of violence that are based on the current infraction do establish the consequential nature of the violent offense, but they are subject to miss-classification if the complete violent offending history is not assessed (Rasmusson et al., 1990). In the present study, violence was assessed using cumulative court records of conviction from age 13 to age 21; age 21 is just past the peak age of risk for violent offending (Elliott, Huizinga & Morse, 1986). The court convictions were complemented by self-report interviews to allow study of the substantial number of repetitively violent individuals in the population who escape official identification (Elliott, Huizinga & Morse, 1986). Thus, to supplement prior clinical samples, this research included violent individuals who have not, as well as who have, been convicted, incarcerated, and treated psychiatrically for their violent behavior.

This research was conducted as part of the Dunedin Multidisciplinary Health and Development Study; the multidisciplinary assessment design of the larger study allowed us systematically to examine a number of variables that might influence the serotonin/violence relation. Men and women were studied, to redress the dearth of studies of serotonin and violence among women (Raine, 1993). Factors known or suspected to influence concentrations of whole blood serotonin were examined, including use of psychiatric medication, illicit drugs or tobacco (Mann et al., 1992), season of venipuncture (Arora, Kregel & Meltzer, 1984; Badcock, Spence & Stern, 1987; Brewerton et al., 1988), blood platelet count (Ritvo, Yuwiler, Geller, Plotkin, Mason, & Saegar, 1971), and available precursor plasma tryptophan concentration (Chamberlain, Ervin, Pihl & Young, 1987; Fernstrom & Wurtman, 1971). We examined body mass because it has been hypothesized that, if larger individuals are more likely to be violent, the serotonin/violence relation may be an artifact of the link between serotonin and somatostatin, a hormone that influences body mass (Raine, 1993). Suicide attempts were examined in an effort to place this research in the context of clinical studies which report that serotonin relates to violence against the self as well as against others (Asberg, Nordstrom & Traskman-Bendz, 1986; Coccaro et al., 1989; Nielson et al., 1994). Psychiatric diagnoses were examined because serotonin has been linked to several disorders (van Kammen, 1987). Alcohol dependence was examined because it has been shown to relate specifically to serotonin within antisocial clinical samples (Coccaro et al., 1989; Linnoila et al., 1983). Two known powerful correlates of violence were also examined, socio-economic status and IQ test scores (Lynam, Moffitt, & Stouthamer-Loeber, 1993) because their relation to serotonin has not been ascertained.

As a final control, we tested the serotonin/violence hypothesis while controlling for frequency of overall criminal offending. Because virtually all violent offenders have lengthy careers of property crime as well, property crime must be controlled as a test to establish the specificity of serotonin as a marker for violentness, per se.

In the last analysis, we asked whether or not a measure of childhood family context might condition the relation between serotonin and violent behavior measures taken in adulthood. Such "bio-social" interactions predicting violent outcomes have been reported previously (Raine, this volume). Moreover, studies of monkeys have documented that subjecting infants to deprivation during early development can bring about lasting changes in the serotonin system and in behavior (Higley, Suomi & Linnoila, 1991; Higley,

Thompson, Champoux, Goldman, Hasert, Kraemer, Scanlan, Suomi & Linnoila, 1993). Therefore, we examined a measure of the quality of each study member's childhood family environment (cohesion, expressiveness, and conflict).

METHODS

Sample

Participants in this study were the 21-year-old members of an unselected cohort, born between April 1 1972 and March 31 1973 in a New Zealand city of approximately 120,000 inhabitants, and studied since birth in the Dunedin Multidisciplinary Health and Development Study. The history of the study has been described by Silva (1990). This report presents findings from the most recent assessment, when the study members were 21 years old, conducted during 1993–1994. The investigation's standard procedure involves bringing participants to the research unit for a full day of individual interviews, examinations, and laboratory studies presented as 50-minute modules in counterbalanced order (e.g., IQ test, NIMH Diagnostic Interview Schedule for Mental Disorders, physical examination, delinquency interview). Study members incarcerated for offending were brought to the research unit under guard or assessed at the unit after parole. Interviews with participants were supplemented by a search of official records. Questionnaires mailed to informants who know each participant well were used to validate self-reports. Because confidentiality has never been violated, study members are by now unusually willing to provide frank interview responses.

The base sample for the study comprises the 1037 individuals originally enrolled (52% male and 48% female). At age 21, 97% of the living study members participated and blood serotonin samples were obtained from 781 participants, a compliance rate of 82% (17 study members had died since age 3, 9 were not located, 19 refused the entire study, 42 were interviewed in the field where blood samples could not be taken, 169 came to the unit but declined venipuncture). The original sample was representative of New Zealand's South Island in family socio-economic status and in caucasian European ethnicity; only 2% of participants were identified by their mothers at birth as Polynesian or Maori. The sample of 781 individuals for the blood serotonin study did not differ from the original representative birth cohort on gender ($Chi\text{-}Square = 1.4, p > .20$), family socio-economic status ($t = 1.2, p > .20$), or criminal conviction ($Chi\text{-}Square = 0.66, p > .40$).

Whole Blood Serotonin Assays

Whole blood serotonin assays were performed on blood samples collected by venipuncture between 4:00 and 5:00 PM at the end of the 8-hour assessment day (with lunch served) that occurred within 60 days of each participant's twenty-first birthday. Thus, although laboratory studies suggest that whole blood serotonin is not affected after puberty by age (Badcock, Spence & Stern, 1987; Ritvo, Yuwiler, Geller, Plotkin, Mason & Saeger, 1971; Siefert, Foxx & Butler, 1980), recent diet (Badcock, Spence & Stern, 1987; Kremer, Goekoop & Van Kempen, 1990; Ritvo, Yuwiler & Geller, 1970), or diurnal variation or activity level (Kremer, Goekoop & Van Kempen, 1990; Badcock, Spence & Stern, 1987), these factors were controlled because not all reports agree (Wirz-Justice, Lichtsteiner & Freer, 1977). The blood was collected into lithium heparin (Vacutainer) tubes and stored in the dark at -20°C until the time of assay, from 1 to 12 months later.

Storage under these conditions does not effect whole blood serotonin concentrations (Badcock, Spence & Stern, 1987).

Blood serotonin (5-HT) concentration was determined using an HPLC procedure (Brammer, unpublished) that bears an essentially 1:1 relation to the fluorometric assay of Yuwiler, Plotkin, Geller & Ritvo (1970). A 100 μl portion of whole blood was diluted with 750 μl 3% freshly prepared ascorbic acid, and the mixture was treated with 100 μl 4 M perchloric acid. The tubes were vortexed for 1 min and then centrifuged (13000 g x 5 min). The supernatant was withdrawn and held on ice until examination by liquid chromatography. A mobile phase was prepared from an aqueous component of 33 mM acetate, 33 mM citrate, 33 mM phosphate, and 0.27 mM EDTA adjusted to pH 4.1 with sodium hydroxide, 0.2 μm filtered, and mixed with acetonitrile to 7%. A flow rate of 1.0 ml/min was maintained, and 50 μl of sample supernatant was injected. The components were separated on an ODS analytical column (4.6 x 150 mm, 5 μm particle) and detected by fluorescence at 283 nm excitation and 330 nm emission. External standards were treated the same as samples. Results are expressed as ng serotonin per ml whole blood. In quality-control tests for this study, the within-day coefficient of variation (CV) was 4.3% (96.6 ± 4.2 ng/ml, n = 20). The between-day CV for a quality-control blood sample was 3.6% (119.8 ± 4.2 ng/ml, n = 5). The average between-day CV for nine patients' blood samples taken on 5 occasions was 5.8% (range 3.1% - 7.8%) and recovery of added serotonin at low and high standards was greater than 97%.

Violence Measures

Two complementary measures of violence were examined.

Court Conviction Criminal Records. for all courts in New Zealand and Australia were obtained by searching the central computer system of the New Zealand police, revealing that 141 study members had been convicted for some offense(s) by age 21. Of the sample members who had provided blood samples, 31 (26 men and 5 women) had been convicted of one or more of the following violent crimes: inciting or threatening violence (n = 17 incidents), using an attack dog on a person (n = 1), presenting an offensive weapon (n = 13), threatening a police officer (n = 5), rape (n = 2), manual assault (n = 25), assault on a police officer (n = 5), assault with a deadly weapon (n = 5), aggravated robbery (n = 1), and homicide (n = 1).

Self-Reports of criminal Offenses. committed during the past 12 months were obtained in private standardized interviews using the Self-Report Delinquency Interview developed for the U.S. National Youth Survey and National Institute of Justice multi-site surveys (Elliott & Huizinga, 1989). The interview assessed 41 different illegal offenses. In the Dunedin sample it yields an internal reliability alpha of .88, a one-month test-retest reliability correlation of .85, and moderate correlations with informant reports and conviction records (Moffitt, Silva, Lynam & Henry, 1994). Serotonin assays and court conviction records were obtained only after all subjects had been interviewed, and thus self-report measures were blind and independent. Of the sample members who had provided blood samples, the 50 (36 men and 14 women) designated as self-reported violent participants had reported at least two of the following in the past 12 months: "attacked someone you lived with, with a weapon or with the idea of seriously hurting or killing them" (n = 11 study members), "attacked someone else with a weapon or with the idea of seriously hurting or killing them" (n = 13), "hit someone you lived with" (n = 33), "hit someone else" (n

= 39), "used a weapon, force, or strongarm methods to rob a person" (n = 2), "were involved in a gang fight" (n = 18).

Potential Intervening Variables

On the basis of prior research, we examined 11 variables that might influence the serotonin/violence relation. *Psychiatric medications* taken in the past 12 months were recorded from study members' self-reports during the age-21 mental health interview. Medication was taken for a mental disorder by 4.1% of study members. *Illicit drug use history* was assessed by age-21 self report of drugs used during the past 12 months: marijuana, stimulants, sedatives, cocaine, heroin, opiates, PCP/MDA, psychedelics, inhalants, or other. Scores ranged from 0 to 10 drugs, with 20% of the sample using two or more types of drugs. *Season* of venipuncture was ascertained from subject's appointment date. *Tryptophan* assays were conducted as part of the assay for serotonin. Quality-control studies paralleled those for serotonin. The within-day CV was 1.3%, the between-day CV was 4.9%, the average between-day CV was 3.6% (range 2.3% - 5.3%), and recovery was greater than 95%. *Platelet counts* were conducted by trained technicians as a routine procedure. *Body mass index* was calculated, using anthropometric measurements taken at the age-21 assessment, as measured weight, divided by measured height squared. *Suicide attempts* in the past 12 months were recorded from study members' self-reports during the age-21 mental health interview. Suicide was attempted by 19 study members (blood could not be sampled from the one study member who has died from suicide). For this study, mental disorders examined were *depressive disorders* (6.7% of this sample had major depressive episode or dysthymia), *anxiety disorders* (9.5% had generalized anxiety, obsessive-compulsive disorder, panic disorder, agoraphobia, simple phobia, or social phobia), manic episode (2%), *alcohol dependence* (9.8%) and *tobacco dependence* (16.5%). Diagnoses of Axis I disorders during the 12 months preceding the age-21 interview were made according to the criteria of the Diagnostic and Statistical Manual (3rd Edition, Revised) of the American Psychiatric Association (1987). Symptoms were privately assessed using the standardized Diagnostic Interview Schedule, designed by the National Institute of Mental Health for the Epidemiological Catchment Area project (Robins & Regier, 1991). Because of the youth and limited size of the sample, disorders with lifetime prevalence rates of 1% or lower were not diagnosed. Epidemiology, reliability, and validity of psychiatric diagnoses in the Dunedin sample have been reported elsewhere (Newman, Moffitt, Caspi, Magdol, Silva & Stanton, in press). *Socio-economic status* (SES) was rated on a six-point scale used to measure the status of occupations in New Zealand, ranging from professional-administrative to unskilled (Elley & Irving, 1985). University students, 20% of the sample, were rated as "highly skilled". Participants who were unemployed at the time of assessment, 14.8% of the sample, were rated on the last job they had held. *Intelligence Quotients* were measured using the Wechsler Intelligence Scales (Revised), the most reliable and valid standardized individual test of intelligence (Wechsler, 1974). Frequency of *nonviolent criminal conviction* and *self-reported nonviolent crime* were assessed as previously described in the sections on measurement of violent conviction and self-reported violence. The *Moos Family Relations Index* (Moos & Moos, 1981) was completed by a parent when study members were 7, 9, and 15 years old (Parnicky, Williams & Silva, 1985). Scores were averaged to form a composite index. The scale measures cohesion ("family members really help and support one another"), expressiveness ("family members keep their feelings to themselves") and conflict ("family members sometimes hit each other").

RESULTS

Whole Blood Serotonin and Measures of Violence

Student's *t*-test was used to test for differences between means of violent versus nonviolent groups. Effect sizes are estimated according to Cohen's (1988) formula as the distance between the mean scores of two groups, in population standard deviation (z) units, where .2 is a small effect, .5 is a moderate effect, and .8 is a large effect. (Degrees of freedom vary in this report because the number of cases with present data varied by analysis, but no analysis had fewer than 98% of the study members who gave blood samples.)

Among Women. Comparisons between violent and nonviolent women revealed that serotonin level was unrelated to women's violence whether measured by self-report ($t = 0.17$, 365*df*, $p > .75$) or conviction ($t = 0.30$, 367*df*, $p > .75$). The effect size among women was very small: .04 for self-reported violence and .13 for violent conviction. To examine whether or not menstrual fluctuations in platelet production might have introduced noise into the women's serotonin measure, we controlled for individual differences in platelet count while comparing violent and nonviolent women, but the insignificant effect sizes were not altered. Because serotonin was unrelated to women's violence in this sample, the women will not be described further in this report.

Among Men. Comparisons between violent and nonviolent men revealed that serotonin level was related to men's violence whether measured by self-report ($t = 2.88$, 407*df*, $p < .01$) or conviction ($t = 2.56$, 410*df*, $p < .01$). The effect size among men was moderate: .50 for self-reported violence and .51 for violent conviction. Self-reported violence and conviction for violence overlapped in the sample of men (*Chi-Square* = 64.27, 1*df*, $p < .001$). The odds of conviction were 12.7 times greater for men who self-reported violence than for men who did not, with a 95% confidence interval (CI) of 6 to 27. Given this overlap, a violent group was constituted as the men who had self-reported violence *or* been convicted for it. Of these 54 violent men, 70% had been convicted for violence by the courts, 13% had been remanded to prison, and 9% had been psychiatrically hospitalized during the past year. With respect to psychiatric diagnoses at age 21, 28% of the violent men were alcohol dependent, 42% were tobacco dependent, 27% had an anxiety disorder, 23% had a depressive disorder, and 4% had a manic episode. The 54 violent men had higher blood serotonin than the nonviolent men ($t = 3.87$, 408*df*, $p < .001$). The effect size was moderate: .56.

Figure 1 shows group means and standard errors. Because the sample is a representative epidemiological sample and the serotonin scores formed an approximate normal distribution, the group mean in standard deviation (z-score) units may be interpreted as the group's deviation from the population norm for young adult males. Figure 2 shows the distributions of serotonin scores for the violent and nonviolent men. Examination of the distribution of serotonin scores shows that the relation between serotonin and violence was not an artifact of cases at the extreme end of the distribution.

Controlling for Potential Intervening Variables

Table 1 shows the mean concentrations of whole blood serotonin for violent and nonviolent men, before and after adjusting for the effects of other study variables. To test for differences between means of violent versus nonviolent groups while controlling for

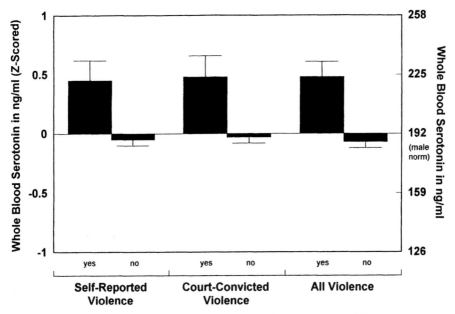

Figure 1. Whole blood serotonin levels (Z-scores) in violent vs. non-violent men.

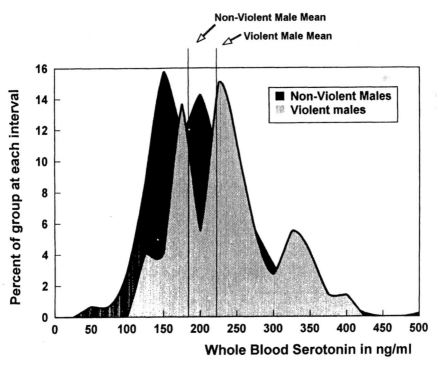

Figure 2. The distribution of serotonin scores for violent and nonviolent men.

potential intervening variables, we used analysis of covariance if the covariate was a continuously distributed variable, or multiple regression if the covariate was a binary variable.

Psychiatric Medications. The 10 men who had taken psychiatric medications in the past year had lower serotonin concentrations than non-medicated men, and although the effect size at .48 was moderate, the difference was not significant because statistical power was limited by the small number of medicated men ($t = 1.5$, $407df$, $p = .13$). Violent men were not more likely to have been medicated than were nonviolent men (3.7% vs. 3.6%, *Chi-Square* = 0.01, $1df$, $p > .90$). Moreover, as shown in Table 1, the serotonin difference between violent and nonviolent men remained after medication was entered as a covariate.

Illicit Drug Use. Serotonin was weakly but significantly correlated with number of drugs used ($r = .11$, $409df$, $p < .05$). When violent participation was statistically controlled, the correlation between serotonin and drug use dropped to nonsignificance (*beta* = .05, $p > .20$); it appeared to be an artifact of disproportionate drug use by violent men; violent men used more than twice as many drugs as nonviolent men ($t = 8.04$, $488df$, $p < .001$). However, the serotonin difference between violent and nonviolent men remained after drug use was entered as a covariate.

Season. Concentrations of serotonin were lower among men assessed during January, February and March (summer in New Zealand) relative to other seasons ($t = 2.48$, $410df$, $p < .01$). The seasonal effect size was small; .29. However, the serotonin difference between violent and nonviolent men remained after season (dummy coded as summer vs. non-summer) was entered as a covariate.

Table 1. Analysis of variance for whole blood serotonin ng/ml by men's violence before and after controlling for potential intervening variables

	Mean serotonin violent men	Nonviolent men's mean serotonin	Effect size	F value	p value
n of subjects	54	356			
Baseline comparison	222.18	185.75	.56	14.97	.000
Adjusted means after controlling for...					
psychiatric medications	222.03	185.79	.55	14.52	.000
illicit drug use	219.41	186.16	.51	11.30	.001
season	221.59	185.83	.54	14.61	.000
tryptophan	220.89	185.94	.53	13.98	.000
platelet count	220.31	186.57	.49	13.38	.000
body mass	221.43	185.81	.54	14.44	.000
suicide attempts	224.56	185.40	.60	17.09	.000
depressive disorder	223.00	185.62	.58	15.40	.000
anxiety disorder	222.37	185.71	.56	14.77	.000
manic episode	219.27	186.56	.50	11.47	.001
alcohol dependence	220.54	185.98	.53	13.25	.000
tobacco dependence	218.21	186.57	.49	9.38	.002
SES	222.24	185.72	.55	13.66	.000
IQ	219.41	186.92	.50	11.29	.001
nonviolent convictions	216.86	186.55	.46	8.81	.003
self-reported nonviolent crime	227.73	184.98	.65	14.81	.000

Tryptophan. Serotonin was weakly but significantly correlated with plasma trypto-phan concentration ($r = .14$, $412df$, $p < .01$). Violent and nonviolent men did not differ sig-nificantly on tryptophan ($t = 1.19$, $408df$, $p > .2$). The serotonin difference between violent and nonviolent men remained after tryptophan was entered as a covariate.

Platelet Count. Serotonin was significantly correlated with platelet count ($r = .20$, $410df$, $p < .001$). Violent and nonviolent men approached a significant difference on plate-let count ($t = 1.92$, $409df$, $p = .06$). The serotonin difference between violent and nonvio-lent men remained after platelet count was entered as a covariate.

Body Mass. Serotonin was weakly but significantly correlated with body mass ($r = -.13$, $410df$, $p < .01$). Violent and nonviolent men did not differ significantly on body mass ($t = 0.19$, $478df$, $p > .75$). The serotonin difference between violent and nonviolent men remained after body mass was entered as a covariate.

Suicide Attempts. The nine men in our sample who had attempted suicide did not differ significantly from non-suicidal men on serotonin level ($t = 1.5$, $409df$, $p = .13$). The mean serotonin level for three attempters who had been treated with psychiatric medica-tions such as 5HT re-uptake inhibitors was 1.05 *SD* below the male norm and the mean for the six unmedicated suicide attempters was .23 *SD* below the male norm. Violent men were 6.16 times more likely to attempt suicide than were nonviolent men (95% CI: 1.73 to 21.88, *Chi-Square* $= 10.14$, $1df$, $p < .001$). None of the violent suicidal men were medi-cated. Moreover, the serotonin difference between violent and nonviolent men remained after suicide was entered as a covariate.

Psychiatric Disorders. Depressed men did not differ from non-depressed men on sero-tonin ($t = .08$, $410df$, $p = .93$). Similarly, anxious men did not differ from non-anxious men on serotonin ($t = .58$, $410df$, $p = .56$). However, manic men did have higher serotonin concentra-tions than non-manic men ($t = 2.16$, $408df$, $p = .03$). Violent men were 2.75 times more likely to have a depressive disorder than were nonviolent men (95% CI: 1.44 to 5.25, *Chi-Square* $= 10.11$, $1df$, $p < .01$), and 3.03 times more likely to have an anxiety disorder than were nonvio-lent men (95% CI: 1.65 to 5.58, *Chi-Square* $= 13.72$, $1df$, $p < .001$). However, violent men were not more likely to be manic than were nonviolent men (*Chi-Square* $= 2.01$, $1df$, $p = .15$). The serotonin difference between violent and nonviolent men remained after depression, anxi-ety and mania were entered separately as covariates.

Alcohol. Alcohol-dependent men had higher blood serotonin concentrations than non-dependent men ($t = 2.03$, $409df$, $p < .05$). Violent men were 3.15 times more likely to be alcohol-dependent than were nonviolent men (95% CI: 1.72 to 5.75, *Chi-Square* $= 15.15$, $1df$, $p < .001$). However, the serotonin difference between violent and nonviolent men remained after alcohol dependence was entered as a covariate.

Tobacco. Tobacco-dependent men had marginally higher blood serotonin concentra-tions than non-dependent men ($t = 1.93$, $390df$, $p = .06$). Violent men were 5.15 times more likely to be tobacco-dependent than were nonviolent men (95% CI: 2.91 to 9.13, *Chi-Square* $= 36.40$, $1df$, $p < .001$). However, the serotonin difference between violent and nonviolent men remained after tobacco dependence was entered as a covariate.

Socio-Economic Status. Serotonin was not correlated with SES, $r = .05$. Violent men had lower SES ratings than nonviolent men ($t = 7.23$, $486df$, p <.001). However, the serotonin difference between violent and nonviolent men remained after SES was entered as a covariate.

IQ. Serotonin was weakly but significantly and negatively correlated with IQ ($r = -.12$, $401df$, $p < .05$). Violent men had lower IQ scores than nonviolent men ($t = 5.24$, $477df$, $p < .001$). However, the serotonin difference between violent and nonviolent men remained after IQ was entered as a covariate.

Controlling for Intervening Variables Simultaneously. Five variables were related to both serotonin and to violence, at either conventional or marginal levels of significance: illicit drug use, platelet count, alcohol dependence, tobacco dependence, and IQ. When these five variables were entered simultaneously as covariates, the serotonin difference between violent and nonviolent men remained significant ($F = 5.92$, $p = .01$). After the mean serotonin scores of violent and nonviolent men were adjusted for these five covariates, the effect size decreased about one quarter, from .56 to .41.

Frequency of Nonviolent Offending. Violent men had much higher rates of all types of crimes than did nonviolent men, whether measured by self-report ($t = 9.56$, $480df$, $p < .01$) or conviction ($t = 9.59$, $487df$, $p < .01$). On average, violent men had been convicted of 7.7 non-violent offenses, but nonviolent men had been convicted of only 0.4 such offenses. However, as shown in Table 1, the serotonin difference between violent and nonviolent men remained after overall frequency of offending was entered as a covariate.

Family Relations Index. As expected, violent men had grown up in more conflicted families than had non-violent men ($t = 4.3$, 484df, $p < .001$). However, serotonin was not related to the family relations index ($r = .09$, $408df$, ns). Because the outcome variable was categorical, "violent" versus "non-violent", logistic regression analysis was used to determine whether or not family and serotonin interacted to predict violent outcome. As anticipated, significant main effects on violence emerged for both serotonin and family. However, the logistic model that included an interaction between serotonin and family made a significantly better fit to the data than a model with main effects alone (*Chi-square* for improvement = 5.5, $1df$, p < .03).

Figure 3 illustrates this interaction. To produce the figure, we cut the distributions of serotonin and the family relations index into thirds: high, medium, and low concentrations of serotonin, and supportive, average, and conflicted family backgrounds. Both the serotonin scores and the family relations scores were normally distributed, and serotonin and family were unrelated, so a three by three crosstabulation produced nine equal-sized cells of approximately 45 men each (range 31–54). Figure 3 shows that when conflicted family background combined with high whole blood serotonin concentration, 37% of men were violent by age 21. The combination of high serotonin and conflicted family accounted for one third of the 54 violent men in this sample. In contrast, only 10%-11% of men were violent if a conflicted family was complemented by low serotonin *or* if high serotonin was complemented by a supportive family.

DISCUSSION

In this study, elevated whole blood serotonin was characteristic of violent men. The violent men's mean serotonin level, 222 ng/ml, was .48 *SD* above the male sample norm

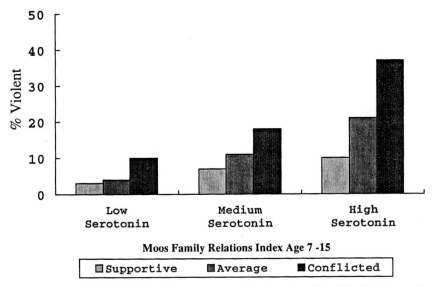

Figure 3. Percent of violent man according to serotonin levels and conflicted family background.

and .56 *SD* above the mean of nonviolent men. The finding was specific to violence as opposed to antisocial activities in general, for it was unaltered when the frequency of overall offending was statistically controlled. The finding was robust across two different methods of measuring violence, self-report and court conviction. To our knowledge, this is the first study to demonstrate that an index of the serotonergic system is related to violence in the general population. The size of the epidemiological serotonin effect was not small, it indicated a moderate effect size in the population. Moreover, the relation between serotonin and violence was not an artifact of cases at the extreme end of the distribution. However, the overlapping distributions shown in Figure 2 revealed that many nonviolent men showed higher concentrations of serotonin than the violent men. As such, although whole blood serotonin is sensitive to violence risk, it does not have enough specificity to serve as a marker for violence risk in the general population.

The sampling design of the present study allowed comparison by gender. Individual differences in serotonin level were linked with violence only among men; the finding did not describe women. This observation agrees with the only prior study of women (Gardner, Lucas & Cowdry, 1990). There is some evidence that whole blood serotonin may fluctuate slightly in the last five days of the menstrual cycle, perhaps from hormonal effects on platelet or serotonin production (Ritvo, Yuwiler, Geller, Plotkin, Mason & Saeger, 1971), although this finding is not consistent (see Ashcroft, Crawford, Binns & MacDougall, 1964). We had not assessed the women's menstrual cycles but we were able to introduce statistical controls for platelet count, which did not alter the null relation between women's violence and serotonin. Nonetheless, it is possible that potential findings among women may have been obscured by noise in the serotonin data. Alternately, it is possible that women's violence differs from men's violence in its psychological meaning, and thereby has different biological correlates. For example, in one comparison of men and women who were matched on levels of violent behaviors against an intimate partner, male perpetrators were significantly more deviant than female perpetrators on several psychiatric and social correlates of violence (Magdol, Moffitt, Caspi, Newman, Fagan, & Silva, in press).

The research procedures of the Dunedin study eliminated potential influences on blood serotonin from diurnal variation, recent diet, recent activity level, or age. The measurement design of the Dunedin study allowed statistical control for potential biasing effects on serotonin by variables suggested in the methodological literature. Although blood serotonin concentrations were indeed lower among study members who had taken psychiatric medications, lower in the austral summer, higher among illicit drug users and heavy smokers, and positively related to plasma tryptophan and platelet count, none of these effects eliminated the relation between serotonin and violence. The serotonin/violence relation also survived statistical control for potential biasing effects of factors hypothesized to be related to violence: body mass, suicide attempts, psychiatric disorders, alcohol dependence, low socio-economic status, and low IQ. In combination, control factors accounted for about one quarter of the relation between serotonin and violence; adjusting for alcohol, tobacco, and drug use, platelet count, and IQ, the violent men's mean serotonin concentration remained .41 *SD* above the mean of nonviolent men. Thus, the whole blood serotonin/violence relation in this sample is neither wholly mediated by these variables nor is it an artifact of them.

The finding of this study is compatible with a growing clinical literature on serotonin and aggression. Elevated blood serotonin has been reported for conduct-disordered patients (Pliszka, Graham, Renner, Sherman & Broussard, 1988) and aggressive depressed patients (Mann, McBride, Anderson, & Mieczkowski, 1992). Consistent with these prior blood findings, low concentrations of serotonin metabolite 5-HIAA in cerebrospinal fluid characterized an aggressive subgroup within conduct-disordered patients (Kruesi et al., 1990) and specific diagnostic subgroups within samples of criminal patients (Brown, Ebert, Goyer, Jimersom, Klein, Bunney & Goodwin, 1982; Linniola, Virkkunen, Scheinin, Nuutila, Rimon, & Goodwin, 1983; Virkkunen et al. 1987, 1989, 1994, 1994). Experimental challenge by a serotonin agonist affected alcoholic, suicidal, and aggressive subgroups within a patient sample (Coccaro et al., 1989). These previous clinical reports, and literature reviews based on them, have speculated that dysfunction in the central serotonergic system will be limited to relatively circumscribed and sometimes rare subtypes of violent offenders (Asberg, Nordstrom & Traskman-Bendz, 1986; Eichelman, 1993; Rogeness, Javors & Pliszka, 1992; Schalling, 1993; Spoont, 1992; Raine, 1993; Virkkunen & Linnoila, 1993). Those subtypes have been variously, and somewhat inconclusively, proposed as personality disordered, pathologically violent, hostile/affective, suicidal, non-alcoholic, early-onset alcoholic, intra-familially violent, impulsive, or mesomorphic in body build. If the serotonin/violence relation were specific to the subgroups we studied here, it should have been reduced to non-significance in the covariance analyses. The relation was not reduced, suggesting the hypothesis that the serotonin/violence relation in the general population may be robust across at least some implicated clinical subgroups. Research that combines brain measures of serotonin with epidemiological sampling frames is needed to determine if the serotonin/violence relation is limited to select subgroups or broadly applicable to most violent offenders.

Observed elevated whole blood serotonin among violent men is consistent with the hypothesis that serotonergic dysfunction is associated with aggression. It must be noted, however, that even though the present results are consistent with a serotonin dysfunction in the central nervous system, the association is certainly indirect and might even be fortuitous. For example, our data cannot distinguish whether the mechanism involved altered serotonin metabolism, altered platelet production, altered serotonin transport into platelets secondary to altered blood flow through the gut, or some other process. The relation between whole blood serotonin and serotonergic function within the brain is not yet understood. Interpretation of our findings will depend on the extent to which central and peripheral indices of the serotonin system are revealed to be influenced by similar kinetic developmental defects or experimental interventions (Cook et al., 1994; Geller, Ritvo, Freeman et al., 1982; Given & Longenecker; 1985; Kremer, Goekoop & VanKempen, 1990; Mann, McBride, Anderson, & Mieczkowski, 1992; Pletscher, 1978; Raleigh, McGuire, Brammer & Yuwiler, 1984; Stahl, 1985; Von Hahn, Honegger & Pletscher, 1980). We also await compelling explanations of why behavior relates to low concentrations of CSF 5-HIAA but high concentrations of whole blood 5-HT. Nonetheless, whatever the mechanism, this study illustrates that whole blood serotonin could become a useful measure for epidemiological research. Recent commentaries on the literature (Berman, Kavoussi, Coccaro, in press; Rasmusson, Riddle, Leckman, Anderson & Cohen, 1990; Stoff, Ieni, Friedman, Bridger, Pollock, & Vitello, 1991) have blamed failures to replicate relations between serotonin and behavior on small, idiosyncratic samples. These commentaries also advise us to address the developmental origins of links between serotonin and aggression by studying representative samples of children longitudinally with repeated behavioral and serotonin measures. If replicated, our findings suggest that whole blood serotonin may be a useful measure for studying larger, more representative samples, and may allow repeated measures in developmental studies. Epidemiological research can complement clinical research by delineating the population parameters of the serotonin-violence relation first revealed by studies of inpatient samples (Rasmusson, Riddle, Leckman, Anderson & Cohen, 1990).

The present investigation was cross-sectional, examining serotonin in only 21-year-olds. Laboratory studies suggest that individual differences in whole blood serotonin are stable over extended periods of a year or more (Badcock, Spence & Stern, 1987; Ritvo, Yuwiler & Geller, 1970; Yuwiler, Brammer, Morley, Raleigh, Flannery, & Geller, 1981). Two previous clinical studies have shown that CSF serotonin indices can predict aggression and violence after two to three years of follow-up (Kruesi et al., 1992; Virkkunen, DeJong, Bartko, Goodwin, & Linnoila, 1989). In addition, in studies of monkeys, experimental manipulations of males' dominance hierarchies (Botchin, Kaplan, Manuck & Mann, 1993; Raleigh, McGuire, Brammer & Yuwiler, 1984) and infants' rearing experiences (Higley, Suomi & Linnoila, 1991; Higley, Thompson, Champoux, Goldman, Hasert, Kraemer, Scanlan, Suomi & Linnoila, 1993) have documented that changes in social experience can bring about changes in the serotonin system. In this study, we found that the strength of the serotonin/violence relation was moderated by prospective measures of family background gathered in childhood; 37% of the high-serotonin men from unsupportive, conflicted families were violent, compared to only 10% of the other men. It is tempting to infer that socialization *experiences* in the men's pasts were vital moderators of the relations between serotonin and violent behavior. However, we caution against the assumption that measures of the family environment tap purely social experience. Indeed, our measure of family context, the Moos Family Relations Index, has shown significant heritability in behavioral genetic studies (Hur & Bouchard, 1995; Plomin, McClearn, Pedersen, Nesselroade, & Bergeman, 1988). This indicates that children's socialization environments are not purely "environmental." Rather, rearing environments are at least partial products of parents' heritable traits. Thus, it remains possible that men are more likely to express serotonin-related violence if they carry a predisposition toward antisocial behavior that they inherited from their unsupportive and highly conflictual (i.e., antisocial) parents. Such a scenario could generate the findings shown in Figure 3. A behavioral genetic research design is the strongest way of isolating what is truly environmental about measures of children's environments, before researchers may proceed to test for interactions between biological and environmental variables.

In future years, a "natural experiment" may present an opportunity to test another, more interpretable, bio-social interaction. Our 21-year-old sample were just at the peak age for violence in New Zealand; in the next ten years many of the violent offenders will desist. Previous research has shown that desistence from antisocial behavior is often accompanied by a "turning point" social experience: pair-bonding with a supportive pro-social partner (Quinton, Pickles, Maughan, & Rutter, 1993; Sampson & Laub, 1993). The bio-social interaction hypothesis would suggest that violence is most likely to persist among men who have (a) abnormal serotonin, but (b) no pro-social partner. Future assessments in the Dunedin longitudinal study may allow us to test the long-term stability of individual differences in whole blood serotonin, to test whether whole blood serotonin concentrations can predict future violent outcomes, and to trace the relations among whole blood serotonin, social experience, and violence across adult development.

ACKNOWLEDGMENTS

This research was supported by the H.I. Romnes Fellowship of the University of Wisconsin, U.S. Public Health Service grants from the Violence and Traumatic Stress Research Branch (MH-45070) and the Personality and Social Processes Research Branch (MH-49414) of the National Institute of Mental Health, the Research Service of the Department of Veterans Affairs, and by the New Zealand Health Research Council. We are grateful to Craig Berridge, HonaLee Harrington, Bob Krueger, Paul Stevenson, David Stoff, Chris Sies and Gavin Tisch of the Canterbury Health Laboratories, the phlebotomists of the Plunkett Society, the New Zealand Police, and to the study members and their families.

REFERENCES

American Psychiatric Association, (1987). Diagnostic and statistical manual of mental disorders, Revised (DSM-III-R). Washington, DC: APA.

Anderson, G.M., Freedman, D.X., & Cohen, D.J. (1987). Whole blood serotonin in autistic and normal subjects. Journal of Child Psychology and Psychiatry, 28, 885–900.

Arora, R.C., Kregel, L., & Meltzer, H.Y. (1984). Seasonal variation of serotonin uptake in normal controls and depressed patients. Biological Psychiatry, 19, 795–804.

Asberg, M., Nordstom, P., & Traskman-Bendz, L. (1986). Cerebrospinal fluid studies in suicide, an overview. Annals of the New York Academy of Sciences, 487, 243–255.

Ashcroft, G.W., Crawford, T.B.B., Binns, J.K., & MacDougall, E.J. (1964). Estimation of 5-hydroxytryptamine in human blood. Clinica Chimica Acta, 9, 364–369.

Badcock, N.R., Spence, J.G., & Stern, L.M. (1987). Blood serotonin levels in adults, autistic and nonautistic children with a comparison of different methodologies. Annuals of Clinical Biochemistry, 24, 625–634.

Berk, R.A. (1983). An introduction to sample selection bias in sociological data. American Sociological Review, 48, 386–398.

Berman, M.E., Kavoussi, R.J. & Coccaro, E.F. (1997). Neurotransmitter correlates of antisocial personality disorder. In D. Stoff, J. Breiling & J. Maser (Eds.). Handbook of antisocial behavior New York: Wiley .

Botchin, M.B., Kaplan, J.R., Manuck, S.B., & Mann, J.J. (1994). Neuroendocrine responses to fenfluramine challenge are influenced by exposure to chronic social stress in adult male cynomolgus macaques. Psychoneuroendrocrinology, 1, 1–11.

Brammer, G., Raleigh, M.J., Ritvo, E.R., Geller, E., McGuire, M.T. & Yuwiler, A. (1991 Fenfluramine effects on serotonergic measures in vervet monkeys. Pharmacology, Biochemistry, and Behavior, 40, 267–271.

Brammer, G.L. (1994) Blood assay procedure for tyrosine, serotonin, and tryptophan. Unpublished manuscript from the Neurobiochemistry Laboratory of the West Los Angeles VA Medical Center.

Brewerton, T.D., Berrettini, W., Nurnberger, J., Linnoila, M. (1988). Analysis of seasonal fluctuations of CSF monoamine metabolites and neuropeptides in normal controls: Findings with 5HIAA and HVA. Psychiatry Research, 23, 257–265.

Brown, G.L., Ebert, M.H., Goyer, P.F., Jimerson, D.C., Klein, W.J., Bunney, W.E., & Goodwin, F.K. (1982). Aggression, suicide, and serotonin: Relationships to CSF amine metabolites. American Journal of Psychiatry, 139, 741–746.

Brown, G.L., Goodwin, F.K., & Bunney, W.E.,Jr. (1982). Human aggression and suicide: Their relationship to neuropsychiatric diagnoses and serotonin metabolism. Advances in Biochemical Psychopharmacology, 34, 287–307.

Brown, C.S., Kent, T.A., Bryant, S.G., Gevedon, R.M., Campbell, J.L., Felthous, A.R., Barratt, E.S., & Rose, R.M. (1989). Blood platelet uptake of serotonin in episodic aggression. Psychiatry Research, 27, 5–12.

Cases, O., Seif, I., Grimsby, J., Gaspar, P., Chen, K., Pournin, S., Muller, U., Aguet, M., Babinet, C., Shih, J.C., & De Maeyer, E. (1995). Aggressive behavior and altered amounts of brain serotonin and norepinephrine in mice lacking MAOA. Science, 268, 1763–1766.

Chamberlain, B., Ervin, F.R., Pihl, R.O., & Young, S.N. (1987). The effect of raising or lowering tryptophan levels on aggression in vervet monkeys. Pharmacology, Biochemistry, and Behavior, 28, 503–510.

Coccaro, E.F. (1989). Central serotonin and impulsive aggression. British Journal of Psychiatry, 155 (supl. 8), 52–62.

Coccaro, E.F., Siever, L.J., Klar, H.M., Maurer, G., Cochrane, K., Cooper, T.B., Mohs, R.C., & Davis, K.L. (1989). Serotonergic studies in patients with affective and personality disorders. Correlates with suicidal and impulsive aggressive behavior. Archives of General Psychiatry, 46, 587–599.

Coccaro, E.F., Silverman, J.M., Klar, H.K., Horvath, H.B., & Siever, L.J. (1994). Familial correlates of reduced central serotonergic system function in patients with personality disorders. Archives of General Psychiatry, 51, 318–324.

Cohen, J. (1988). Statistical power analysis for the behavioral sciences. Hillsdale, NJ: Erlbaum .

Cohen, P. & Cohen, J. (1984). The clinician's illusion. Archives of General Psychiatry, 41, 1178–1182.

Cook, E.H.,Jr., Arora, R.C., Anderson, G.M., Berry-Kravis, E.M., Yan, S., Yeoh, H.C., Sklena, P.J., Charak, D.A., & Leventhal, B.L. (1993). Platelet serotonin studies in hyperserotonemic relatives of children with autistic disorder. Life Sciences, 52, 2005–2015.

Cook, E.H.,Jr., Fletcher, K.E., Wainwright, M., Marks, N., Yan, S., & Leventhal, B.L. (1994). Primary structure of the human platelet serotonin 5-HT2a receptor: Identity with frontal cortex serotonin 5-HT2a receptor. Journal of Neurochemistry, 63, 465–469.

Depue, R.A. & Spoont, M.R. (1987). Conceptualizing a serotonin trait: A dimension of behavioral constraint. In J. Mann & M. Stanley (Eds.). The psychobiology of suicidal behavior New York: Academy of Sciences.

Eichelman, B. (1993). Bridges from the animal laboratory to the study of violent or criminal individuals. In S. Hodgins (Ed.). Mental disorder and crime (pp. 194–207). Newbury Park, CA: Sage.

Elley, W.B. & Irving, J.B. (1985). The Elley-Irving socio-economic index 1981 census revision. New Zealand Journal of Educational Studies, 20, 115–128.

Elliott, D.S., Huizinga, D., & Morse, B. (1986). Self-reported violent offending: A descriptive analysis of juvenile violent offenders and their offending careers. Journal of Interpersonal Violence, 1, 472–514.

Elliott, D.S. & Huizinga, D. (1989). Improving self-reported measures of delinquency. In M.W. Klein (Ed.). Cross-national research in self-reported crime and delinquency (pp. 155–186). Dordrecht: Kluwer Academic Publisher.

Fernstrom, J.D. & Wurtman, R.J. (1971). Brain serotonin content: Physiological dependence on plasma tryptophan levels. Science, 173, 149–152.

Gardner, D.L., Lucas, P.B., & Cowdry, R.W. (1990). CSF metabolites in borderline personality disorder compared with normal controls. Biological Psychiatry, 28, 247–254.

Geller, E., Ritvo, E.R., Freeman, B.J., & Yuwiler, A. (1982). Preliminary observations on the effect of fenfluramine on blood serotonin and symptoms in three autistic boys. New England Journal of Medicine, 307, 165–169.

Given, M.B. & Longenecker, G.L. (1985). Characteristics of serotonin uptake and release by platelets. In G.L. Longenecker (Ed.). The platelets: Physiology and pharmacology (pp. 463–479). NY: Academic Press.

Hanna, G., Yuwiler, A., & Cantwell, D.P. (1991). Whole blood serotonin in juvenile obsessive-compulsive disorder. Biological Psychiatry, 29, 738–744.

Higley, J.D., Suomi, S.J., & Linnoila, M. (1991). CSF monoamine metabolite concentrations vary according to age, rearing, and sex, and are influenced by the stressor of social separation in rhesus monkeys. Psychopharmacology, 103, 551–556.

Higley, J.D., Thompson, W.W., Champoux, M., Goldman, D., Hasert, M.F., Kraemer, G.W., Scanlan, J.M., Suomi, S.J., & Linnoila, M. (1993). Paternal and maternal genetic and environmental contributions to cerebrospinal fluid monoamine metabolites in Rhesus monkeys (Macaca mulatta). Archives of General Psychiatry, 50, 615–623.

Hur, Y. & Bouchard, T. (1995). Genetic influences on perceptions of childhood family environment: A reared apart twin study. Child Development, 66, 330–345.

Kremer, H.P.H., Goekoop, J.G., & Van Kempen, G.M.J. (1990). Clinical use of the determination of serotonin in whole blood. Journal of Clinical Psychopharmacology, 10, 83–87.

Kruesi, M.J.P., Rapoport, J.L., Hamburger, S., Hibbs, E., Potter, W.Z., Lenane, M., & Brown, G.L. (1990). CSF monoamine metabolites, aggression, and impulsivity in disruptive behavior disorders of children and adolescents. Archives of General Psychiatry, 47, 419–426.

Kruesi, M., Hibbs, E., Zahn, T., Keysor, C.S., Hamburger, S.D., Bartko, J.J. Rapaport, J.L. (1992). A 2-year prospective follow-up study of children and adolescents with disruptive behavior disorders: Prediction by CSF 5HIAA, HVA, and autonomic measures? Archives of General Psychiatry, 49, 429–435.

Linnoila, M., Virkkunen, M., Scheinin, M., Nuutila, A., Rimon, R., & Goodwin, F.K. (1983). Low CSF 5HIAA concentration differentiates impulsive from nonimpulsive violent behavior. Life Sciences, 33, 2609–2614.

Lynam, D., Moffitt, T.E., & Stouthamer-Loeber, M. (1993). Explaining the relation between IQ and delinquency: Class, race, test motivation, school failure or self-control. Journal of Abnormal Psychology, 102, 187–196.

Magdol, L., Moffitt, T.E., Caspi, A., Newman, D.L., Fagan, J., & Silva, P.A. (In press). Gender differences in partner violence in a birth-cohort of 21-year-olds: Bridging the gap between clinical and epidemiological approaches. Journal of Consulting and Clinical Psychology.

Mann, J.J., McBride, P.A., Anderson, G.M., & Mieczkowski, T.A. (1992). Platelet and whole blood serotonin content in depressed inpatients: Correlations with acute and life-time psychopathology. Biological Psychiatry, 32, 243–257.

Mednick, S.A. (1978). Berkson's fallacy and high-risk research in schizophrenia. In L.C. Wynne & R.L. Cromwell (Eds.). The nature of schizophrenia New York, NY: Wiley

Moos, R. & Moos, B. (1981). Family environment scale manual. Palo Alto, CA: Consulting Psychologists Press.

Moss, H.B., Yao, J.K., & Panzak, G.L. (1990). Serotonergic responsivity and behavioral dimensions in antisocial personality disorder with substance abuse. Biological Psychiatry, 28, 325–338.

Newman, D.L., Moffitt, T.E., Caspi, A., Magdol, L., Silva, P.A., & Stanton, W. (1996). Psychiatric disorder in a birth cohort of young adults: Prevalence, comorbidity, clinical significance, and new cases incidence from age 11 to 21. Journal of Consulting and Clinical Psychology, 64, 552–562.

Nielson, D.A., Goldman, D., Virkkunen, M., Tokola, R., Rawlings, R., & Linnoila, M. (1994). Suicidality and 5-Hydroxyindoleacetic acid concentration associated with a tryptophan hydroxylase polymorphism. Archives of General Psychiatry, 51, 34–38.

O'Keane, V., Moloney, E., O'Neill, H., O'Conner, A., Smith, C., & Dinan, T.G. (1992). Blunted prolactin responses to d-Fenfluramine in sociopathy: Evidence for subsensitivity of central serotonergic function. British Journal of Psychiatry, 160, 643–646.

Parnicky, J.J., Williams, S., & Silva, P.A. (1985). Family environment scale: A Dunedin (New Zealand) pilot study. Australian Psychologist, 20, 195–204.

Pletscher, A. (1978). Platelets as models for monoaminergic neurons. In M.B.H. Youdim, W. Lovenberg, D.F. Sharman & J.R. Lagnado (Eds.). Essays in neurochemistry and neuropharmacology, Vol. 3 (pp. 49–101). New York: Wiley.

Pliszka, S.R., Graham, A.R., Rogeness, G.A., Renner, P., Sherman, J., & Broussard, T. (1988). Plasma neurochemistry in juvenile offenders. Journal of the American Academy of Child and Adolescent Psychiatry, 27, 588–594.

Plomin, R., McClearn, G.E., Pedersen, N.L., Nesselroade, J.R., & Bergeman, C.S. (1988). Genetic influence on childhood family environment perceived retrospectively from the last half of the life span. Developmental Psychology, 24, 738–745.

Plutchik, R. & Van Praag, H. (1989). The measurement of suicidality, aggressivity and impulsivity. Progress in Neuro-Psychopharmacology and Biological Psychiatry, 13, S23-S34.

Pucilowski, O. & Kostowski, W. (1983). Aggressive behavior and the central serotonergic systems. Behavior & Brain Research, 9, 33–48.

Quinton, D., Pickles, A., Maughan, B., & Rutter, M. (1993). Partners, peers, and pathways: Assortative pairing and continuities in conduct disorder. Development and Psychopathology, 5, 763–783.

Raine, A. (1993). The psychopathology of crime. NY: Academic Press.

Raleigh, M.J., McGuire, M., Brammer, G.L., & Yuwiler, A. (1984). Social and enviromental influences on blood serotonin concetrations in monkeys. Archives of General Psychiatry, 41, 405–410.

Ransohoff, D.F. & Feinstein, A.R. (1978). Problems of spectrum and bias in evaluation of the efficacy of diagnostic tests. New England Journal of Medicine, 299, 926–930.

Rasmusson, A.M., Riddle, M.A., Leckman, J.F., Anderson, G.M. & Cohen, D.J. (1990). Neurotransmitter assessment in neuropsychiatric disorders of childhood. In S.I. Deutsch, A. Weizman & R. Weizman (Eds.). Application of basic neuroscience to child psychiatry (pp. 33–59). NY: Plenum.

Ritvo, E.R., Yuwiler, A., Geller, E., Ornitz, E.M., Saeger, K., & Plotkin, S. (1970). Increased blood serotonin and platelets in early infantile autism. Archives of General Psychiatry, 23, 566–572.

Ritvo, E.R., Yuwiler, A., Geller, E., Plotkin, S., Mason, A., & Saegar, K. (1971). Maturational changes in blood serotonin levels and platelet counts. Biochemical Medicine, 5, 90–96.

Robins, L.N. & Regier, D.A. (1991). Psychiatric disorders in America. New York, NY: The Free Press.

Rogeness, G.A., Hernandez, J.M., Macedo, C.A., & Mitchell, E.L. (1982). Biochemical differences in children with CD socialized and undersocialized. American Journal of Psychiatry, 139, 307–311.

Rogeness, G.A., Javors, M.A., & Pliska, S.R. (1992). Neurochemistry and child and adolescent psychiatry. Journal of the American Academy of Child and Adolescent Psychiatry, 31, 765–781.

Sampson, R.J. & Laub, J.H. (1993). Crime in the making. Cambridge: Harvard University Press.

Schalling, D. (1993). Neurochemical correlates of personality, impulsivity, and disinhibitory suicidality. In S. Hodgins (Ed.). Mental disorder and crime (pp. 208–226). Newbury Park, CA: Sage.

Siefert, W.E., Foxx, J.L., & Butler, I.J. (1980). Age effects on dopamine and serotonin metabolite levels in CSF. Annals of Neurology, 8, 38–42.

Silva, P.A. (1990). The Dunedin multidisciplinary health and development study: A fifteen year longitudinal study. Paediatric and Perinatal Epidemiology, 4, 96–127.

Soubrie, P. (1986). Reconciling the role of central serotonin neurons in human and animal behavior. The Behavioral and Brain Sciences, 9, 319–365.

Spoont, M. (1992). Modulatory role of serotonin in neural information processing: Implications for human psychopathology. Psychological Bulletin, 112, 330–350.

Stahl, S.M. (1985). Platelets as pharmacologic models for the receptors and biochemistry of monoaminergic neurons. In G.L. Longenecker (Ed.). The platelets: Physiology and pharmacology (pp. 307–340). NY: Academic Press.

Stoff, D.M., Ieni, J., Friedman, E., Bridger, W.H., Pollock, L., & Vitiello, B. (1991). Platelet 3H-Imipramine binding, serotonin uptake, and plasma alpha1 acid glycoprotein in disruptive behavior disorders. Biological Psychiatry, 29, 494–498.

van Kammen, D.P. (1987). 5-HT, a neurotransmitter for all seasons? Biological Psychiatry, 22, 1–3.

Virkkunen, M., Nuutila, A., Goodwin, F.K., & Linnoila, M. (1987). CSF monoamine metabolite levels in male arsonists. Archives of General Psychiatry, 44, 241–247.

Virkkunen, M., De Jong, J., Bartko, J., Goodwin, F.K., & Linnoila, M. (1989). Relationship of psychobiological variables to recidivism in violent offenders and impulsive fire setters. A follow-up study. Archives of General Psychiatry, 46, 600–603.

Virkkunen, M. & Linnoila, M. (1993). Serotonin in personality disorders with habitual violence and impulsivity. In S. Hodgins (Ed.). Mental disorder and crime (pp. 227–243). Newbury Park, CA: Sage.

Virkkunen, M., Kallio, E., Rawlings, R., Tokola, R., Poland, R.E., Guidotti, A., Nemeroff, C., Bissette, G., Kalogeras, K., Karonen, S., & Linnoila, M. (1994a). Personality profiles and state aggressiveness in Finnish alcoholic, violent offenders, fire starters, and healthy volunteers. Archives of General Psychiatry, 51, 28–33.

Virkkunen, M., Rawlings, R., Tokola, R., Poland, R.E., Guidotti, A., Nemeroff, C., Bissette, G., Kalogeras, K., Karonen, S., & Linnoila, M. (1994b). CSF biochemistries, glucose metabolism, and diurnal activity rhythms in alcoholic, violent offenders, fire setters, and healthy volunteers. Archives of General Psychiatry, 51, 20–27.

Von Hahn, H.P., Honegger, C.G., & Pletscher, A. (1980). Similar kinetic characteristics of 5-hydroxytryptamine binding in blood platelets and brain membranes of rats. Neuroscience Letters, 20, 319–322.

Wechsler, D. (1974). Manual for the Wechsler Intelligence Scale for Children - Revised. New York, NY: Psychological Corporation.

Wirz-Justice, A., Lichtsteiner, M., & Freer, H. (1977). Diurnal and seasonal variations in human platelet serotonin in man. Journal of Neural Transmission, 41, 7–15.

Yuwiler, A., Brammer, G.L., Morley, J.E., Raleigh, M.J., Flannery, J.W., & Geller, E. (1981). Short-term and repetitive administration of oral tryptophan in normal men. Archives of General Psychiatry, 38, 619–626.

Yuwiler, A., Plotkin, S., Geller, E., & Ritvo, E.R. (1970). A rapid accurate procedure for the determination of serotonin in whole human blood. Biochemical Medicine, 2, 426–436.

HORMONES–CONTEXT INTERACTIONS AND ANTISOCIAL BEHAVIOR IN YOUTH[*]

Elizabeth J. Susman and Angelo Ponirakis

Department of Biobehavioral Health
The Pennsylvania State University
University Park, Pennsylvania 16802

1. INTRODUCTION

The notion that hormones affect behavior has been around for centuries. Berthold's 1849 famous experiment with roosters showed that removing the source of testosterone (T) decreased fighting, crowing, and mating behavior (McEwen & Schmeck, 1994). When T was restored, precastration behaviors and interests were restored leading to the conclusion that hormones have powerful effects on behavior. Unfortunately, since Berthold's famous experiment, the specific mechanisms whereby hormones affect behavior remain unknown. What is known is that hormone concentrations, sex steroids in particular, are related to a diverse array of emotions and antisocial and aggressive behaviors. In addition to the growing body of empirical literature, theoretical models for considering the effects of hormones on behavior and the moderators of hormones and behavior in youth now are beginning to be articulated (Brooks-Gunn, Graber & Paikoff, 1994; Susman, Worrall, Murowchick, Frobose, & Schwab, in press). Hormones of both gonadal and adrenal origin are considered in these models. Earlier studies assessed only the associations between hormones and behavior whereas studies now consider the importance of the social context in which hormone-behavior interactions occur, as well as the developmental status of the individuals.

The purpose here is to present findings linking hormones of the hypothalamic-pituitary-adrenal (HPA) and hypothalamic-pituitary-gonadal (HPG) axes to antisocial behavior in youth. Our perspective is that hormone-behavior links can be meaningfully interpreted only in relation to the context in which hormones are assessed. Contexts of development are defined broadly ranging from exogenous macro-sociocultural influences (e.g., neigh-

[*] The research reported here was supported in part by grants RO1 HD26004 and P20 HD29356, National Institute of Child Health and Human Development, and the John D and Catherine T. McArthur Foundation and the National Institute of Justice and the intramural programs of the National Institute of Child Health and Human Development and the National Institute of Mental Health.

Biosocial Bases of Violence, edited by Raine *et al.*
Plenum Press, New York, 1997

borhoods) to endogenous physiological states (e.g., reproductive status and gender). These physiological, psychological, and social contexts may moderate or mediate the effects of hormone-behavior interactions. The discussion that follows covers three issues inherent to the integration of biological, psychological, and contextual processes in antisocial behavior. The chapter begins with a brief overview of hormone action and models for conceptualizing hormone-behavior research. Empirical findings connecting hormones of the adrenal and gonadal axes and antisocial behavior in various contexts are then presented primarily for children and youth. The manuscript concludes with challenges and directions for future research.

2. MECHANISMS AND MODELS OF HORMONE ACTION

Hormones are chemical messengers that act on the cells that produce them (autocrine), act on adjacent cells (paracrine), or are released from one type of cell and act on distant receptors on different types of target cells. The most general of cellular hormone processes is the initial step of binding of the hormone to a receptor in the target tissue (Hedge, Colby, & Goodman, 1987). Hormones may bind to their receptor by either passing through the cell membrane to receptors located in the cytoplasm or nucleus of the cell (lipophilic hormones), or by binding to receptors on the cell surface (protein hormones). After binding, a series of events follows resulting in the activation of messenger ribonucleic acid (mRNA) and enzymes that either decrease or increase a physiological response. Termination of a hormone action usually requires dissociation of the hormone from its receptor.

With regard to hormone physiology and behavior, hormone receptors activate genes that control neurotransmitters with the potential for affecting behavior. Additional explanations of hormone action include (Beach, 1975; Buchanan, Eccles & Becker, 1992): (1) Hormone concentrations can alter structures essential for carrying out certain behaviors. Increases in physical size, strength, and endurance at puberty, for instance, can increase the probability that aggressive behavior will occur. (2) Hormone concentrations can alter flexible peripheral systems, such as the sensory systems, thereby increasing or decreasing the potential for instigating a behavior. (3) Hormone concentrations may influence central nervous system (CNS) processes, such as autonomic nervous system and neurotransmitter action. These actions can potentiate or dampen receptor binding. (4) Hormone concentrations may affect CNS processes involved in emotional regulation. Hormone receptor sites for steroid hormones are located in many brain regions: hypothalamus, amygdala, septal nucleus, and hippocampus (Buchanan et al., 1992). Given the diverse sites and actions of hormones in the CNS, hormones can be viewed as both causes and consequences of antisocial behavior.

2.1. Models of Hormones and Behavior

Models recently have been formulated for considering hormone, behavior, and contextual interactions. Specifically, four general models (Susman, Worrall, et al., in press; Susman, 1996) and specific models of hormones and affective expression (Brooks-Gunn et al., 1994) exist for linking hormones, behavior and contextual factors. The models appear in Figure 1.

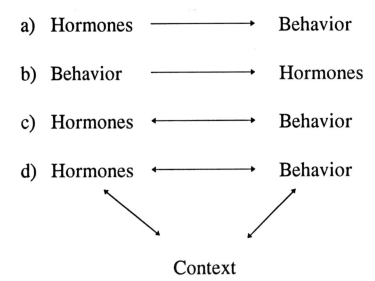

Figure 1. Models for considering hormone, behavior, and contextual interactions.

a. *Hormones as influences on behavior.* The research supporting hormones as causal influences on behavior is based on early animal model studies showing organizational and activational effects of hormones on behavior at different periods of development in subhuman species (Phoenix, Goy, Gerall, & Young, 1959). Higher levels of prenatal androgens are presumed to organize the brain in such a way so as to affect later behavior. High concentrations of androgens in males are considered to be a contributing factor to greater aggressive behavior in males than in females. Evidence for organizational influences of hormones on humans was derived from studies of children born to mothers with known alterations in the prenatal endocrine milieu (Money & Ehrhardt, 1972). Specifically, Money's work showed that prenatal exposure to high concentrations of androgens secondary to congenital adrenal hyperplasia (CAH) shifts behavior in children in a masculine direction. Recent well-designed studies (Berenbaum & Snyder, 1995) extend these early findings by showing that girls with CAH prefer to play with boys' toys.

b. *Behavior as influences on hormones.* The perspective that behavior affects hormone concentrations is illustrative of modern biological perspectives on development. The environment in which organisms develop affects both biological structures and functions (see Mayr, 1988). Studies showing hormone changes as a consequence of behavior or experiences in adult males began in the early 1980s. Mazur and Lamb (1980) showed that T concentrations rise in males before a tennis match, as if in anticipation of the competition. Booth and colleagues (Booth, Shelley, Mazur, Tharp, & Kittok, 1989) expanded this paradigm by showing that the experience of winning a tennis match elevated concentrations of T in the winners but not in the losers. Cortisol concentrations were not affected by winning or losing but higher seeded players had higher cortisol concentrations than unseeded players, which declined as the season progressed. These two sets of observations provided the first convincing findings in humans that the context of interactions, the experience of winning or losing, affects hormone concentrations.

c. *Hormone-behavior interactions.* The concept of hormone-behavior interactions assumes that change in one domain influences change in the other domain. This perspective is consistent with Raine and colleagues' (Raine, Brennan, Farrington & Mednick, in this volume) sequential biosocial effect. Increases or decreases in hormones may precipitate increases in aggression. Aggressive behavior then may further increase or decrease hormone concentrations, which further increases aggressive behavior. Sequential biosocial effects of hormones and behavior illustrate the phenomenon that behavior and biology are fused in ontogeny (Magnusson & Cairns, 1992). Development in one behavioral domain influences development in a physiological domain and vice versa. Sapolsky's work (1982) demonstrated these interaction influences when he reported that a high ranking baboon, after certain experiences, has a different T response than a submissive animal. Higher ranked baboons increased in T concentrations after a challenge experience. Higher concentrations of T may then instigate even greater social dominance. Lower ranked males reacted with lower concentrations of T that may instigate more social submission. In this instance, experiences of social dominance interact with hormone responses, that then change behavior. The mutual hormone-behavior interaction model has not been applied to humans.

d. *Hormones, behavior, and context.* The role of physiological and social contextual influences on both behavior and hormones addresses the critical issue of mechanisms whereby events in the endogenous or exogenous environment affect development. The contexts of development are proposed to influence secretion by the adrenals, gonads, and thyroid by interacting with genes to alter the expression of genetic makeup thereby affecting CNS and peripheral cells (McEwen et al., 1992). Events in the environment may influence secretion of hormones through arousal tendencies associated with novel and challenging experiences. In brief, contextual factors may both moderate or mediate hormone-behavior interactions.

2.2. Hormones and Antisocial Behavior in Human Adolescents

Studies of hormones and behavior in human adolescents have historical roots in animal model studies of aggression, primarily, studies with rodents showing the influence of androgens on aggressive behavior. Higher concentrations of testosterone, for instance, lead to increases in attach behavior in rodents (Brain, 1994). The implicit model for studies of hormones and aggressive behavior in youth is based on the assumption that changes in hormone concentrations at puberty increase the likelihood of increases in antisocial behavior (Susman & Petersen, 1992). Testosterone is the hormone most often associated with aggressive and antisocial behavior. Estrogens may also be involved in aggressive behavior since a portion of T is aromatized to estrogen. In rodents, the estrogen produced from aromatization of T to estrogen mediates androgenic activity but its role in humans is less well understood.

The association between T concentrations and aggressive behavior in adolescents first was reported for Swedish, 15 to 17-year-old males (Olweus, Mattson, Schalling, & Low, 1980). Higher concentrations of T were related to self reports of physical and verbal aggression resulting from provocation. The causal role of T in provoked and unprovoked aggression also was examined (Olweus, Mattson, Schalling, & Low, 1988). The path model findings indicated that from grade six to grade nine, there is a direct causal effect of T on provoked aggressive behavior. Unprovoked aggressive behavior was less directly influenced by T. The indirect effect of T on unprovoked aggressive behavior appeared to be mediated by low frustration tolerance at grade nine. The overall findings indicated that T has both direct and indirect effects on aggressive behavior.

The links between T and antisocial behavior have been less consistently reported in recent studies of adolescents. Specifically, testosterone and aggressive behavior associations are less consistently reported in studies of healthy pubertal age adolescents (Brooks-Gunn & Warren, 1989; Inoff-Germain et al., 1988; Nottelmann et al., 1987; Nottelmann, Inoff-Germain, Susman, & Chrousos, 1990; Susman et al., 1985, 1987, Susman, Dorn, & Chrousos, 1991). A collaborative NIMH-NICHD study examining hormones and behavior at puberty will be used to illustrate testosterone-aggressive behavior findings. The sample consisted of 108 healthy adolescents: 56 boys, ages 10–14 years, and 52 girls, ages 9–14 years. There were three assessments, at 6-month intervals. The biological measures include a physical examination to assess stage of pubertal maturation based on genital and pubic hair development in boys and breast and pubic hair development in girls (Tanner criterion) (Marshall & Tanner, 1969, 1970) and hormone concentration assessments of luteinizing hormone (LH), follicle stimulating hormone (FSH), testosterone (T), estradiol (E_2), dehydroepiandrosterone (DHEA), dehydroepiandrosterone sulphate (DHEAS), androstenedione (Δ 4-A), testosterone-estradiol binding globulin (TeBG), cortisol, and cortisol binding globulin (CBG). A full range of gonadal and adrenal hormones and their binding proteins were included since the study was exploratory and at that time it was not yet known which hormones were related to psychological processes at puberty. The psychological domains assessed included: self and parent reports and observations of dominance and aggressive behavior, self-image, moods, cognition, behavior problems, and psychiatric disorders.

One aspect of the study focused on relations between adrenal and gonadal axis hormones, emotional dispositions, and aggressive attributes (Susman et al., 1987). The hypothesis was that higher concentrations of androgens (primarily T, and to a lesser extent the adrenal androgens and estrogen) are related to aggressive behavior. The general pattern of findings, for boys, for hormones and aggressive attributes, was that higher scores on delinquent behavior problems were related to lower concentrations of E_2 and higher concentrations of Δ4-A. Higher scores on rebellious attitude were related to higher concentrations of LH, lower concentrations of FSH, and higher concentrations of DHEA. When emotional dispositions and aggressive attributes were considered together with hormones, the following pattern of findings emerged: Higher scores on delinquent behavior problems were related to lower T/E2 ratios (indicating lower T), lower concentrations of TeBG, and lower concentrations of DHEAS. There were no significant findings for T. There also were no significant findings for girls. This pattern of findings appears in Figure 2.

In a more extensive analysis assessing psychosocial functioning and hormone concentrations, chronological age, pubertal status, and height and weight, similar findings were reported (Nottelmann et al., 1990). Adjustment was assessed on the Offer Self-Image Questionnaire for Adolescents (Offer, Ostrov, & Howard, 1977), which assesses both emotions and behavior problems. For boys, a profile of higher Δ4-A concentrations (HPA axis) and either lower T or lower T to E_2 ratio (HPG axis) was associated with greater degrees of adjustment problems. The relationships held when hormone concentrations were controlled for chronological age, pubertal state or both. For girls, the findings were relatively weak and somewhat inconsistent across hormones and self-image subscales.

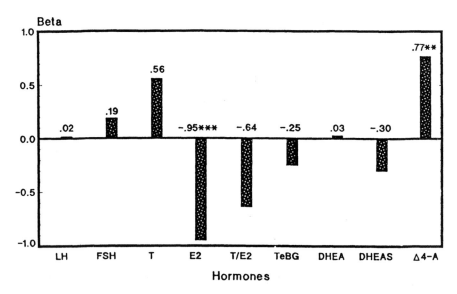

Figure 2. Relations between adrenal and gonadal axes hormones and aggressive attributes.

The findings showing that lower concentrations of gonadal steroids and higher adrenal androgens were related to antisocial behavior were interpreted in relation to two different pubertal processes. During the pubertal period, there may be heightened sensitivity of the gonadal axis to stress-related secretion of hormones from the adrenal axis, resulting in suppression of the gonadal axis (Nottelmann et al., 1987; Nottelmann et al., 1990). A second interpretation was that higher adrenal and lower gonadal hormone concentrations may be a reflection of a predisposition, in some individuals, to heightened physiological reactivity to environmental challenges, which undoubtedly are plentiful during adolescence.

2.2.1. Interaction of HPA-HPG Axes. The pattern of lower concentrations of sex steroids and higher concentrations of adrenal androgens, in conjunction with lower pubertal stage for age is consistent with a profile of hormones found under conditions of stress. Endocrine glands interact or cross talk with one another (Brain, 1989; Brain & Susman, in press) such that reproductive hormones interact with stress hormones. Hormones of the HPA axis secreted under stressful conditions can suppress the HPG axis through both central and peripheral mechanisms (Chrousos & Gold, 1992; Johnson, Kamilaris, Chrousos, & Gold, 1992). Corticotropin releasing hormone (CRH) released in the hypothalamus during stressful experiences inhibits gonadotropin secretion (Sapolsky, 1991) resulting in the suppression of gonadal steroids. These effects can occur as a result of a direct effect on the brain or through the mediation of the opiate β-endorphin. Glucocorticoids released in the periphery from the adrenal cortex during stress can affect the HPG axis by suppressing gonadotropin releasing hormone (GnRH) from the hypothalamus. Glucocorticoids can have an inhibitory effect directly at the gonadal level as well (Johnson et al., 1992; Sapolsky, 1991). These same stress-related processes may suppress the gonadal axis during puberty even under minor stressful circumstances of adolescence (See Nottelmann et al., 1990; Susman, Nottelmann, Dorn, Gold, & Chrousos, 1989). The links between stressful life experiences of adolescents and their potential effects on pubertal timing illustrate the importance of considering interactions between endocrine axes and the contexts of development.

Contexts that involve interpersonal dominance and competition are associated with gonadal and adrenal hormone concentrations in both animals and humans. Systematic observational ratings of dominance (attempts to control) and aggressive behavior of young adolescents interacting with their parents was the focus of another aspect of the NIMH-NICHD study (Inoff-Germain et al., 1988). For boys, *expression of modulated anger* was associated with higher concentrations of DHEA and a lower level of DHEAS. *Shows no sign of anger when aggressed against* was related to a lower T/E_2 ratio and a lower level of TeBG. For girls, there were more significant findings than for boys. *Expresses modulated anger, acts defiantly to mother and father*, and *shows anger toward mother* were related to higher concentrations of E_2. *Tries to dominate mother, tries to dominate father*, and *shows anger toward father* were related to high concentrations of $\Delta4$-A. *Is explosive* was related to higher concentrations of FSH. In brief, concentrations of both adrenal and gonadal hormones that increase at puberty were associated with adolescent behavioral expressions of anger and power while interacting with their parents. The findings were stronger for girls than for boys. Noteworthy is that E_2 and not T was linked to interpersonal dominance. The interpretation, especially with regard to E_2, was that pubertal-age girls may be sensitive to changes in estrogen levels as generally is thought to be true for changes in T in boys. Testosterone was not associated with anger or dominance behavior in either boys or girls.

In another study of pubertal age boys, T was related to peer rating of dominance. Boys were rated by unfamiliar same-age peers for their toughness and leadership. Boys rated as tough had significantly higher concentrations of T than the not tough or leader boys (Schaal, Tremblay, Soussignan & Susman, in press). Peers appeared to be sensitive to a dominant style of leadership in the dominant boys.

A measurement issue may help to explain the relative absence of findings for T in some studies but not in others. Total T containing both bound and free T was measured in plasma in the NIMH-NICHD study whereas T was measured in saliva in the Schaal et al. (in press) study. Saliva T is free and is the physiologically active component of T. A more consistent pattern of findings across studies is likely to be identified if saliva is assayed for free T.

2.3. The Stress System: A Potential Mechanism Involved in Hormone-Behavior and Context Relations

In a pioneering study, Rose (1980) showed that short- and-longer term stressors appear to affect hormone concentrations. Since these early studies, the effects of stressors on adrenal androgens and gonadal steroid hormone concentrations in nonclinical studies received only minimal research attention and these studies were done primarily in adult males (e.g., Booth et al., 1989). Adrenal hormones as indices of stress reactivity are known to be associated with infant distress (See Stansbury & Gunnar, 1994), childhood conduct disorder (McBurnett et al., 1991; Scarbo & Kolko), adolescent behavioral distress (Susman, Dorn et al., in press) and older adult adaptation to new living arrangements (Preville et al., 1995). Cortisol and other components of the HPA axis are exquisitely sensitive to environmental perturbations and can be considered mediators of other endogenous hormone systems.

Subsequent to identifying a profile of hormones, in the NIMH-NICHD study, showing that gonadal steroids and adrenal androgens may reflect stressful experiences or stress-related disorders, the physiology of stress and emotions and antisocial behavior has been examined in a number of studies in our lab (Dorn, Susman, & Petersen, 1993; McCool & Susman, 1994; McCool, Dorn, Susman, 1994; Preville et al., in press; Susman, Dorn, et al., in press; Ponirakis, Susman, & Stifter, 1996). The effect of stressful experiences on cortisol concentrations is an instance of the "Behavior-influences on hormones" model mentioned earlier. In one study, individual differences in HPA reactivity to a known stressor, reflected in cortisol concentrations, were examined (Susman, Dorn et al., in press). Changes in cortisol level during a novel and challenging situation, having blood drawn and a physical examination, at the first time of measurement in the NIMH-NICHD study, were examined in relation to concurrent and longitudinal, distress behaviors, behavior problems, and symptoms of anxiety and depression. For the assessment of cortisol reactivity, the first occasion of measurement was considered a novel and challenging situation. Novelty is considered a major precipitant of HPA axis activation (Levine & Weiner, 1989).

Individual differences in cortisol reactivity, assessed by changes in cortisol concentrations across three samples obtained at 40-minute intervals, were found among the adolescents. Some adolescents increased, some did not change, and others decreased in cortisol level during the novel and challenging situation. These three patterns of change were used to construct three reactivity groups. Adolescents were categorized as those who: (1) increased in cortisol concentrations (increase) (2) decreased in concentrations (decrease), and (3) remained stable across samples (stable). The prediction was that adolescents who increased in cortisol level would be more prone to develop affective and behavior problems than adolescents in the stable or decrease groups. The hypothesis was based on the literature on depression and other affective disorders showing that higher concentrations of cortisol are an inherent aspect of the pathophysiology of depression (See Gold, Loriaux & Chrousos, 1988a & b). Symptoms of depression, anxiety and conduct disorder were assessed by the Diagnostic Interview for Screening Children (DISC) (Costello, Edelbrock, Kalas, Dulcan, & Klaric, 1983). Adolescents who increased in cortisol level, the highly reactive adolescents, at the first time of measurement, reported more behavior problems and symptoms of depression one year later than adolescents who did not change or who decreased in cortisol level. Specifically, there was a significant effect for the reactivity group for one of the two categories of conduct disorder behavior problems, (nonaggressive behavior problems) and for the total conduct behavior problems (sum of aggressive and nonaggressive behavior problems). There were no differences between the groups on aggressive behavior problems.

Similar patterns were found for depression. Adolescents who showed increases in cortisol during the novel and challenging situation at the first time of measurement reported significantly more depression symptoms (affective, cognitive, and vegetative) one year later than adolescents in the other groups. The group of adolescents who increased in cortisol had low concentrations of cortisol at the beginning of the lab visit which may reflect low arousal tendencies when confronted with challenging situations. Low arousal in the adolescents who later displayed behavior problems one year later may parallel the low arousal reflected in autonomic nervous system (ANS) reactivity and behavior problems in adolescents (See Raine in this volume). The group of adolescents who decreased in cortisol levels appeared to experience anticipatory anxiety about coming to the laboratory as reflected by decreases in higher cortisol levels, which declined after arrival at the lab. The adolescents who did not change on cortisol levels had higher cortisol levels than the other two groups of adolescents. These adolescents are similar to the inhibited children, described by Kagan and colleagues (Kagan, Resnick & Snidman, 1987) who had high and

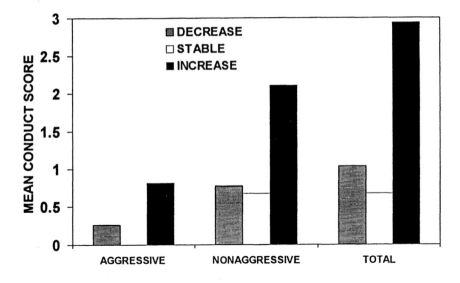

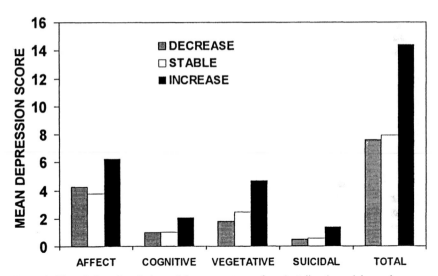

Figure 3. The relation of cortisol reactivity to symptoms of conduct disorder and depression.

stable arousal even in minimally stressful contexts. Difficulties in regulating physiological reactivity in novel and challenging situations appear in the current study to be a risk factor for symptoms of conduct disorder and depression symptoms.

2.4. Reproductive Status as a Context for Hormone and Antisocial Behavior

Reproductive transitions provide periods of development in which to examine hormone-behavior relations because there are rapid changes in both hormones (e.g., sex steroids), emotions, and occasionally mental health problems (Dorn & Chrousos, in press).

Reproductive transitions provide an excellent paradigm for showing the interactive effects of hormones and behavior since hormones tend to rise during some reproductive transitions (i.e., puberty and pregnancy) and fall in others (i.e., menopause) (Dorn & Chrousos, in press). Charting behavior changes in relation to these hormone transitions is a needed effort in research.

The first question to be addressed here regarding reproductive transitions and antisocial behavior focused on the issue of stability of hormone effects across pregnancy. Specifically, the question was: Does antisocial behavior have effects on hormone concentrations over time and how do these effects become translated into behavior? Earlier in this chapter the sequential biosocial effects model (Raine et al., this volume) was discussed, which states that changes in a social factor cause changes in biological functioning, which in turn influences violence. To demonstrate the validity of this model, one must first demonstrate that the biological or social factor or changes in these factors predict across time to the hormone or antisocial outcome. We and others have begun to address this issue by examining the predictive power of emotions and antisocial behavior at one point in time to hormones at a later point in time in adolescents in the puberty and pregnancy reproductive transitions.

Tremblay and colleagues (Schaal et al., in press) addressed this issue in boys with disruptive behavior problems. Disruptive behavior patterns since age six were hypothesized to affect T concentrations at the reproductive transition of puberty. Boys who were stable in their disruptive behavior and were anxious from age six to age 13 were significantly lower in T concentrations than nondisruptive boys. Similarly, Susman and colleagues (Susman, Granger, et al., in press) showed that emotional regulation (anxiety), behavior problems and poor impulse control predicted T, E_2, dehydroepiandrosterone (DHEA), dehydroepiandrosterone sulphate (DHEAS), androstenedione, and cortisol levels predicted behavior problems across a one year period during puberty. Cortisol was the hormone most often predicted by behavior problems or emotional dysregulation. Specifically, in pubertal age girls, externalizing behavior problems on the Child Behavior Checklist (CBCL) (Achenbach, 1991) and anxiety both predicted higher concentrations of cortisol either six or 12 months later. Two things are noteworthy in these findings. First, the direction of the relationship between behavior problems and cortisol was not consistent across occasions of measurement. In some instances the relationship was positive whereas in other instances the relationship was negative. The inconsistence in directionality may reflect the current-day experiences of the adolescents which vary from day to day (Eccles et al., 1988). Second, lower anxiety, as opposed to higher anxiety, was related to higher concentrations of cortisol. In two samples of pregnant adolescents (McCool & Susman, 1994; Susman, Granger et al., in press), depression and anxiety and cortisol were negatively related. We propose that the physiological context of the individual, the reproductive transition of pregnancy in this case, may alter the links between hormones and emotions.

These findings in pubertal-age and pregnant adolescents linking hormones and emotional regulation led us to consider further the HPA axis as a possible mechanism whereby antisocial behavior may affect hormone concentrations or hormone concentrations may affect antisocial behavior. As discussed above, HPA axis functioning is considered an aspect of the pathophysiology of depression (Chrousos & Gold, 1992) but the role of the HPA axis in conduct disorder is unknown. Since there is comorbidity in depression and conduct disorder in some adolescents, HPA axis functioning may be implicated in conduct disorder as well. To assess the possible longitudinal interactive effects of antisocial behavior on hormone concentrations and hormone concentration on antisocial behavior, we chose two groups of adolescent girls. The sample (referred to above) consisted of 78 pregnant adolescents who were enrolled in a longitudinal study prior to 16 weeks gestation. An age and socioeconomic status matched group of 57 nonpregnant adolescents constituted a comparison group. The pregnant adolescents were assessed at ≤16 gestation (Time 1), 30–32 weeks gestation (Time 2), and two weeks postpartum (Time 3). The comparison adolescents were seen at comparable time periods. The measures were the Diagnostic Interview Schedule for Children (Schwab-Stone et al., 1993) and saliva cortisol. The index of cortisol concentrations used in the analysis was the mean of five saliva samples obtained at 20 minute intervals beginning at 8:30 a.m.

2.4.1. Conduct Disorder Symptoms: Predictors of Cortisol. To assess the longitudinal effects of conduct disorder symptoms on cortisol we used regression analysis. *Conduct disorder symptoms* at Time 1, Time 2, and Time 3 of measurement in the pregnant and comparison adolescents were entered into a regression model to predict cortisol concentrations at Time 3 of measurement (two weeks postpartum). The overall model was significant, $F(3, 77) = 3.23$, $p = .02$, $R^2 = .12$. The findings appear in Figure 4. Conduct disorder symptoms at Time 3 were positively and significantly related to cortisol at Time 3, $p = .02$. The direction of the relation between conduct symptoms and cortisol was not consistent across pregnancy. Although not significant, fewer conduct symptoms at Time 1 and more symptoms at Time 2 predicted cortisol at Time 3 suggesting that experiences of adolescents that vary across pregnancy may have different effects on later hormone concentrations.

2.4.2. Cortisol: Predictors of Conduct Disorder Symptoms. This analysis tested the hypothesis that hormone concentrations influence a behavior outcome. Cortisol concentra-

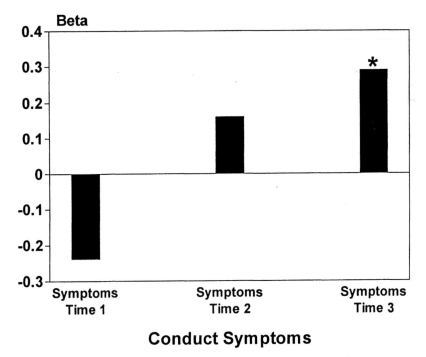

Figure 4. Conduct disorder symptoms predicting cortisol concentrations.

tions at Time 1, Time 2, and Time 3 of measurement were entered into a regression model to predict *number of conduct disorder symptoms* at Time 3 of measurement. The overall model was significant, $F(3, 95) = 2.27$, $p = .08$, $R^2 = .07$. Cortisol at Time 3 was significantly related to number of conduct symptoms, $p = .05$. Higher cortisol was related to a greater number of conduct symptoms. The findings appear in Figure 5.

The same regression model was done with cortisol concentrations at Time 1, Time 2, and Time 3 of measurement predicting *severity of conduct disorder symptoms* at Time 3 of measurement. Severity of conduct disorder symptoms was established by weighting the DISC conduct disorder symptom scores. DISC items are scored 0, 1 or 2. The items scored as 2 were multiplied by 2 to establish the severity of symptoms score. The overall model was significant, $F(3,95) = 4.63$, $p = .004$, $R^2 = .13$. The findings appear in Figure 6. Cortisol at Time 3 was significantly associated with severity of conduct symptoms. Noteworthy is that cortisol concentrations were consistently positively associated with severity of conduct disorder symptoms at all three times of measurement. This positive association is different from the negative association between cortisol and antisocial behavior reported in other studies (see later citations and discussion). The difference in these and the current findings may reflect the pregnancy status of the adolescents in the current sample. Dorn and colleagues (1993) reported cortisol associations that were different during pregnancy than during the nonpregnant state.

To assess the effect of the physiological context of pregnancy on severity of conduct disorder symptoms, a blocking factor of physiological status, pregnant or comparison, was added to test the interactive effects of cortisol and reproductive status and severity of symptoms. This analysis addressed the question posed by Raine et al. (Raine et al., in this volume) regarding the interactions between biological factors and moderating factors in

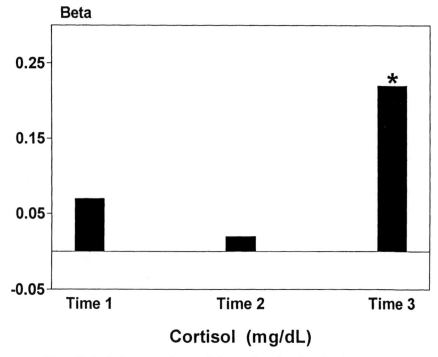

Figure 5. Cortisol concentrations predicting number of conduct disorder symptoms.

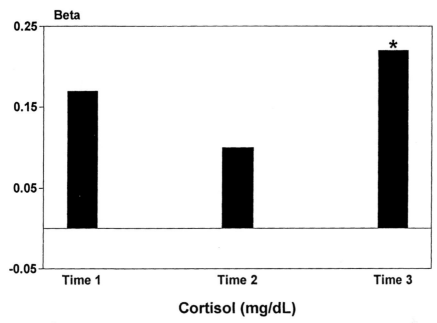

Figure 6. Cortisol concentrations predicting severity of conduct disorder symptoms.

predicting violence. In the current instance, physiological status (pregnant or comparison) was the moderating factor. The overall model was significant, $F (4, 76)= 2.87$, $p = .03$. There was a significant effect for reproductive status, $p = .03$, and an interaction between reproductive status and severity of symptoms, $p = .01$. Pregnant adolescents had more severe symptoms than comparison adolescents. The interaction showed that both the pregnant and comparison groups increase in antisocial behavior from Time 1 to Time 2 and decrease at Time 3 but the comparison adolescents decrease more than the pregnant adolescents. The context of physiological status moderated the effect cortisol had on conduct symptoms such that severity of symptoms declined less slowly in the pregnant than in the comparison group. The physiological context of pregnancy interacted with cortisol resulting in a greater risk for severity of conduct symptoms.

The relatively small N in our current sample precludes using causal models for sorting out the causal effect of cortisol on conduct disorder *symptoms* and the effect of conduct disorder on cortisol concentrations. Nonetheless, statements can be made about causality from the findings cited above. Conduct disorder *symptoms* were more strongly predicted by cortisol than cortisol by conduct disorder symptoms. In addition, cortisol was a stronger predictor of conduct disorder *severity* than cortisol by conduct disorder severity. In adolescents with more severe conduct disorder symptoms, the interaction between cortisol and behavior may have become stable over time as the behavior and the biology of the disorder may have been in place for sometime. Overall the pregnant and comparison adolescents were comparable in the effects of cortisol on *number of conduct disorder symptoms* and the effects of conduct disorder symptoms on cortisol. In contrast, *severity of symptoms* was moderated by reproductive status. The physiological context of pregnancy appears to be an important mediator of hormone-behavior interactions. The consistency of the pattern of findings across the different studies summarized here suggests that HPA

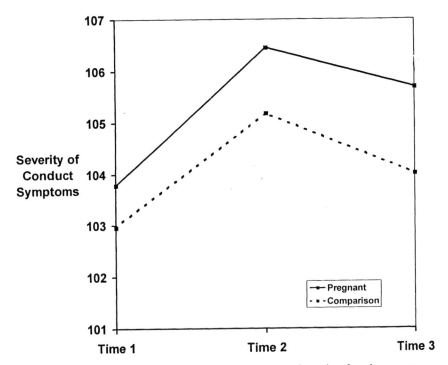

Figure 7. Interactive effect of cortisol and reproductive status and severity of conduct symptoms.

axis dysfunction may be an aspect of the pathophysiology of conduct disorder as well as an aspect of the pathophysiology of depression (Gold, Goodwin, & Chrousos, 1988a & b).

2.5. Corticotropin Releasing Hormone (CRH) and Antisocial Behavior

Human model studies of hormones and behavior rely on peripherally circulating hormone concentrations secreted in the gonads or adrenals as indices of the internal endocrine milieu. Hormone concentrations measured in the periphery may not be an accurate indicator of the actual exposure of the brain to that hormone. The hormones discussed in the preceding chapters are secreted from the pituitary or from target organs (gonads and adrenal glands). Peripherally circulating hormones measured in saliva or blood may undergo multiple conversions prior to or after reaching the brain. These conversions make it difficult to establish the link between a specific hormone concentration in the periphery and an actual brain process. A partial solution to this problem is to measure substances in the periphery that are secreted in the brain and directly released into the periphery. CRH is secreted directly into the periphery from the hypothalamus following a stressful event. CRH is a 41 amino acid peptide (Vale, Spiess, Rivier, & Rivier, 1981) that induces the pituitary corticotropin to release ACTH, which regulates the synthesis of glucocorticoids (e.g., cortisol). CRH also is secreted from the placenta during pregnancy but it does not respond to negative feedback from cortisol allowing concentrations of CRH to rise over the course of pregnancy. Pregnancy is one of the few times during development when CRH can be measured in peripherally circulating blood with current assay technologies. If an association between CRH and behavior problems exists, then we will have additional evidence that the HPA axis is either causally or consequentially related to the development of antisocial behavior.

In the sample of pregnant adolescent described above, CRH was measured in plasma in late first trimester or early second trimester. Conduct disorder symptoms were assessed concurrently with CRH, during the third trimester and 2 weeks postpartum. Lower concentrations of CRH for gestational age were predictive of greater numbers of conduct disorder problems in the postpartum period (Susman, Schmeelk, Worrall, Granger, & Chrousos, 1996). The associations also were negative between CRH and conduct disorder symptoms during late first or early second trimester and during the postpartum period, although they did not reach conventional levels of significance.

These findings demonstrate for the first time that individual differences in CRH concentrations are linked to problem behaviors in a nonclinical sample of adolescents. Past research on CRH included primarily clinically depressed adults (Gold et al., 1988) or adolescents (Dorn et al., in press) so our results are difficult to compare with other findings. The negative association between CRH and conduct disorder symptoms in pregnant adolescents is consistent with a number of studies showing that low cortisol and corticotropin levels are associated with aggressive, antisocial, or violent behavior. Urinary cortisol was low in habitually violent offenders (Virkkunen, 1985) and in males with antisocial personality (Bergman & Brismar, 1994) when contrasted with comparisons. Similarly, cerebral spinal fluid (CSF) corticotropin concentrations were low in alcoholic, impulsive offenders with antisocial personality disorder (Virkkunen et al., 1994). In preadolescent boys, saliva cortisol was associated with greater conduct disorder symptoms counts (Vanyukov et al., 1993). Cortisol levels were lower in sons whose fathers had conduct disorder as children and who subsequently developed antisocial personality compared to cortisol level in sons whose fathers either did not have an Axis I psychiatric disorder or who did not develop antisocial behavior (Vanyukov et al., 1993). In both pubertal-age boys and girls, lower levels of plasma cortisol at the beginning of a stressful situation (a phlebotomy procedure) that increased across the time of the procedure had significantly more symptoms of conduct disorder than adolescents who either did not change in cortisol levels or who decreased in levels across the time of the phlebotomy procedure (Susman, Dorn, et al., in press). Hypothalamic-pituitary-adrenal (HPA) axis hyporesponsivity and aggressive, antisocial, and violent behavior correspond to the cardiovascular system under arousal and aggressive behavior observed in adolescents and adults (Raine, 1996). The collective findings indicate that hyporesponsivity of the HPA axis may be a significant risk factor for antisocial social and aggressive behavior in children, adolescents, and adults.

3. SUMMARY AND CONCLUSIONS

Variations in physiological and social contexts of development are key components to understanding hormonal contributions to violence. Contexts of development appear to influence concentrations of adrenal androgens, gonadal steroids, glucocorticoids (cortisol) and the releasing factor CRH. Changes in these products of the HPA and HPG endocrine axes, in turn, may instigate additional violence. The demonstrated links reviewed above between hormones of adrenal origin, CRH and antisocial behavior suggest that the neuroendocrinology of stress is an inherent aspect of antisocial behavior.

In spite of the consistency of hormone-behavior relations, the specific role of hormones as biosocial influences on violence is extremely difficult to assess as the forms of aggressive, antisocial behavior and violence are difficult concepts to conceptualize and measure. Added to this complexity are the complex integrated endocrine system and neurotransmitters that may predispose toward aggressive behavior (Brain & Haug, 1992). Investigators are further hampered from identifying specific physiological mechanisms and pathways involved in violence since the classic experimental approaches to assessing the effects of hormones on aggressive behavior and the effects of aggressive behavior on hormone concentrations are ruled out on ethical reasons in human model studies. As in Berthold's roosters, the classic experimental endocrine paradigm is to remove the endogenous source of the hormone, observe what transpires, and replace the hormone with exogenous hormones and observe again for changes in behavior. This ablatement-replacement approach has obvious advantages for establishing hormone effects but cannot be used with nonclinical samples of humans. Similarly, the classic approach in experimental psychology is to assess hormone change in highly structured settings such as during experimentally induced stressors (see Preville et al., 1996; Susman, Dorn, et al., in press) Advocated here is the assessment of hormone behavior interactions in naturalistic settings. The probability of being able to assess the interactive influence of hormones and behavior in contexts where adolescents commit violent acts is highly unlikely. Nonetheless, in simulated-natural laboratory settings it is possible to observe the effect of conflict or aggressive behavior on hormone concentrations under different conditions and to observe the effect of different types of contextual rewards/sanctions for aggressive behavior and conflict resolution.

The theoretical models for considering the bidirectional effects of hormones, antisocial behavior, and context now are articulated to the point where they can be useful in sorting out the complex factors involved in the etiology of aggression (See Brooks-Gunn, Graber, Paikoff, 1994; Susman, Worrall et al., in press). Statistical methods and user friendly software also are available for determining hormone-behavior interactions. A popular statistical method for deciphering cause effect relations is the use of structural equation models. LISREL, AMOS and other software programs make these analyses fairly easy to compute. Use of causal statistical models have been used sparingly primarily because of the lack of a well-articulated model of antisocial behavior as well as the cost of accruing a sufficient number of subjects in any one biosocial study of antisocial behavior. Endocrine assays are expensive but the costs of assays can now be brought within a range of cost that is affordable in most studies.

In conclusion, two areas of integration are proposed for future research. First, the integration of endocrine physiological indices with studies of behavior in naturalistic settings is key to understanding the effects of aggressive behavior on hormones and the effects of hormones on aggressive behavior. This integration will require the close collaboration of behavioral and biological scientists. Second, the integration of physiological systems known to be involved in aggressive and violent behavior and the combined influence of these systems is one of the most conceptually and empirically difficult tasks confronting the field. The endocrine system, autonomic nervous system, and neurotransmitter systems have evolved together as well integrated systems. These systems also are known to be involved in aggression or violence in animal and human primates. There is a great need to work collaboratively to integrate these physiological systems in studies of human aggression.

REFERENCES

Achenbach, T. M. (1991). *Manual for the Child Behavior Checklist/4–18 and 1991 Profile*. Burlington, VT: University of Vermont.

Beach, F. A. (1975). Behavioral endocrinology: An emerging discipline. *American Science, 63*, 178. (1993). Longitudinal study of delinquency, drug use, sexual activity, and pregnancy among children and youth in three cities. *Public Health Reports, 108*, 90–96.

Bergman, B., & Brismar, B. (1994). Hormone levels and personality traits in abusive and suicidal male alcoholics. *Alcoholism: Clinical and Experimental Research, 18*, 311–316

Berenbaum, S. A., & Snyder, E. 1995). Early hormonal influences on childhood sex-typed activity and playmate preferences: Implications for the development of sexual orientation. *Developmental Psychology, 31*, 31–42.

Booth, A., Shelley, G., Mazur, A., Tharp, G., & Kittok, R. (1989). Testosterone, winning and losing in human competition. *Hormones and Behavior, 23*, 556–571.

Brain, P. F. (1989). An ethoexperimental approach to behavioral endocrinology. In R. J. Blanchard, P. F. Brain, D. C. Blanchard, & S. Parmigiani (Eds.), *Ethoexperimental Approaches to the Study of Behavior* (pp. 539–557). Dordrecht: Kluwer Academic Publishers.

Brain, P. F. (1994). Hormonal aspects of aggression and violence. In A. J. Reiss Jr., K. A., Miczek, & J. I. Roth (Eds.), *Understanding and Preventing Violence Vol. 2 Biobehavioral Influences* (pp. 173–244). Washington, DC: National Academy Press.

Brain, P. F., & Haug, M. (1992). Hormonal and neurochemical correlates of various forms of animal aggression. *Psychoneuroendocrinology, 17*, 537–551.

Brain, P., & Susman, E. J. (in press). Hormonal aspects of antisocial behavior and violence. In D. M. Stoff, J. Breiling, & J. Maser (Eds.), *Handbook of Antisocial Behavior*. Hillsdale, NJ: Lawrence Erlbaum Associates.

Brooks-Gunn, J., Graber, J., & Paikoff, R. (1994). Studying links between hormones and negative affect: Models and measures. *Journal of Research on Adolescence, 4*, 469–486.

Brooks-Gunn, J., & Warren, M. (1989). Biological and social contributions to negative affect in young adolescent girls. *Child Development, 60*, 40–55.

Buchanan, C., Eccles, J. S., & Becker, J. B. (1992). Are adolescents the victims of raging hormones: Evidence for activational effects of hormones on moods and behavior at adolescence. *Psychological Bulletin, 111*, 62–107.

Chrousos, G. P., & Gold, P. W. (1992). The concepts of stress and stress system disorders. *Journal of the American Medical Association, 267*, 1244–1252.

Costello, A. J., Edelbrock, C., Kalas, R., Dulcan, M. K., & Klaric, S. H. (1983). Development and testing of the NIMH Diagnostic Interview Schedule for Children in a clinical population. National Institute of Mental Health.

Dorn, L. D., Susman, E. J., & Petersen, A. C. (1993). Cortisol reactivity and anxiety and depression in pregnant adolescents: A longitudinal perspective. *Journal of Psychoneuroendocrinology, 18*, 219–239.

Dorn, L. D. & Chrousos, G. P. (In press). The neurobiology of stress: Understanding regulation of affect during female biological transitions. In S. Berga (Ed.). *Seminars in reproductive endocrinology*. New York, NY: Thieme Stratton.

Dorn, L. D., Burgess, E. S., Susman, E. J., von Eye, A., DeBellis, M. D., Gold, P. W., & Chrousos, G. P. (in press). Response to °CRH in depressed and non-depressed adolescents: Does gender make a difference? *Journal of the American Academy of Child and Adolescent Psychiatry*.

Eccles, J. S., Miller, C. L., Tucker, M. L., Becker, J., Schramm, W., Midgley, R., Holmes, W., Pasch, L., & Miller, M. (1988). Hormones and affect at early adolescence. In J. Brooks-Gunn (Chair), *Hormonal contributions to adolescent behavior*. A symposium conducted at the second biennial meeting of Society for Research on Adolescence, Alexandria, VA.

Gold, P. W., Goodwin, & Chrousos, G. P. (1988a). Clinical and biochemical manifestations of depression: Relation to the neurobiology of stress (First of two parts). *New England Journal of Medicine, 319*, 348–353.

Gold, P. W., Goodwin, F. K., & Chrousos, G. P. (1988b). Clinical and biochemical manifestations of depression: Relation to the neurobiology of stress (Second of two parts). *New England Journal of Medicine, 319*, 413–420.

Hedge, G. A., Colby, H. A., & Goodman, R. L. (1987). *Clinical Endocrine Physiology*. W. B. Saunders Company.

Huizinga, D., Loeber, R., & Thornberry, T. P. (1993). Longitudinal study of delinquency, drug use, sexual activity, and pregnancy among children and youth in three cities. *Public Health Reports, 108*, 90–96.

Inoff-Germain, G. E., Arnold, G. S., Nottelmann, E. D., Susman, E. J., Cutler, G. B., Jr., & Chrousos, G. P. (1988). Relations between hormone levels and observational measures of aggressive behavior of early adolescents in family interactions. *Developmental Psychology, 24*, 129–139.

Johnson, E. O., Kamilaris, T. C., Chrousos, G. P., & Gold, P. W. (1992). Mechanisms of stress: A dynamic overview of hormonal and behavioral homeostasis. *Neuroscience and Biobehavioral Reviews, 16*, 115–130.

Kagan, J., Resnick, S. & Snidman, N. (1987). The physiology and psychology of behavioral inhibition in children. *Child Development, 58*, 1459–1473.

Kovacs, M., Krol, R. S., & Voti, L. (1994). Early onset psychopathology and the risk for teenage pregnancy among clinically referred girls. *Journal of the American Academy of Child & Adolescent Psychiatry, 33,* 106–113.

Levine, S., & Weiner, S. G. (1989). Coping with uncertainty: A paradox. In D. S. Palermo (Ed.), *Coping with uncertainty: Behavioral and developmental perspectives* (pp. 1–16). Hillsdale, NJ: Lawrence Erlbaum Associates, Inc.

Magnusson, D., & Cairns, R. B. (1992). Developmental science: Toward a unitary framework for the investigator of psychological phenomena. Unpublished manuscript. University of Stockholm.

Marshall, W. A., & Tanner, J. (1969). Variations in the pattern of pubertal change in girls. *Archives of the Disabled Child, 44,* 291–303.

Marshall, W. A., & Tanner, J. (1970). Variations in the pattern of pubertal change in boys. *Archives of the Disabled Child, 45,* 13–23.

Mayr, E. (1988). *Towards a new philosophy of biology: Observations of an evolutionist.* Cambridge, MA: Harvard University Press.

Mazur, A., & Lamb, T. A. (1980). Testosterone, status, and mood in human males. *Hormones and Behavior, 14,* 236–246.

McBurnett, K., Lahey, B. B., Frick, P. J., Risch, C., Loeber, R., Hart, E. L., Christ, M. A., & Hanson, K. A. (1991). Anxiety, inhibition, and conduct disorder in children: II. Relation to salivary cortisol. *Journal of the American Academy of Child and Adolescent Psychiatry, 30,* 192–196.

McCool, W., & Susman, E. J. (1994). Cortisol reactivity and self-report anxiety in antepartum: Predictors of maternal intrapartal outcomes in gravid adolescents. *Journal of Psychosocial Obstetrics and Gynecology, 15,* 9–18.

McCool, W., Dorn, L. D., & Susman, E. J. (1994). Relations of cortisol reactivity and anxiety to perinatal outcome in primiparous adolescents. *Research in Nursing and Health, 17,* 411–420.

McEwen, B. S., & Schmeck, H. M. Jr. (1994). *The Hostage Brain.* The Rockefeller University Press: Rockefeller, NY.

McEwen, B. S., Angulo, J., Cameron, H., Chao, H. M., Daniels, D., Gannon, M. N., Gould, E., Mendelson, S., Sakai, R., Spencer, R., & Woolley, C. (1992). Paradoxical effects of adrenal steroids on the brain: Protection versus degeneration. *Biological Psychiatry, 31,* 177–199.

Money, J., & Ehrhardt, A. A. (1972). *Man and woman, boy and girl.* Baltimore, MD: Johns Hopkins University Press.

Nottelmann, E. D., Inoff-Germain, G. E., Susman, E. J., & Chrousos, G. P. (1990). Hormones and behavior at puberty. In J. Bancroft & J. M. Reinisch (Eds.), *Adolescence and puberty,* (pp. 513–517). New York, NY: Garland Press.

Nottelmann, E. D., Susman, E. J., Dorn, L. D., Inoff-Germain, G. E., Cutler, G. B., Jr., Loriaux, D. L., & Chrousos, G. P. (1987). Developmental processes in American early adolescents: Relationships between adolescent adjustment problems and chronological pubertal stage and puberty-related serum hormone levels. *Journal of Pediatrics, 110,* 473–480.

Offer, D., Ostrov, E., & Howard, K. I., (1977). *The Offer Self-Image Questionnaire for Adolescents: A manual.* Chicago: Michael Reese Hospital.

Olweus, D., Mattsson, A., Schalling, D., & Low, H. (1980). Testosterone, aggression, physical, and personality dimensions in normal adolescent males. *Psychosomatic Medicine, 42,* 253–269.

Olweus, D., Mattsson, A., Schalling, D., & Low, H. (1988). Circulating testosterone levels and aggression in adolescent males: A causal analysis. *Psychosomatic Medicine, 50,* 261–272.

Phoenix, C. H., Goy, R. W., Gerall, A. A., & Young, W. C. (1959). Organizing action of prenatally administered testosterone propionate on the tissues mediating mating behavior in the female guinea pig. *Endocrinology, 65,* 369–382.

Ponirakis, A., Susman, E. J., & Stifter, C. (1996). *Emotionality during adolescent pregnancy and its effects on infant health and autonomic nervous system reactivity.* Manuscript in preparation.

Preville, M., Susman, E. J., Zarit, S., Smyer, M., Bosworth, H., Smyer, M., Bosworth, H. B., & Reid, J. (in press). A measurement model of cortisol reactivity of healthy older adults during relocation to a retirement home. *Journal of Gerontology.*

Rose, R. (1980). Endocrine responses to stressful psychological events. In E. J. Sachar (Ed.), *Advances in psychoneuroendocrinolgy* (pp. 251–276). The Psychiatric Clinics of North America 3.

Sapolsky, R. M. (1982). The endocrine stress-response and social status in the wild baboon. *Hormones and Behavior, 16,* 279–292.

Sapolsky, R. M. (1991). Testicular function, social rank and personality among wild baboons. *Psychoneuroendocrinology, 16,* 281–293.

Schaal, B., Tremblay, R. E., Souissignan, R., & Susman, E. J. (1995). Male pubertal testosterone linked to high social dominance but low physical aggression: A 7 year longitudinal study. *Journal of the American Academy of Child Psychiatry.*

Schwab-Stone, M., Fisher, M. S., Piacentini, J., Shaffer, D., Davies, M., & Briggs, M. (1993). Diagnostic Interview Schedule for Children-Revised Version (DISC-R): II. Test-retest reliability. *Journal of the American Academy of Child and Adolescent Psychiatry, 32,* 651–657.

Susman, E. J. (1993). Psychological contextual and psychobiological interactions: A developmental perspective on conduct disorder. *Development and Psychopathology, 5,* 181–189.

Susman, E. J., Dorn, L. D., & Chrousos, G. P. (1991). Negative affect and hormone levels in young adolescents: Concurrent and longitudinal perspectives. *Journal of Youth and Adolescence, 20,* 167–190.

Susman, E. J., Dorn, L. D., Inoff-Germain, G. E., Nottelmann, E. D., & Chrousos, G. P. (in press). Cortisol reactivity, distress behavior, behavior problems, and emotionality in young adolescents: A longitudinal perspective. *Journal of Research on Adolescence.*

Susman, E. J., Granger, D. A., Murowchick, E., Ponirakis, A., & Worrall, B. K. (in press). Gonadal and adrenal hormones: Developmental transitions and aggressive behavior. *New York Academy of Sciences.*

Susman, E. J., Nottelmann, E. D., Dorn, L. D., Gold, P., & Chrousos, G. P. (1989). The physiology of stress and behavioral development. In D. S. Palermo (Ed.), *Coping with Uncertainty: Behavioral and Developmental Perspectives* (pp. 17–37). Hillsdale, NJ: Lawrence Erlbaum Associates.

Susman, E. J., & Petersen, A. C. (1992). Hormones and behavior in adolescence. In E. R. McAnarney, R. E. Kreipe, D. P. Orr, & G. D. Comerci (Eds.), *Textbook of Adolescent Medicine* (pp. 125–130), New York: Saunders Publ.

Susman, E. J., Nottelmann, E. D., Inoff, G. E., Dorn, L. D., Cutler, G. B., Loriaux, D. L., & Chrousos, G. P. (1985). The relation of relative hormonal levels and social-emotional behavior in young adolescents. *Journal of Youth and Adolescence, 14,* 245–252.

Susman, E. J., Inoff-Germain, G. E., Nottelmann, E. D., Cutler, G. B., Loriaux, D. L., & Chrousos, G. P. (1987). Hormones, emotional dispositions and aggressive attributes in young adolescents. *Child Development, 58,* 1114–1134.

Susman, E. J., Worrall, B. K., Murowchick, E., Frobose, C., & Schwab, J. E. (in press). Experience and neuroendocrine parameters of development: Aggressive behavior and competencies. In D. Stoff & R. Cairns (Eds.). *Neurobiological approaches to clinical aggression research.* Hillsdale, NJ: Lawrence Erlbaum Associates.

Susman, E. J., Dorn, L. D., Schwab, J. E., Frobose, C. A., Murowchick, E., & Murray, D. D. (1996). *Cortisol reactivity, environmental processes and emotionality in pregnant and nonpregnant adolescents.* Manuscript submitted for publication.

Susman, E. J., Schmeelk, K. H., Worrall, B. K., & Heaton, J. A. (1996, March). *Corticotropin releasing hormone and emotional regulation in pregnant adolescents.* Paper presented at the Society for Research on Adolescence, Boston, MA.

Vale, W., Spiess, J., Rivier, C., & Rivier, J. (1981). Characterization of a 41 residue ovine hypothalamic peptide that stimulates the secretion of corticotropin and β-endorphin. *Science, 213,* 1394–1397.

Virkkunen, M., Rawlings, R., Tokola, R., Poland, R. E., Guidotti, A., Nemeroff, C., Bissette, G., Kalogeras, K., Karonen, S. L. & Linnoila, M. (1994). CSF biochemistries, glucose metabolism, and diurnal activity rhythms in alcoholic, violent offenders, impulsive fire setters, and healthy volunteers. *Archives of General Psychiatry, 51,* 20–27.

16

MALE PHYSICAL AGGRESSION, SOCIAL DOMINANCE AND TESTOSTERONE LEVELS AT PUBERTY

A Developmental Perspective

Richard E. Tremblay, Benoist Schaal, Bernard Boulerice, Louise Arseneault, Robert Soussignan, and Daniel Pérusse

University of Montréal
Montréal, Québec H2C 1A6
Canada

"On June 1, 1889, Charles Édouard Brown-Séquard, a prominent French physiologist, announced at the Société de Biologie in Paris that he had devised a rejuvenating therapy for the body and mind. The 72-year-old professor reported that he had drastically reversed his own decline by injecting himself with a liquid extract derived from the testicles of dogs and guinea pigs. These injections, he told his audience, had increased his physical strength and intellectual energy, relieved his constipation and even lengthened the arc of his urine" (Hoberman & Yesali, 1995, p.77).

1. OVERVIEW OF THE COURSE AND FUNCTION OF TESTOSTERONE FROM FETAL LIFE TO ADULTHOOD

Testosterone (T) is an androgen hormone produced mainly by the testes. Its production is regulated by the hypothalamic-pituitary-gonads (HPG) axis. Towards the end of the first trimester of pregnancy (10–12 weeks) male fetuses have higher plasmatic levels of T than female fetuses. Peak levels of T are obtained by the middle of the second trimester. During the third trimester no sex differences in T level have been detected, although there is evidence of some testicular activity (Forest, 1989).

In many species T secretion during fetal life influences the development of male primary sexual characteristics and has long term organizational effects on brain and behavior (Goy & McEwen, 1980; Hines, 1982; Wilson, George, & Griffin, 1981). For example, adult female rhesus monkeys whose mothers were administered T during pregnancy had masculinized behaviors, including more aggressive behaviors than normal females (Eaton, Goy, & Phoenix, 1973; Goy, Bercovitch, & McBrair, 1988). Human females exposed to high levels of androgens during fetal life (because of adrenal disfunction, or because of a

Biosocial Bases of Violence, edited by Raine *et al.*
Plenum Press, New York, 1997

drug treatment given to the mother) have masculinized genitals; behavioral studies have shown that during childhood they are more active, they prefer activities and toys which are traditionally masculine, they prefer to play with boys, they have higher spatial abilities, and during adolescence and adulthood they have higher rates of homosexual or bisexual fantasies (Berenbaum & Snyder, 1995; Dittmann, Kappes, & Kappes, 1992; Hines & Kaufman, 1994; Money & Schwartz, 1976; Reinisch, 1981; Resnick, Berenbaum, Gottesman, & Bouchard, 1986).

At birth females have plasma T levels similar to adult females, but these levels decrease rapidly. Two weeks after birth the T level of females has reached the low level it will maintain until puberty. The course of plasma T levels is quite different in males. At birth they have higher peripheral plasma T levels than females. The male levels decrease rapidly during the first week after birth, and increase again starting from the second week after birth. By the second month after birth they have reached half the T levels of adults, and T levels similar to adolescents. These male T levels then decrease until the seventh month after birth, when no differences can be observed between males and females. Differences between males and females will reappear only with puberty (Forest, 1989). It has been suggested that the sex differences in T levels during the months after birth play a role in hypothalamic sexual differentiation, and could explain later sex differences in behavior (Archer, 1991, 1994; Forest, Sizonenko, Cathiard, & Bertand, 1974).

The changes in male salivary T levels during the first year after birth has been linked with changes in testicular volume. Cho, Sanayama, Sasaki, & Nakajima (1985) showed that between one and four months after birth males had larger testicular volume than between six to twelve month after birth. In one of the few longitudinal studies of infants' T levels, Huhtaniemi et al. (1986) measured the salivary T levels of 22 boys every 14 days from birth to 6 months. They found that salivary T was highest after birth and decreased linearly over the next six months; but they failed to replicate with salivary T the pattern described above of rapid decrease of plasma T during the first postnatal week followed by an increase up to the second postnatal month. Because salivary T reflects the fraction of plasma T which is not bound to protein, it is considered a better indicator of the T level which is biologically active (Riad-Fahmy, Read, Walker, & Griffiths, 1982). Huhtaniemi et al. (1986) suggested that the different developmental patterns of plasma and saliva T could be due to an increase of plasma T binding to protein over time and a decline of unbound T. Unfortunately, we are not aware of any study which has compared the developmental course of male and female salivary T during the first year after birth.

During adrenarche, the period between age 7 to 10–12 years when the adrenal glands start producing higher levels of androgens (mainly DHA and DHAS), T rises slightly in males and females and is mainly produced by the adrenal glands (Forest, 1989).

With puberty, male levels of plasma T rise sharply from a mean of 0.28 nmol/l at age 8 years to 4 nmol/l around age 13 years, 15 nmol/l by age 16 years, and 20 nmol/l at the end of adolescence. There is also an increase in female T levels during puberty, but the rise is much less spectacular. At 8 years of age male and female levels are similar, around 0.28 nmol/l. By the end of adolescence females will have attained levels of approximately 0.75 nmol/l during their follicular phase, and 1.28 nmol/l during their luteal phase. At the end of pregnancy, T levels will reach 2.35 nmol/l (Forest, 1989).

The increase in T levels during adolescence is highly correlated with bone growth, increase in muscle mass, growth of the genital organs, and the development of secondary sexual characteristics (Ducharme et al., 1976; Forest, 1981; Forest, 1989; Marshall & Tanner, 1969; Winter & Faiman, 1973). Sexual motivation and sexual behavior during adolescence have been shown to increase with age, in parallel with the increase in T level.

However, there is no clear evidence that the increase in T causes the changes in sexual motivation and behavior (Halpern, Udry, Campbell, & Suchindran, 1993a).

In 1939, Adolf Butenandt and Leopold Ruzicka received the Nobel Prize for Chemistry following their successful attempts to create a synthetic T product. Since that discovery, synthetic T has been used to treat different conditions in men (e. g., initiation of puberty for boys with important delays, hypogonadism, impotence, homosexuality, lack of libido, aging),and women (e. g., breast tumors, lack of libido, aging). It has also been extensively used by males and females to increase muscularity for athletic performance or simply for appearance (Hoberman & Yesali, 1995).

2. TESTOSTERONE AND PHYSICAL AGGRESSION

For decades T has been linked to the activation of dominance and aggressive behaviors in male primates, including humans (Archer, 1988, 1991, 1994; Brain, 1979; Dabbs, 1992b; Reiss & Roth, 1993). Comparisons between groups of adult humans selected for their high or low aggressiveness (or antisocial behavior) indicate higher levels of T in the more aggressive groups (Dabbs & Morris, 1990; Mattson, Schalling, Olweus, Low, & Svenson, 1980; Virkkunen, et al., 1994; Windle & Windle, 1995). Correlations between T and behavioral measures of aggressiveness are low but generally positive (Christiansen & Knussmann, 1987; Dabbs, 1992a; Gladue, 1991; Gray, Jackson, & McKinlay, 1991).

Archer (1991, 1994) has suggested that the link between T and overt aggression and, in humans, to aggression-related traits or attitudes, fits a general model according to which behavior and gonadal androgen production interact in a regulation loop, being alternately causal and consequential. Dabbs (1992b) depicts T as a trait and a state. T can be considered a trait variable because individual differences are large and appear relatively stable over time. It can also be considered a state because relatively rapid changes in T levels occur in response to environmental factors. The behavior-to-hormone causal influences in humans are far better understood than the reverse. For example, T levels change following nonaggressive social challenges, such as a chess match (Booth, Shelley, Mazur, Tharp, & Kittok, 1989; Elias, 1981; Mazur, Booth, & Dabbs, 1992; Mazur & Lamb, 1980; McCaul, Gladue, & Joppa, 1992).

The mechanisms by which such events have an impact on circulating T levels involve both direct modulation of the HPG axis, and indirect modulation through the Hypothalamic-Pituitary-Adrenal (HPA) axis. Studies on non-human primates and adult humans have shown the links between the HPG and HPA hormonal systems in conditions of environmental stress (Bambino & Hsueh, 1981; Collu, Gibb, & Ducharme, 1984; Gladue, Boechler, & McCaul, 1989; Higley et al., 1992; Keverne, 1992; Sapolsky, 1991). Both pathways, but primarily the HPA axis, are sensitive to physical and psychological stress.

The evidence for this regulation model comes from studies conducted mainly with adult males (Booth et al., 1989; Elias, 1981; Mazur et al., 1992; Mazur & Lamb, 1980; McCaul et al., 1992; Virkkunen et al., 1994). Few studies have focused on the interactions between the hormonal systems and behavior from childhood to adolescence and adulthood. One would expect that longitudinal studies from childhood to adolescence, when gonadal androgens rise progressively from undetectable to mature levels, would provide crucial data on the interactions between hormones and behavior (Archer, 1991, 1994). Adult studies suggest that the increase in androgens during adolescence will act as a strong regulator of motivations and behavior. One study of older adolescent males has provided a confirmation that T is associated with self-reports of aggressive response to provocation (Olweus, Mattsson, Schalling, & Low, 1980; Olweus, Mattsson, Schalling, & Low, 1988). However, the few available studies of pre- and early adolescent males provide no evidence of an association between T and aggressive behavior.

Inoff-Germain et al. (1988) and Susman et al. (1987) found no relationship between plasma T level and concurrent self-reported negative affect, maternal reports of delinquency and opposition, or direct observation of irritability and assertiveness in pre- and early adolescents. In a small-scale study, Constantino et al. (1993) did not find any consistent association between high T concentration in serum blood and high level of aggression in pre-adolescents. Having followed a sample of 100 males from age 12–13 years to 15–16 years, Halpern, Udry, Campbell, & Suchindan (1993b) failed to show any clear relationship between self-reported aggression and antecedent or concurrent T levels. Finklestein, von Eye, & Preece (1994) followed a sample of 106 boys and girls from ages 9 to 15 years, assessing aggressive behaviour with the Olweus (1988) self-report inventory, and pubertal development with the Tanner (1962) method. They found no significant association between aggressive behavior and pubertal development.

It has been suggested that the rise in T levels with puberty could explain the rise in criminal behavior at the same age period (Ellis & Coontz, 1990; Eysenck & Gudjonsson, 1989). However, although there is an increase in T for all males during puberty, most adolescents do not display increased aggression (Finkelstein et al., 1994; Halpern et al., 1993b; Tremblay & Schaal, 1996). In contrast, in some clinical syndromes inducing very low T levels at puberty, adolescents can behave in a highly aggressive way to compensate for physical stigma (Johnson, Myhre, & Ruvalcaba, 1970). Thus the evidence for the increase in T leading to aggression during adolescence remains controversial (Eichelman, 1992; Reiss & Roth, 1993).

3. AIMS OF THIS CHAPTER

The USA National Research Council panel on the understanding and control of violent behavior (Reiss & Roth, 1993) called for more systematic investigation of the relationship between violent behavior and sex hormones. They specifically suggested that longitudinal data, from population based samples, linking peer interactions, behavior, and levels of gonadal hormones, would help predict violent behavior (pp. 160 & 339).

This chapter describes results from such a longitudinal study. A sample of males was followed from kindergarten to middle adolescence with a particular focus on physically aggressive behavior. Annual measurements of T levels between 13 and 16 years of age have provided the opportunity to study the developmental patterns of aggressive behavior and T from early to middle adolescence.

Three related questions were addressed. First we investigated to what extent T levels during early adolescence were related to concurrent social dominance. T levels have been shown to correlate with social dominance in a wide range of social mammals. For example, when male monkey hierarchies are changed, those who dominate experience increases in T levels (Rose, Bernstein, & Gordon, 1975; Sapolsky, 1983). Similarly, when humans win a tennis match (Mazur & Lamb, 1980) or a chess match (Mazur et al., 1992), their increase in T level is greater than their opponents'. We studied to what extent T levels were associated with social dominance as soon as they become reliably measurable from saliva samples in early adolescence. The model summarized above would predict that T concentration is higher in young adolescent males who succeed in making themselves perceived by their peers as dominant, that is those who are most able to impose their will on others. Because perception of social status is essential to the maintenance of dominance hierarchies in most social species, it can be assumed that adolescent human males can quickly assess their social status in a group (Coie, Dodge, & Kupersmidt, 1990; de Waal, 1982; Sapolsky, 1992b).

Second, we investigated whether long-term stable physically aggressive behavior during childhood might influence T levels in early adolescence. Childhood experiences, specifically those deriving from social interaction strategies and coping styles integrated in a stable behavioral repertoire, may indeed influence the endocrine processes controlling pubertal biological change, and make a significant contribution to individual differences in adolescent psychobiological functioning (Belsky, Steinberg, & Draper, 1991). According to previous work (Olweus et al., 1980, 1988) the expected direction of influence would be that children who had a history of physical aggressivity would have higher T levels at puberty. Alternatively, physically aggressive children who tend to be rejected by their social environment, would have been exposed to negative, highly stressful social encounters. The HPA axis of these subjects might then have been much more activated than in less physically aggressive boys, with the consequence of inhibiting the HPG axis, thus limiting T production (Nottelman, Inoff-Germain, Susman, & Chrousos, 1990; Susman, Dorn, & Chrousos, 1991; Susman, this volume).

Finally, we studied to what extent the association between childhood patterns of aggressive behavior and adolescent T levels was stable over time. In other words, if aggressive children have high levels of T at the start of puberty, do they still have high levels of T as they grow older? After having followed olive baboons living freely in the Serengeti Ecosystem of East Africa over many years, Sapolsky (1992a; 1992b) concluded that the relation between hormone levels and behavior at one point in time will not necessarily be the same at another point in time. Over time individuals change social environments, they change status, and their relative hormone levels are likely to be altered. It is a challenge for the investigators to monitor the important changes over time, whether the subjects are free ranging baboons or adolescents in modern cities.

4. THE MONTREAL LONGITUDINAL AND EXPERIMENTAL STUDY

The subjects involved in the present study were drawn from a large sample (n = 1161) of boys followed from kindergarten in a longitudinal-experimental study of social development (Tremblay, Kurtz, Mâsse, Vitaro, & Pihl, 1995; Tremblay, Pihl, Vitaro, & Dobkin, 1994). To control for cultural and socio-economic biases, the boys were recruited according to the following criteria: (i) attending school in low socio-economic areas of Montréal; (ii) born to Caucasian French-speaking parents themselves born in Canada; (iii) living with parents having medium to low educational status. The boys all attended a kindergarten class in one of the 53 schools with the lowest socio-economic catchment areas in Montreal. Their social behavior was rated by their kindergarten teachers. At 10, 11, and 12 years of age, their social behavior was rated by their teachers and their peers.

Subsamples of subjects were brought to our university laboratory during the summer holidays to obtain a variety of cognitive, behavioral, neuropsychological, and psychophysiological measurements (see Kindlon et al., 1995; Mezzacappa et al., in press; Séguin, Pihl, Boulerice, Tremblay, & Harden, 1996; Séguin, Pihl, Harden, Tremblay, & Boulerice,

1995; Soussignan et al., 1992). Boys came to the laboratory by groups of four unfamiliar peers.

Pubertal status and physical development (height, weight, wrist robustness, and skin fold measurements at triceps, shoulder and abdomen) were assessed during the laboratory visit. Pubertal status was self-assessed using the Pubertal Development Scale (Petersen, Crockett, Richards, & Boxer, 1988). This scale assesses self-reported growth spurt, body and facial hair development, skin and voice changes on a 4-point scale. Classification into one of the five pubertal status categories (pre-, early, mid-, late and post-pubertal) is based on the level of development of the three most salient indices of pubertal change (i.e.; body hair, facial hair and voice alterations)

T was measured each summer from the time the boys were 11 years of age to the time they were 16 years of age. T levels were assayed from multiple saliva samples collected at approximately 8:30 am, 10:00 am, 11:30 am, and 3:30 pm during the one-day visit to the laboratory (for details concerning the dosage methodology see Schaal, Tremblay, Soussignan, & Susman, 1996). The titration of T from saliva was preferred to any other way of obtaining similar data. Its application being unobtrusive, and the handling of saliva additionally being uncomplicated relatively to blood or urine. Salivary T level being highly correlated with the unbound fraction of circulating T, it is assumed to be a precise indicant of the behaviorally active fraction of T (Riad-Fahmy et al., 1982; Wang et al., 1981). Because one study (Halpern & Udry, 1992) has shown saliva measures to be less reliable compared to plasma measures, using multiple saliva samples during the same day should improve the reliability of the measures in the present study.

At 11 years of age only a few boys had measurable T levels. At 12 years of age most boys had a measurable T level, but data was obtained on a smaller sample than in the following years (n=76). The age 12 data indicated that T levels were moderately correlated with height (r=.33), weight (r=.26), wrist circumference (r=.29), and shoulder skinfold (r=.27), but not with concurrent, or antecedent (age 6, 10, and 11 years) measures of self-reported delinquency, physical aggression rated by teachers, by peers, and the boys themselves.

At age 13, ratings of toughness and leadership were obtained from individual interviews during which every subject from a peer group was asked to nominate the leader ("who would you choose as leader?") and identify the toughest boy ("who was the toughest?"). The interviews were made at approximately 10:30 a.m.. This was after a 3-hr period during which they were picked up at home and driven together to the lab in a van, individually seen to assess personality and cognitive functioning, and finally took part in a competitive group task to provide an opportunity to measure social interactions. This task consisted of a 15-minute competitive game during which they threw sand bags in holes to win points which were exchanged for money. No rules were set except that bags had to be thrown from a fixed distance. When the task was completed, the winner received two dollars, but all the other boys were told they had done well and were given a dollar each.

Each subject received a toughness and a leadership score ranging from 0 to 5 depending on the number of nominations he received (including self-nominations). The crossing of both toughness and leadership scores, using the median, yielded four groups defined as: (i) tough-leader (n = 52), (ii) tough-not leader (n = 27), (iii) not tough-leader (n = 44), and (iv) not tough-not leader (n = 48). These four groups were not different on a number of demographic and socio-economic variables (family status; both parents' age at birth of first child and at birth of subject; both parents' educational status; both parents' job socio-economic level; and total family adversity).

Childhood aggression was assessed through teacher ratings using the Social Behavior Questionnaire (Tremblay et al., 1991). For this study, a *fighting score* was derived by using 3 items ("fights"; "kicks, bites, hits"; and "bullies or intimidates other children"). The range of possible values of the fighting score was 0–6. Cronbach's alpha varied between .78 and .87 from ages 6 to 12 years. From the total sample of subjects with complete behavioral data (n = 948), two extreme groups of subjects were identified on the basis of the intensity and stability of their physical aggressivity score as assessed by teachers during the kindergarten year (age 5–6), and during the three last primary school years (ages 9 to 12). Boys were entered into the following categories according to their behavior rating scores on at least 3 of 4 assessments: (i) stable high fighter(SHF; n = 64): fighting scores at the 70th percentile or above; (ii) stable low fighter (SLF; n = 65): less than the 70th percentile on the fighting scale.

5. DO SOCIALLY DOMINANT EARLY ADOLESCENT BOYS HAVE HIGHER TESTOSTERONE LEVELS?

As explained above, previous studies on the relationship between social dominance and testosterone level led us to expect that socially dominant early adolescent boys would have higher levels of testosterone than their less socially dominant peers. The results which follow were published in Schaal et al. (1996). We first examined to what extent the dominance ratings by the unfamiliar peers in the laboratory were reflecting the perception of more familiar peers by comparing the results of the age 13 assessment in the laboratory with the age 12 assessments by classroom peers. Results showed that the boys rated tough by unfamiliar peers in the lab (especially the tough not leaders) had also been rated by familiar peers as more aggressive than those rated not tough. The boys rated as leaders had been rated by peers as more popular than those rated as not leaders. These results indicated that the sociometric status obtained after having met for only three hours corresponded to perceptions by familiar peers at school during the preceding 15 months.

We then assessed to what extent these perceptions of toughness and leadership by unfamiliar peers after a brief encounter were associated with concurrent testosterone level. As anticipated from previous research (Dabbs, 1991; Nieschlag, 1974), average T levels at each of the four sampling times during the visit to the laboratory decreased over time from morning to afternoon. Boys who were rated tough (regardless of concurrent leadership ratings) produced a higher average level of salivary T as compared to not tough subjects. But a toughness by leadership by time interaction indicated that tough-leaders produced higher levels of T than any other category, at the sampling times 1, 2 and 3, but not at 4.

Because T was found to correlate with pubertal stage and somatic growth (see section 4), these variables were controlled by entering them as covariates in an analysis of covariance (MANCOVA) with T level as the dependent variable. A significant toughness by leadership by time interaction confirmed that the tough-leader group had higher levels of T at the first three sampling times. Interestingly, the main effect for toughness was no more significant after having included these covariates, indicating that this main effect was explained by pubertal maturity and somatic growth.

Thus, the link between aggression and T appeared to be modified by leadership. The subjects who had the highest T levels were rated tough and leaders by their lab peers, but they were not rated highest on physical aggression in everyday life by classmates. It was the tough-not leader boys who were rated highest on physical aggression by their classmates; their T level was lower than that of the tough-leader boys, and similar to that of both not tough groups (not tough-leader and not tough-not leader). From this perspective, boys with a history of aggression who are not socially successful (e.g., not designated as a leader) do not appear to have high testicular activity before or after a competitive group task in a novel situation. This hypothesis was tested more directly with the following analyses.

6. DO PHYSICALLY AGGRESSIVE BOYS HAVE HIGHER LEVELS OF TESTOSTERONE WHEN THEY REACH PUBERTY?

Results showing that aggressive adolescents and aggressive adults have higher levels of T (Dabbs, Frady, Carr, & Besch, 1987; Olweus et al., 1980; Virkkunen et al., 1994) would lead us to expect that boys who were rated by four different teachers as highly physically aggressive between age 6 and 12 years would have higher levels of T at the start of puberty, compared to boys who were never rated physically aggressive.

Classroom peer ratings at age 12 were used to verify to what extent the classification of subjects based on teacher ratings of behavior from age 6 to 12 (described in section 4) corresponded to familiar peer perception. From Figure 1 it can be seen that the teacher rated physically aggressive boys (SHF) compared to the not physically aggressive boys (SLF) were rated higher on proactive aggression ($F(1,119) = 93.17$; $p < .001$) by their classroom peers. They were also rated less popular ($F(1,119) = 33.14$; $p < .001$).

To assess to what extent long-term stable aggressivity before puberty might influence T level in early adolescence, an analysis was conducted in which antecedent fighting and T sampling time were the factors, and T level was the dependent variable. To control for the potential effect of pubertal stage and somatic growth, MANCOVA was used introducing the

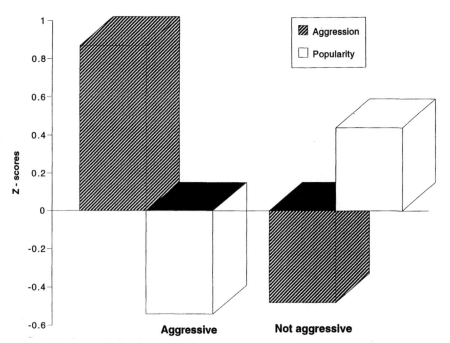

Figure 1. Classroom peer ratings of proactive aggression and popularity at age 12 years for the aggressive and not aggressive groups.

variables described earlier as covariates. As expected, the two groups showed an overall steep decline of salivary T (figure 2) as a function of sampling time, $F(2,125) = 49.64$, $p<.0001$. The main effect for stable fighting was also significant [$F(1,122) = 4.34$; $p = .039$]. ANCOVAs performed at each T sampling time also indicated significant main effects for the fighting factor at t1, t2 and t3. Thus boys who were stable high fighters from ages 6 through 12 had lower salivary T levels at age 13 compared to boys who were stable low fighters.

It is important to note that from the standpoint of another social adjustment index, school status,(Power, Manor, & Fox, 1991; Wadsworth, 1991) the stable fighters also had more problems. Only 34.4% of the stable high fighters were in a regular classroom at the level appropriate for their age (age 13) compared to 80% of the stable low fighters ($c^2(1) = 27.45$, $p<.0001$).

7. SUMMARY OF FINDINGS AT THE START OF PUBERTY (AGE 13 YEARS

The analyses in sections 5 and 6 approached the relation of T and behavior by measuring different aspects of aggressiveness in two very different time periods of child and adolescent development. The first analysis (section 5) showed high salivary T concentration in boys concurrently nominated by unfamiliar peers as tough and a leader after having

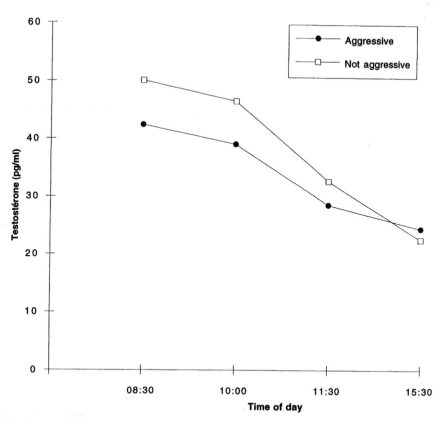

Figure 2. Testosterone levels at four points in time during the laboratory visit for the aggressive and not aggressive groups.

met for only 3 hours. In the second analysis (section 6), the pattern was reversed when aggression was considered in terms of stable frequent physical aggression throughout childhood. Boys who were rated consistently high on physical aggressiveness between ages 6 and 12 had lower T levels at age 13, as compared to stable non-aggressive boys. Clearly, both characterizations of aggressiveness differ. The subjects nominated as tough-leaders, who had the highest morning T levels, were perceived by their unfamiliar peers as socially dominant; they were also perceived by their familiar peers (classmates) 15 months earlier as more popular and less aggressive than the tough-not leader group, but as more aggressive than the not tough groups (not tough-leaders, not tough-not leaders). Thus, the high T boys at age 13 appear to succeed in imposing their will on peers, sometimes by aggressive behavior, but they remain socially attractive over time (Savin-Williams, 1979; Strayer & Trudel, 1984). On the other hand, the boys rated by their teachers as physically aggressive from kindergarten to age 12, who had the lowest T levels, were not socially dominant. They were perceived by their classroom peers as more aggressive than the not stable fighter group, but they were also rated less popular and were failing in school. These boys who were physically aggressive since kindergarten appear to have been rejected by their peers and more generally by the school system. Their T levels seem to reflect this status.

This interpretation fits well with the McBurnett et al. (1991) data indicating that 8–13 year old boys had higher levels of salivary cortisol only when they were diagnosed as concurrently anxious and conduct disordered. The high chronic activation of the HPA axis in these children could explain the relative suppression of the HPG axis. It should be noted that the significant results discussed above were obtained before and after having controlled for pubertal status and physical growth. Thus the T-aggression link in early adolescence appears to be dependent on the way aggression is measured and on the social context from where it is derived. Our results suggest that high levels of T are associated with an aggressive style only if the latter confers a dominant status. Higher salivary T level might thus be considered a marker of social well-being and success. Low T concentration in adolescents who were previously highly aggressive may be explained either by increased social rejection, by stable anxiety throughout childhood, or by the conjunction of both influences.

8. IS THE AGGRESSION-TESTOSTERONE ASSOCIATION STABLE FROM EARLY TO MIDDLE ADOLESCENCE?

Sapolsky (1992a, 1992b) has shown that the neuroendocrine mechanisms of a given individual change over time. For example, low-ranking sub adult baboons with elevated basal cortisol can have a better working adrenocortical axis as they grow older and rise in rank. Adolescence is a period of rapid change, both at the biological and the psychosocial level. The 12 year old boys of our sample were all in elementary school. At age 13, those who were in an age-appropriate classroom had moved to high school, while the others had been held back in elementary school. By 16 years of age 15% of our sample had dropped out of school. Over these years almost half (48.6%) of the boys had started having sexual intercourse, and increased the frequency of their delinquent behaviour (Haapasalo & Tremblay, 1994; Malo & Tremblay, 1997; Tremblay et al., 1995). These are important contextual changes which could have an impact on social dominance and hormone levels.

An analysis of variance procedure (BMDP 5V) was used to compare the mean T levels of the SHF and SLF groups from age 13 to age 16 years. Raw salivary T scores were transformed to Z-scores within each year to correct for variations in assay sensitivity over the years; this transformation obviously screens the large increase in T levels during that period. Figure 3 shows the two curves based on the mean Z-scores of the two groups over time. An important group by age interaction effect can be observed (c2 = 10.41, p = .001). As described in the previous section, the boys who were physically aggressive between age 6 and 12 years had, at age 13, a lower mean T level compared to those who had never been physically aggressive. From 14 to 15 years of age the difference between the two groups decreased, and then increased, but in reversed order, at 16 years of age; the physically aggressive boys during childhood now had higher levels of T compared to those who were never physically aggressive.

These results are still more intriguing than the unexpected results at 13 years of age. Can the logic used to explain the results at age 13 (i.e., reduced T due to highly stressful life conditions) be used to explain the reversed results at age 16 years? Well, the curves on Figure 3 appear to indicate a steady pattern over the years. Apparently, we are not observing differences which appear and disappear randomly. Data for the following years would be helpful to confirm if the age 16 differences are maintained. However, in the absence of these data we can only look at other variables to test the hypothesis that the gradual reversal of mean differences over the years can be explained by changes in the boys' social adjustment. An alternative hypothesis, which we cannot test, is that these differences are partly or completely due to some form of genetic or perinatal programming.

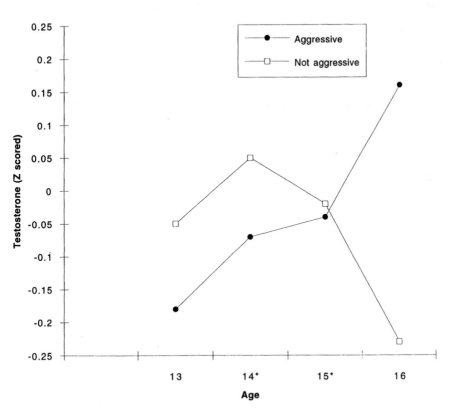

Figure 3. Mean testosterone levels (Z-scored) for the aggressive and not aggressive groups between age 13 and 16.

Four factors were identified as possible explanations for the aggressiveness by age interaction effect on T: 1) Pubertal maturation could have accelerated more quickly in the aggressive group after age 13; 2) more aggressive boys may have started sexual intercourse, stimulating the production of T; 3) more aggressive boys may have been getting involved in delinquent gangs in which aggression enables them to become socially dominant, or in which they receive support to feel more socially dominant; 4) more aggressive boys dropped out of school, thus reducing the stress linked to the daily experience of failing in school and being rejected by peers and teachers.

No significant differences between the aggressives and not aggressives were observed from 13 to 16 years of age for mean body surface and pubertal stage. More aggressive boys had started sexual intercourse at age 13 and 14, more were involved in delinquent gangs at age 13 and 16; the aggressives were committing significantly more delinquent acts, and more aggressive delinquent acts, than the never aggressives in each year between 13 and 16 years; also, more aggressive boys were not in an age-appropriate regular classroom from 13 to 16 years. However, for all of these variables, approximately the same magnitude of differences were present from age 13 to age 16 years. In other words, there was no interaction by age effect which could have explained the interaction effect for T. The only variable that appeared to mirror the T pattern was dropping out of school (see Figure 4). None of the never aggressive boys had dropped out of school by age 16, while the percentage of aggressive boys who had dropped out of school went from 0% at age 13 to 30% at 16 years of age. However, when school drop out was entered as a covariate in the analysis of variance which indicated an age by group interaction effect for T, the interaction remained. Thus, none of the hypothesized variables explain the important relative change in T level between the aggressive and not aggressive boys from early to middle puberty.

9. CONCLUSION

Most studies of aggression and T in humans have been cross-sectional, and focused on adult antisocial behavior. The data presented in this chapter underscore the complexity of the behavior-hormone interactions at the developmental period when the largest increase in T level occurs, and the incidence of serious violent offences is peaking (Elliott, 1994). It would be difficult to conclude from these data that increases in T levels at puberty explain the increase in violent offending, as suggested by Ellis and Coontz (1990).

Frequent physical aggression can be observed from infancy onwards (Cummings, Iannotti, & Zahn-Waxler, 1989; Noël, Leclerc, & Strayer, 1990; Tremblay, Mâsse, Pagani, & Vitaro, 1996). In fact, the period in life when humans are most often physically aggressive towards their peers seems to be between 18 and 30 months after birth (Finkelstein et al., 1994; Kingston & Prior, 1995; Noël et al., 1990; Restoin et al., 1985; Tremblay et al., 1996). By the time children go to kindergarten, most have learned to control their physically aggressive actions or reactions. Those who have not tend to be rejected by their peers and their teachers (Boivin & Vitaro, 1995; Coie et al., 1990; Vitaro, Tremblay, Gagnon, & Boivin, 1992). Thus, they are not socially dominant in the school peer group, which may be the main reference group for children from kindergarten to high school.

When such a developmental perspective is considered, one would expect that early adolescents with a history of frequent physical aggression will have low levels of T. This is what we observed at 13 years of age. However, by age 16, the physically aggressive boys who had been rejected by their school peers, and had failed in school, now had

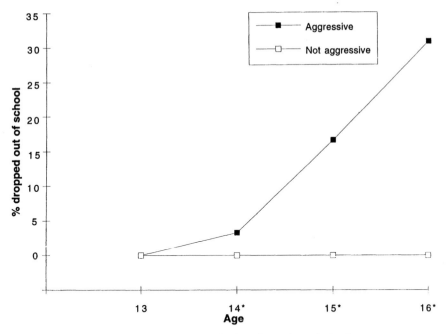

Figure 4. Percent of boys who have dropped out of school.

higher levels of T than those who were never physically aggressive. Also, they were apparently better socially adjusted, i.e. they were still in school and were less delinquent. These results are puzzling.

We obviously need to continue to monitor the development of these subjects to obtain a better understanding of the dynamic relations between change in social adjustment over time and change in T level. Physical aggression is a very small proportion of an aggressive adolescent's behaviors. To understand the association between T and physical aggression we need to know the context in which he is living, and his social adjustment in that context. In some contexts physical aggression will increase social dominance, while in others it will decrease social dominance. In the school context, physical aggression will most likely not lead to social dominance; but in the street life context, physical aggression may well lead to social dominance. The boys who showed a pattern of chronic physical aggression throughout their school years may be joining a reference group (e.g., street gangs) which makes them feel more socially dominant. Our measures of gang involvement and everyday activities may have been inadequate to capture the context in which they were living.

However, from the social dominance perspective, it remains puzzling to observe that boys who are failing to meet the requirements needed to be successful young adults have higher T levels than those who meet these requirements. A number of studies have shown that social status is related to illness and mortality (Marmot, Kogevinas, & Elston, 1987). With a large sample of British civil servants, Marmot (1994) showed that the grade of employment (1. administrative, 2. professional/executive, 3. clerical, 4. others) was strongly related to morbidity and age adjusted mortality rates. The civil servants who had the highest social position were the less at risk of illness and early death. Interestingly, the risk level formed a perfect gradient with grade of employment. The results of these studies fit well with the animal data showing that low social dominance has an impact on the physiological processes which lead to illness and aging through the activation of the HPA axis and inhibition of the HPG axis (Sapolsky, 1992a, 1992b).

Thus, if socio-economic position as a civil servant is related to illness and death, one would expect that it is also related to levels of T. Marmot and his colleagues do not appear to have T data on their civil servants, but such data is available in a study of 4462 US military veterans (Dabbs, 1992a; see also Windle & Windle, 1995). Surprisingly, Dabbs showed that the mean level of T was highest for men who were unemployed, and that blue collar workers had higher T levels than white collar workers. These results are exactly the opposite of those expected from a social dominance perspective.

We need to understand the mechanisms which lead to these unexpected findings. Because T tends to be positively associated with social dominance, one would expect that adolescent boys who fail in school and adult males with lower socio-economic status will have lower T levels. Why would Dabbs observe the reverse? And why would our non-aggressive subjects who continue their studies have, by age 16, lower levels of T than the physically aggressive boys who are failing in school? Could prolonged schooling have an impact on T levels? Or do low levels of T help adolescents to concentrate on school work rather than invest in courtship and competing for a job?

To explain his results, Dabbs (1992a) suggested that because T levels have a genetic component (Meikle, Bishop, Stringham, & West, 1987; Meikle, Stringham, Bishop, & West, 1988; Miller & Martin, 1995; Sakai, Baker, Jacklin, & Shulman, 1991) they affect in utero brain development, social behavior, cognitive development, success in school and finally occupational status. Individuals who inherit high T levels would be more likely to have low verbal intelligence, to show wild and excessive behavior, to react aggressively to frustration, to be antisocial, to fail in school, and thus end up in a low socio-economic status job or unemployed. From an evolutionary perspective Dabbs (1992a, 1992b) concludes that the positive association of T and aggression brings high status to animals because social dominance is based on physical aggression, but it brings low status to humans because aggression generally does not lead to social dominance.

If we accept that T levels are mostly inherited and determine behavior from early childhood onwards, then the interpretation given by Dabbs for his results is plausible[1]. This interpretation could apply to the differences we observed when our subjects were 16 years of age, but they do not explain the changes in relative T levels we observed from 13 to 16 years of age between the aggressive and not aggressive boys. Why would inherited T levels show up at 16 years of age, but not at 13 years of age? It cannot be excluded that inherited levels become apparent only when full pubertal maturation has occurred. But this line of argument rules out the well documented influence of social interaction on T levels. Given that many animal and human studies have shown that T levels change dramatically with the outcome of social encounters, it would be surprising if T levels were not determined by both an inherited disposition and the social context.

Surprisingly, Dabbs (1992a) ruled out the possibility that the military veterans' T levels he observed were the partial results of their social context at the time of sampling. These men (mean age 38 years) were required to go to a site in Albuquerque and live there for 5 days during which they were subjected to medical, physical, psychological, neuropsychological, and clinical laboratory testing. It would be difficult to argue that these men were living in a natural environment. It is not clear what the living conditions were for these five days. To what extent did the men interact with each other? One can imagine that a professional man who leaves his family and business for five days to live in less comfortable conditions and interact with strangers who are mostly "blue collar" workers or unemployed will react differently to the situation, compared to an unemployed, single, and antisocial man. Context-dependent responses of the HPG axis could bring about effects that are at least equivalent to the small effect size (0.01) which was observed between T and occupational status[2].

Future studies need to assess T levels in different contexts. Ideally, one would need to obtain T level measurements in a socially neutral situation, after social success, and after social failure. Studies with repeated sampling over many days have been done with athletes (Cook, Read, Walker, Harris, & Riad-Fahmy, 1992). Similar studies could help to understand the causes of violent behavior. We were successful in obtaining four saliva samples during a single day, but a measure of T taken at home in the morning on a non-stressful day would have been a better indicator of base rate T levels. There is a possibility that at age 13 years, the non-aggressive boys felt more at ease than the aggressive boys in the laboratory situation, while at age 16, the aggressive boys may have been successful in dominating the non aggressive boys, thus creating the reverse differences in T levels.

Although we have gone beyond the still photograph provided by cross-sectional studies, the movement we are observing clearly suggests more questions than it gives answers. To really understand why mean levels of T were higher in physically aggressive and lower social class male adult samples, we may need to monitor hormone, behavior, and social context from fetal development to adult status. Studies of T levels during the first five months after birth, with follow-ups of impulsive, aggressive, and antisocial behavior over the following two or three years, should help understand to what extent the differences in hormone behavior observed during adulthood are present in infancy and interact with environmental conditions.

NOTES

1. There are problems with an evolutionary interpretation which would lead to the conclusion that T would be negatively associated with occupational status because the traditional "masculine" behaviors needed to win "physically-based dominance disputes" are maladaptive in societies which reward academic success, but adaptive in lower socio-economic groups (see Archer, 1994).

    First, it minimises the effect of social dominance on T levels. It suggests that T levels of males are determined by processes (inherited and learned) which bring them to adopt traditional male behaviors, independently of the value of these behaviors in achieving success in our modern societies. It can be argued that males with lower socio-economic status can achieve dominance in their narrow reference group. But the same argument would apply to males with higher socio-economic status. Some are more dominant than others in their narrow reference group. For example, Dabbs (1992b) reports that trial lawyers have higher T levels than non-trial lawyers. Why would the highly dominant males in the high socio-economic status groups have lower T levels than the highly dominant males in the low socio-economic groups?

    The second problem is that the "masculinity" hypothesis implies that in working class environments social dominance is achieved by physical aggression. Although most studies of social class and physical aggression have shown that there are more aggressive individuals in lower socio-economic classes, most males in lower socio-economic classes are not highly physically aggressive (Haapasalo & Tremblay, 1994; Tremblay et al., in press). It should be remembered that the subjects in the present study were all from low socio-economic status schools and families. At age 13 years the most dominant boys were not those with a history of physical aggression.

To support the "masculinity" hypothesis as an explanation of the class differences in T level, one needs to argue that the socially dominant males in lower socio-economic environments are physically aggressive, impulsive, and antisocial. The data from the present study with a male sample who were all from low socio-economic environments shows that the physically aggressive males do not become socially dominant. From school entry onwards they are rejected by the majority. This phenomenon has been observed in most studies of aggression and peer rejection across socio-economic classes, cultures, and races (Coie et al., 1990).

Drift towards low socio-economic status is an alternative hypothesis to explain social class differences in physical aggression, impulsivity, antisocial behavior, and T. To the extent that high levels of physical aggression and antisocial behavior lead to rejection by the peer group and failure in school, one would expect that individuals with these traits will be unable to integrate in the work force, and will thus remain or drift into low socio-economic status. If extremely high levels of T are, for one reason or another, associated with highly aggressive and antisocial behavior (Virkkunen et al., 1994), one would then expect that high T individuals would remain or drift into a lower socio-economic status. Over a number of generations it would make sense to observe differences in levels of aggressive behavior, antisocial behavior and T levels among social classes. But this would not mean that aggression and high T levels are adaptive in lower socio-economic environments. The physically aggressive individuals will tend to be rejected even in lower socio-economic environments.

2. It is important to note that Dabbs (1992a) did not control for race and diet of the subjects. There have been reports of racial differences in T levels and diet effects on T levels (Bishop et al., 1988; Ellis & Nyborg, 1992; Hill, Wynder, Garbaczewski, Garnes, & Walker, 1979; Ross et al., 1986). There are also important differences between blacks and whites on occupational status and on most of the variables Dabbs reports as being associated with T. To test if there are cumulative effects over generations, one needs to show that high levels of physical aggression and T are transmitted from one generation to another, through genetic and/or environmental processes. One would then expect family effects for physical aggression and T. With a random sample (N = 10,287) of Canadian families having children between the ages of 4 and 11 years, we have recently shown that physically aggressive children are concentrated in certain families, and this concentration is greater in lower socio-economic status families compared with higher socio-economic status families (Tremblay et al., in press). Familial factors explained 53% of the variance in physical aggression within the lower social class and only 3% within the higher social class. Thus, higher levels of physical aggression in low socio-economic environments are not due to generalized physical aggression among the less economically disadvantaged, but rather to a concentration of physically aggressive children in a restricted number of families. Data from twin studies also show that familial factors (including genetic and intrauterine factors) account for more than 50% of the variation in T levels (Meikle et al., 1987; Meikle et al., 1988; Miller & Martin, 1995; Sakai et al., 1991).

The social class differences in T levels could thus be the result of a greater concentration of individuals with abnormally high levels of T in lower social classes. To the extent that high T levels would be part of a syndrome which includes impulsive behavior, aggression, and antisocial behavior (Archer, 1994), these same individuals could account for the mean differences in impulsivity, physical aggression, and antisocial behavior between social classes.

ACKNOWLEDGMENTS

This work was made possible by grants from the following sources: Program on Human Development and Criminal Behavior at the Harvard School of Public Health, Cambridge, MA, Quebec's CQRS, FRSQ and FCAR funding programs, the National Health Research and Development Program of Canada (NHRDP),Ottawa, and the Social Sciences and Humanities Research Council of Canada (SSHRC), Ottawa. We are grateful to D. Laurent, R. Soussignan and E.J. Susman for their contributions to planning and analysing parts of this work, to L. Arseneault, H. Beauchesne, L. David, J. Séguin & I.M.-C Tremblay for data collection; to L. Desmarais-Gervais, M. Rosa for data management and analyses; to L. Leblanc for manuscript typing.

REFERENCES

Archer, J. (1988). *The behavioural biology of aggression.* Cambridge: Cambridge University Press.
Archer, J. (1991). The influence of testosterone on human aggression. *British Journal of Psychology, 82,* 1–28.
Archer, J. (1994). Testosterone and aggression. *Journal of Offender Rehabilitation, 21*(3–4), 3–39.
Bambino, T. H., & Hsueh, A. J. (1981). Direct inhibitory effect of glucocorticoids upon testicular luteinizing hormone receptor and steroidogenesis in vivo and in vitro. *Endocrinology, 108*(6), 2142–2148.
Belsky, J., Steinberg, L., & Draper, P. (1991). Childhood experience, interpersonal development and reproductive strategy: An evolutionary theory of socialization. *Child Development, 62,* 647–670.
Berenbaum, S. A., & Snyder, E. (1995). Early hormonal influences on childhood sex-typed activity and playmate preferences: Implications for the development of sexual orientation. *Developmental Psychology, 31*(1), 31–42.
Bishop, D. T., Meikle, A. W., Slattery, M. L., Stringham, J. D., Ford, M. H., & West, D. W. (1988). The effect of nutritional factors on sex hormone levels in male twins. *Genetics and Epidemiology, 5,* 43–59.
Boivin, M., & Vitaro, F. (1995). The impact of peer relationships on aggression in childhood: Inhibition through coercion or promotion through peer support. In J. McCord (Ed.), *Coercion and punishment in long-term perspectives.* (pp. 183–197). New York: Cambridge Press.
Booth, A., Shelley, G., Mazur, A., Tharp, G., & Kittok, R. (1989). Testosterone, and winning and losing in human competition. *Hormones and Behavior, 23*(2), 556–571.
Brain, P. F. (1979). *Hormones and Aggression.* Montréal: Eden Press.
Cho, H., Sanayama, K., Sasaki, N., & Nakajima, H. (1985). Salivary testosterone concentration and testicular volume in male infants. *Endocrinology, 32*(1), 135–140.
Christiansen, K., & Knussmann, R. (1987). Androgen levels and components of aggressive behavior in men. *Hormones and Behavior, 21,* 170–180.
Coie, J. D., Dodge, K. A., & Kupersmidt, J. B. (1990). Peer group behavior and social status. In S. R. Asher & J. D. Coie (Eds.), *Peer rejection in childhood.* Cambridge: Cambridge University Press.
Collu, R., Gibb, W., & Ducharme, J. R. (1984). Effects of stress on the gonadal function. *Journal of Endocrinological Investigation, 7*(5), 529.
Constantino, J. N., Grosz, D., Saenger, P., Chandler, D. W., Nandi, R., & Earls, F., J. (1993). Testosterone and aggression in children. *Journal of the American Academy of Child and Adolescent Psychiatry, 32*(6), 1217–1222.

Cook, N. J., Read, G. F., Walker, R. F., Harris, B., & Riad-Fahmy, D. (1992). Salivary cortisol and testosterone as markers of stress in normal subjects in abnormal situations. In C. Kirschbaum, G. F. Read, & D. H. Hellhammer (Eds.), *Assessment of hormones and drugs in saliva in biobehavioral research.* (pp. 147–162). Göttingen, Germany: Hogrefe & Huber Publishers.

Cummings, E. M., Iannotti, R. J., & Zahn-Waxler, C. (1989). Aggression between peers in early childhood: Individual continuity and developmental change. *Child Development*, 60(4), 887–895.

Dabbs, J. M. J. (1991). Salivary testosterone measurements: Reliability across hours, days, and weeks. *Physiology and Behavior*, 48, 83–86.

Dabbs, J. M. J. (1992a). Testosterone and occupational achievement. *Social Forces*, 70, 813–824.

Dabbs, J. M. J. (1992b). Testosterone measurements in social and clinical psychology. *Journal of Social and Clinical Psychology*, 11, 302–321.

Dabbs, J. M. J., Frady, R. L., Carr, T. S., & Besch, N. F. (1987). Saliva testosterone and criminal violence in young adult prison inmates. *Psychosomatic Medicine*, 49, 174–182.

Dabbs, J. M. J., & Morris, R. (1990). Testosterone, social class, and antisocial behavior in a sample of 4,462 men. *Psychological Science*, 1(3), 209–211.

de Waal, F. (1982). *Chimpanzee Politics*. New York: Harper Colophon.

Dittmann, R. W., Kappes, M. E., & Kappes, M. H. (1992). Sexual behavior in adolescent and adult females with congenital adrenal hyperplasia. *Psychoneuroendocrinology*, 17, 153–170.

Ducharme, J. R., Forest, M. G., de Peretti, E., Sempé, M., Collu, R., & Bertrand, J. (1976). Plasma adrenal and gonadal sex steroids in human pubertal development. *Journal of Clinical Endocrinology and Metabolism*, 42, 468–476.

Eaton, G., Goy, R., & Phoenix, C. (1973). Effects of testosterone treatment in adulthood on sexual behavior of female pseudohermophrodite rhesus monkeys. *Nature*, 242, 119–120.

Eichelman, B. (1992). Aggressive behavior: From laboratory to clinic: Quo Vadis? *Archives of General Psychiatry*, 49(5), 488–492.

Elias, M. (1981). Serum cortisol, testosterone, and testosterone-binding globulin responses to competitive fighting in human males. *Aggressive Behavior*, 7, 215–224.

Elliott, D. S. (1994). Serious violent offenders: Onset, developmental course and termination: The American Society of Criminology 1993 Presidential Address. *Criminology*, 32(1), 1–21.

Ellis, L., & Coontz, P. D. (1990). Androgens, brain functionning, and criminality: The neurohormonal foundations of antisociality. In L. Ellis & H. Hoffman (Eds.), *Crime in biological, social and moral contexts.* (pp. 162–193). New York: Praeger.

Ellis, L., & Nyborg, H. (1992). Racial/ethnic variations in male testosterone levels. *Steroids*, 57, 72–75.

Eysenck, H. J., & Gudjonsson, G. H. (1989). *The causes and cures of criminality*. New York: Plenum.

Finkelstein, J. W., von Eye, A., & Preece, M. A. (1994). The relationship between aggressive behavior and puberty in normal adolescents: A longitudinal study. *Journal of Adolescent Health*, 15, 319–326.

Forest, M. G. (1981). Control of the onset of puberty. In R. Crosignani (Ed.), *Endocrinology of human infertility: New aspects.* (pp. 267–306). London: Academic Press.

Forest, M. G. (1989). Androgens in childhood. *Pediatric and Adolescent Endocrinology*, 19, 104–129.

Forest, M. G., Sizonenko, P. C., Cathiard, A. M., & Bertand, J. (1974). Hypophyso-gonadal function in humans during the first year of life. I. Evidence for testicular activity in early infancy. *Journal of Clinical Investigation*, 53, 819–828.

Gladue, B. A. (1991). Qualitative and quantitative sex differences in self-reported aggressive behavioral characteristics. *Psychological Reports*, 68(2), 675–684.

Gladue, B. A., Boechler, M., & McCaul, K. D. (1989). Hormonal response to competition in human males. *Aggressive Behavior*, 15(6), 409–422.

Goy, R. W., Bercovitch, F. B., & McBrair, M. C. (1988). Behavioral masculinization is independent of genital masculinization in prenatally androgenized female rhesus macaques. *Hormones and Behavior*, 22, 552–571.

Goy, R. W., & McEwen, B. S. (1980). *Sexual differentiation of the brain*. Cambridge, MA: MIT Press.

Gray, A., Jackson, D. N., & McKinlay, J. B. (1991). The relation between dominance, anger, and hormones in normally aging men: Results from the Massachusetts male aging study. *Psychosomatic Medicine*, 53, 375–385.

Haapasalo, J., & Tremblay, R. E. (1994). Physically aggressive boys from ages 6 to 12: Family background, parenting behavior, and prediction of delinquency. *Journal of Consulting and Clinical Psychology*, 62(5), 1044–1052.

Halpern, C. T., & Udry, J. R. (1992). Variation in adolescent hormone measures and implications for behavioral research. *Journal of Research on Adolescence*, 2, 103.

Halpern. C. T., Udry, J. R., Campbell, B., & Suchindran, C. (1993a). Relationships between aggression and pubertal increases in testosterone: A panel analysis of adolescent males. *Social Biology, 40*(1–2), 8–24.

Halpern. C. T., Udry, J. R., Campbell, B., & Suchindran, C. (1993b). Testosterone and pubertal development as predictors of sexual activity: A panel analysis of adolescent males. *Psychosomatic Medicine, 55*(5), 436–447.

Higley, J. D., Mehlman, P. T., Taub, D. M., Higley, S. B., Suomi, S. J., Linnoila, M., & Vickers, J. H. (1992). Cerebrospinal fluid monoamine and adrenal correlates of aggression in free-ranging Rhesus monkeys. *Archives of General Psychiatry, 49*(6).

Hill, P., Wynder, E. L., Garbaczewski, L., Garnes, H., & Walker, A. R. P. (1979). Diet and urinary steroids in black and white North American men and black South African men. *American Journal of Clinical Nutrition, 42*, 127–134.

Hines, M. (1982). Prenatal gonadal hormones and sex differences in human behavior. *Psychological Bulletin, 92*, 56–80.

Hines, M., & Kaufman, F. R. (1994). Androgen and the development of human sex-typical behavior: Rough-and-tumble play and sex of preferred playmates in children with congenital adrenal hyperplasia (CAH). *Child Development, 65*, 1042–1053.

Hoberman, J. M., & Yesali, C. E. (1995). The history of synthetic testosterone. *Scientific American, 272*, 76–81.

Huhtaniemi, I., Dunkel, L., & Perheentupa, J. (1986). Transient increase in postnatal testicular activity is not revealed by longitudinal measurements of salivary testosterone. *Pediatric Research, 20*(12), 1324–1327.

Inoff-Germain, G., Arnold, G. S., Nottelmann, E. D., Susman, E. J., Cutler, G. B., & Chrousos, G. P. (1988). Relations between hormone levels and observational measures of aggressive behavior of young adolescents in family interactions. *Developmental Psychology, 24*, 129–139.

Johnson, H. R., Myhre, S. A., & Ruvalcaba, R. H. (1970). Effects of testosterone on body image and behavior in Klinefelter's syndrome: A pilot study. *Developmental Medicine and Child Neurology, 12*(4), 454–460.

Keverne, E. B. (1992). Primate social relationships: Their determinants and consequences. *Advance Study Behavior, 21*(1).

Kindlon, D. J., Tremblay, R. E., Mezzacappa, E., Earls, F., Laurent, D., & Schaal, B. (1995). Longitudinal patterns of heart rate and fighting behavior in 9 through 12 year old boys. *Journal of the American Academy of Child and Adolescent Psychiatry, 34*(3), 371–377.

Kingston, L., & Prior, M. (1995). The development of patterns of stable, transient, and school-age onset aggressive behavior in young children. *Journal of the American Academy of Child and Adolescent Psychiatry, 34*, 348–358.

Malo, J., & Tremblay, R. E. (1997). The impact of paternal alcoholism and maternal social position on boy's school adjustment, pubertal maturation and sexual behaviour: A test of two competing hypotheses. *Journal of Child Psychology and Psychiatry and allied disciplines, 38*(2).

Marmot, M. G. (1994). Social differentials in health within and between populations. *Journal of the American Academy of Arts and Sciences, 123*(4), 197–216.

Marmot, M. G., Kogevinas, M., & Elston, M. A. (1987). Social/economic status and disease. *Annual Revue of Public Health, 8*, 111–137.

Marshall. W. A., & Tanner, J. M. (1969). Variations in pattern of pubertal changes in boys. *Archives of Disease in Childhood, 44*, 13–23.

Mattson, A., Schalling, D., Olweus, D., Low, H., & Svenson, J. (1980). Plasma testosterone, aggressive behavior, and personality dimensions in young male delinquents. *Journal of the American Academy for Child Psychiatry, 19*, 476–490.

Mazur, A., Booth, A., & Dabbs Jr., J. M. (1992). Testosterone and chess competition. *Social Psychology Quarterly, 55*(1), 70–77.

Mazur, A., & Lamb, T. A. (1980). Testosterone, status, and mood in human males. *Hormones and Behavior, 14*, 236–246.

McBurnett, K., Lahey, B. B., Frick, P. J., Risch, S. C., Loeber, R., Hart, E. L., Christ, M. A. G., & Hanson, K. S. (1991). Anxiety, inhibition, and conduct disorder in children: II. Relation to salivary cortisol. *Journal of the American Academy of Child and Adolescent Psychiatry, 30*(2), 192–196.

McCaul, K. D., Gladue, B. A., & Joppa, M. (1992). Winning, losing, mood, and testosterone. *Hormones and Behavior, 26*, 486–504.

Meikle, A. K., Bishop, D. T., Stringham, J. D., & West, D. W. (1987). Quantitating genetic and nongenetic factors that determine plasma sex steroid variation in normal male twins. *Metabolism, 35*, 1090–1095.

Meikle, A. K., Stringham, J. D., Bishop, D. T., & West, D. W. (1988). Quantitating genetic and nongenetic factors influencing androgen production and clearance rates in men. *Journal of Clinical Endocrinology and Metabolism, 67*, 104–109.

Mezzacappa, E., Tremblay, R. E., Kindlon, D., Saul, J. P., Arseneault, L., Séguin, J., Pihl, R. O., & Earls, F. (in press). Anxiety, antisocial behavior and heart rate regulation in adolescent males. *Journal of Child Psychology and Psychiatry*.

Miller, E. M., & Martin, N. (1995). Analysis of the effect of hormones on opposite-sex twin attitudes. *Acta Geneticae Medicae et Gemmellologiae, 44*(1), 41–52.

Money, J., & Schwartz, M. (1976). Fetal androgens in the early treated adrenogenital syndrome of 46XX hermaphroditism: Influence on assertive and aggressive types of behavior. *Aggressive Behavior, 2,* 19–30.

Nieschlag, E. (1974). Circadian rhythms of plasma testosterone. In J. Aschoff, F. Ceresa, & F. Halberg (Eds.), *Chronobiological aspects of endocrinology*. (pp. 117–120). Stuttgart: Schattauer Verlag.

Noël, J. M., Leclerc, D., & Strayer, F. F. (1990). Une analyse fonctionnelle du répertoire social des enfants d'âge pré-scolaire en groupe de pairs. *Enfance, 45*(4), 405–421.

Nottelman, E. D., Inoff-Germain, G., Susman, E. J., & Chrousos, G. P. (1990). Hormones and behavior at puberty. In J. Bancroft & J. M. Reinisch (Eds.), *Adolescence and Puberty*. New York: Oxford University Press.

Olweus, D. (1988). *Development of a multifaceted aggression inventory for boys*. Reports from the Institute of Psychology (No. 6). Bergen: University of Bergen.

Olweus, D., Mattsson, A., Schalling, D., & Low, H. (1980). Testosterone, aggression, physical, and personality dimensions in normal adolescent males. *Psychosomatic Medicine, 42*, 253–269.

Olweus, D., Mattsson, A., Schalling, D., & Low, H. (1988). Circulating testosterone levels and aggression in adolescent males: A causal analysis. *Psychosomatic Medicine, 50*, 261–272.

Petersen, A. C., Crockett, L., Richards, M., & Boxer, A. (1988). A self-report measure of pubertal status: Reliability, validity, and initial norms. *Journal of Youth and Adolescence, 17*(2), 117–133.

Power, C., Manor, O., & Fox, J. (1991). *Health and class: The early years*. London: Chapman & Hall.

Reinisch, J. M. (1981). Prenatal exposure to synthetic progestins increases potential for aggression in humans. *Science, 211*, 1171–1173.

Reiss, A. J., & Roth, J. A. (Ed.). (1993). *Understanding and preventing violence*. Washington, D.C.: National Academy Press.

Resnick, S. M., Berenbaum, S. A., Gottesman, I. I., & Bouchard, T. (1986). Early hormonal influences on cognitive functioning in congenital adrenal hyperplasia. *Developmental Psychology, 22*, 191–198.

Restoin, A., Montagner, H., Rodriguez, D., Girardot, J. J., Laurent, D., Kontar, F., Ullmann, V., Casagrande, C., & Talpain, B. (1985). Chronologie des comportements de communication et profils de comportement chez le jeune enfant. In R. E. Tremblay, M. A. Provost, & F. F. Strayer (Eds.), *Ethologie et développement de l'enfant*. (pp. 93–130). Paris: Editions Stock/Laurence Pernoud.

Riad-Fahmy, D., Read, G. F., Walker, R. F., & Griffiths, K. (1982). Steroids in saliva for assessing endocrine function. *Endocrine Reviews, 3*, 367–395.

Rose, R. M., Bernstein, I. S., & Gordon, T. P. (1975). Consequences of social conflict on plasma testosterone levels in Rhesus Monkeys. *Psychosomatic Medicine, 37*, 50.

Ross, R. Bernstein, L., Judd, H., Hanisch, R., Pike, M., & Henderson, B. (1986). Serum testosterone levels in healthy young black and white men. *Journal of the National Cancer Institute, 76*, 45–48.

Sakai, L. M., Baker, L. A., Jacklin, C. N., & Shulman, I. (1991). Sex steroids at birth: Genetic and environmental variation and covariation. *Developmental Psychobiology, 24*(8), 559–570.

Sapolsky, R. (1983). Individual differences in cortisol secretory patterns in the wild baboon: Role of negative-feedback sensitivity. *Endocrinology, 113*, 2263.

Sapolsky, R. M. (1991). Testicular function, social rank and personality among wild baboons. *Psychoneuroendocrinology, 16*(4), 281–293.

Sapolsky, R. M. (1992a). Neuroendocrinology of the stress-response. In J. B. Becker, S. M. Breedlove, & D. Crews (Eds.), *Behavioral Endocrinology*. (pp. 287–324). Cambridge, MA: A Bradford Book, The MIT Press.

Sapolsky, R. M. (Ed.). (1992b). *Stress, the aging brain, and the mechanisms of neuron death*. Cambridge, MA: A Bradford Book, The MIT Press.

Savin-Williams, R. C. (1979). Dominance hierarchies in groups of early adolescents. *Child Development, 50*, 923–935.

Schaal, B., Tremblay, R. E., Soussignan, R. G., Paquette, D., & Laurent, D. (1996). Aggression, testosterone and physical development in early adolescence: A longitudinal perspective. *Aggressive Behavior*.

Schaal, B., Tremblay, R. E., Soussignan, R., & Susman, E. J. (1996). Male testosterone linked to high social dominance but low physical aggression in early adolescence. *Journal of the American Academy of Child and Adolescent Psychiatry, 34*.

Séguin, J. R., Pihl, R. O., Boulerice, B., Tremblay, R. E., & Harden, P. W. (1996). Low pain sensitivity and stability of physical aggression in boys. *Journal of Child Psychology and Psychiatry*, 671–673.

Séguin, J. R., Pihl, R. O., Harden, P. W., Tremblay, R. E., & Boulerice, B. (1995). Cognitive and neuropsychological characteristics of physically aggressive boys. *Journal of Abnormal Psychology, 104*(4), 614–624.

Soussignan, R., Tremblay, R. E., Schaal, B., Laurent, D., Larivée, S., Gagnon, C., LeBlanc, M., & Charlebois, P. (1992). Behavioural and cognitive characteristics of conduct disordered-hyperactive boys from age 6 to 11: A multiple informant perspective. *Journal of Child Psychology and Psychiatry, 33*(8), 1333–1346.

Strayer, F. F., & Trudel, M. (1984). Developmental changes in the nature and function of social dominance among young children. *Ethology and Sociobiology, 5*, 279–295.

Susman, E. J., Dorn, L. D., & Chrousos, G. P. (1991). Negative affect and hormone levels in young adolescents: Concurrent and predictive perspectives. *Journal of Youth and Adolescence, 20*(2), 167–190.

Susman, E. J., Inoff-Germain, G., Nottelmann, E. D., Loriaux, D. L., Cutler, G. B., & Chrousos, G. P. (1987). Hormones, emotional dispositions, and aggressive attributes in young adolescents. *Child Development, 58*, 1114–1134.

Tanner, J. M. (1962). *Growth at adolescence.* Oxford: Blackwell.

Tremblay, R. E., Boulerice, B., Harden, P. W., McDuff, P., Pérusse, D., Pihl, R. O., & Zoccolillo, M. (in press). *Do Canadian children become more aggressive as they approach adolescence?* Ottawa: Statistics Canada.

Tremblay, R. E., Kurtz, L., Mâsse, L. C., Vitaro, F., & Pihl, R. O. (1995). A bimodal preventive intervention for disruptive kindergarten boys: Its impact through mid-adolescence. *Journal of Consulting and Clinical Psychology, 63*(4), 560–568.

Tremblay, R. E., Loeber, R., Gagnon, C., Charlebois, P., Larivée, S., & LeBlanc, M. (1991). Disruptive boys with stable and unstable high fighting behavior patterns during junior elementary school. *Journal of Abnormal Child Psychology, 19*, 285–300.

Tremblay, R. E., Mâsse, L. C., Pagani, L., & Vitaro, F. (1996). From childhood physical aggression to adolescent maladjustment: The Montréal Prevention Experiment. In R. D. Peters & R. J. McMahon (Eds.), *Preventing childhood disorders, substance abuse and delinquency.* (pp. 268–298). Thousand Oaks, CA: Sage.

Tremblay, R. E., Pihl, R. O., Vitaro, F., & Dobkin, P. L. (1994). Predicting early onset of male antisocial behavior from preschool behavior. *Archives of General Psychiatry, 51*, 732–738.

Tremblay, R. E., & Schaal, B. (1996). Physically aggressive boys from age 6 to 12 years: Their biopsychosocial status at puberty. In G. Ferris & T. Grisso (Eds.), *Understanding aggressive behavior in children. 794* (pp. 192–208). New York: Annals of the New York Academy of Sciences.

Virkkunen, M., Kallio, E., Rawlings, R., Tokola, R., Poland, R., E., Guidotti, A., Nemeroff, C., Bissette, G., Kalogeras, K., Karonen, S.-L., & Linoila, M. (1994). Personality profiles and state aggressiveness in Finnish alcoholic, violent offenders, fire setters, and healthy volunteers. *Archives of General Psychiatry, 51*, 20–27.

Vitaro, F., Tremblay, R. E., Gagnon, C., & Boivin, M. (1992). Peer rejection from kindergarten to grade 2: Outcomes, correlates, and prediction. *Merrill-Palmer Quarterly, 38*(3), 382–400.

Wadsworth, M. E. J. (1991). *The imprint of time: Childhood, history, and adult life.* Oxford: Clarendon Press.

Wang, C., Plymate, S., Nieschlag, E., & Paulsen, C. (1981). Salivary testosterone in men: Further evidence of a direct correlation with free serum testosterone. *Journal of Clinical Endocrinology and Metabolism, 53*, 1021–1024.

Wilson, J. D., George, F. W., & Griffin, J. E. (1981). The hormonal control of sexual development. *Science, 211*, 1278–1284.

Windle, R. C., & Windle, M. (1995). Longitudinal patterns of physical aggression: Associations with adult social, psychiatric, and personality functioning and testosterone levels. *Development and Psychopathology, 7*, 563–585.

Winter, J. S. D., & Faiman, C. (1973). The development of cyclic pituitary-gonadal function in adolescent females. *Journal of Clinical Endocrinology and Metabolism, 37*, 714–718.

17

KEY ISSUES IN STUDYING THE BIOSOCIAL BASES OF VIOLENCE

David P. Farrington

Institute of Criminology
University of Cambridge
Cambridge, United Kingdom

In this chapter, I will attempt to summarize some of the key issues arising in the NATO Advanced Study Institute on Biosocial Bases of Violence. However, it is not possible to convey in graphic detail the many vigorous yet constructive exchanges that took place. The Institute was very successful in bringing together biological and psychosocial researchers, including both leading scholars and young scientists. The multi-disciplinary nature of the meeting was important in exposing biological and psychosocial approaches to productive critiques and thoughtful points from other disciplines. I hope that this resulting volume will be seen as a showcase of the best current biosocial research and that it will stimulate even more fruitful collaborative efforts in the future.

1. A GENERAL MODEL OF VIOLENCE

I find it useful to organize the key issues using a general model of influences on violence, shown in Figure 1. This model is similar to one presented at the Institute by Per-Olof Wikström. It is assumed that biological and psychosocial factors can influence each other, and that they can interact in specified ways to produce an individual potential for violence, as well as influencing this potential directly. It is further assumed that this individual potential and situational factors can influence each other, and that they can interact in specified ways to produce violent behavior, as well as influencing this behavior directly. The model is, of course, highly simplified. However, it is presented in response to several of the more social researchers at the Institute, who felt that existing biosocial models paid insufficient attention to situational factors and to how the individual potential for violence was translated into violent behavior in specific situations.

2. VIOLENCE

As pointed out in Chapter 1, the Institute was concerned with antisocial, criminal, and psychopathic behavior as well as with violence. A key issue is how far all of these

Biosocial Bases of Violence, edited by Raine *et al.*
Plenum Press, New York, 1997

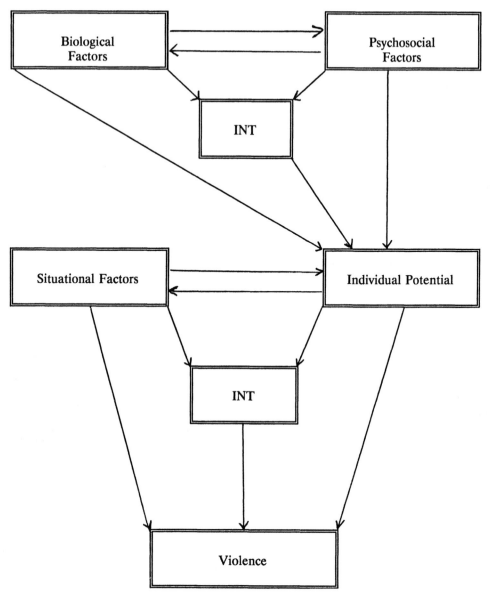

Figure 1. A model of influences on violence.

types of behavior reflect the same underlying individual potential. Is it necessary to postulate a specific potential for violence or a more general potential for antisocial behavior? In
my own paper on heart rate and violence, I discussed how far violent offenders were basically frequent offenders. It is clear that there is comorbidity between violence and nonviolent offending, conduct disorder, and antisocial personality disorder. How far are different
theories required for these different outcome variables? In some cases, different situational factors may interact with the same general underlying potential to produce different observed behaviors.

Another key issue concerns how far it is possible to distinguish unambiguously between risk factors and comorbid conditions. For example, are substance use and risky sex behavior risk factors for violence, comorbid conditions, or both? One possible solution is to distinguish between long-term between-individual differences in antisocial potential and short-term within-individual variations in antisocial potential. For example, heavy drinking could be a symptom of long-term between-individual differences in antisocial potential, and it could also be a cause of short-term within-individual variations in antisocial potential.

Focussing on violence, many key issues of definition and measurement were raised at the Institute. The main types of violence that are studied consist of murder, rape, robbery, assault, and domestic violence. However, researchers such as Rowell Huesmann stressed the need to distinguish between proactive and reactive aggression, direct and indirect aggression, hostile and instrumental aggression, and so on. It may be a mistake to treat violence as a homogeneous category. The definition of violence can have important implications for theories and conclusions. For example, Ernest Barratt pointed out that impulsivity was likely to be a more important risk factor for impulsive as opposed to planned violence. Malcolm Klein argued that it was essential to distinguish between group and individual violence; obviously, the influence of groups or gangs is likely to be greater on group violence.

Several researchers called for more use of self-report measures of violence as opposed to the use of official records (e.g. samples defined by arrest, conviction, or incarceration), but others were more doubtful about the value of self-reports. Arguably, many acts reported in self-report surveys (e.g. gang fights) may not constitute unambiguously violent crimes, and many of the most violent people in the population may be missing from interviewed samples. Strictly speaking, it is only acts leading to convictions that have passed the legal criteria for violence (including intentions and consequences). However, many violent acts can only be measured in self-reports, because they never lead to a police record or a conviction. More research is needed on the validity and reliability of self-report as opposed to official record measures of violence (Huizinga, 1991; Farrington *et al.*, 1996).

3. BIOLOGICAL AND PSYCHOSOCIAL VARIABLES

Several scholars were concerned about the problems of defining biological, psychological, and social variables. It seemed necessary to distinguish clearly between the nature of the construct, the method of measurement, and the nature of the mechanism of effect (or process). Where all of these were clearly biological, a variable could be defined unambiguously as biological (and similarly for psychological or social). Where the construct, measure, and mechanism were not all of the same type, it was difficult to classify a variable. The general conclusion was that it was more important to specify constructs, measures and mechanisms than to place variables into categories.

While recognising the classification problems, the most important biological variables that were related to violence and antisocial behavior were resting heart rate (e.g. by Adrian Raine and myself), serotonin (e.g. by Markus Kruesi and Terrie Moffitt), and testosterone (e.g. by Elizabeth Susman and Richard Tremblay). Perinatal complications were studied by Patricia Brennan and constitutional factors by Robert Sampson. Several researchers emphasized the importance of genetic factors. For example, Laura Baker argued that behavior genetic research designs were needed to disentangle genetic from other factors, such as in the link between violent parents and violent children. Numerous other biological variables are also studied in this volume, including skin conductance, cortisol, and MAO.

The main psychological variables that were related to violence and antisocial behavior were personality factors (see especially the papers by Hans Eysenck and Robert Cloninger), hyperactivity and impulsivity, intelligence and self-esteem. The main social variables were socio-economic status, a single parent family, child-rearing factors such as poor parental supervision, parental violence, a convicted parent, and school performance. Arguably, more efforts should be made by biological researchers to measure psychosocial variables, and by psychosocial researchers to measure biological variables. Neighborhood, community, situational, or larger societal factors were rarely studied, except in the paper by Michael Rutter. Unchanging variables such as gender and ethnicity were studied as interacting rather than as influencing factors; for example, there was a great deal of interest in whether relationships differed between males and females, although much of the research was conducted on males.

Responding to chapter 1 (which was circulated to lecturers well in advance of the Institute), several scholars attempted to study protective factors as well as risk factors, although the emphasis in all the papers was more on risk factors. A key challenge for the future is for researchers to discover protective factors that have no opposite risk factor (i.e. that are not at the opposite end of the scale from a risk factor). Interactive protective factors will be discussed later.

A key measurement problem for the future is how far it is possible to measure biological variables in large-scale representative community surveys. Because of the perceived need for carefully controlled laboratory or hospital conditions, much past biological research has been based on small, selected samples and has involved largely cross-sectional or retrospective case-control studies. There is a pressing need to measure more biological variables in large-scale longitudinal surveys, which often have numerous psychological, social and behavioral (e.g. violence) measures but few biological measures. More longitudinal researchers should be inspired in the future by surveys such as that conducted in Mauritius by Peter Venables and Adrian Raine, in Montreal by Richard Tremblay, and in Dunedin (New Zealand) by Terrie Moffitt.

4. CAUSAL INFLUENCES AND INDIVIDUAL POTENTIAL

It is important to investigate biological influences on biological and psychosocial variables, and psychosocial influences on biological and psychosocial variables. The aim should be to establish causal sequences, where changes in one variable X are followed by changes in another Y, and then changes in Y are followed by changes in another variable Z. The challenge for researchers is to convert a general model, such as that shown in Figure 1, into a specific testable theory. More explicitly biosocial theories are needed.

It was clear in the Institute that different researchers favored different key underlying constructs to describe individual potential, including arousal, fearlessness, impulsivity, disinhibition, novelty seeking, self-control, and conditionability. Some of these variables could be risk factors rather than indicators of individual potential. It would be desirable if researchers could agree on what to call the key underlying construct of an individual potential for violence or antisocial behavior. Some of the above terms may be different names for essentially the same construct.

Rowell Huesmann argued that the effects of biological and psychosocial factors were translated into behavior through mediating cognitive processes. Similarly, Hans Eysenck argued that personality factors mediated between biological and psychosocial factors and behavior. It might be desirable to incorporate these kinds of mediating factors, along with other key processes such as emotional, attachment and social learning processes, into the model.

As already mentioned, biological and psychosocial researchers should investigate influences of and interactions with situational or opportunity factors. For example, in the Cambridge Study, reported motives for physical fights depended on whether the boy fought alone or with others (Farrington, Berkowitz, & West, 1982). In individual fights, the boy was usually provoked, became angry, and hit out to hurt his opponent and to discharge his own internal feelings of tension. In group fights, the boy often said that he became involved to help a friend or because he was attacked, and rarely said that he was angry. The group fights were more serious, occurring in bars or streets, and they were more likely to involve weapons, produce injuries, and lead to police intervention. Fights often occurred when minor incidents quickly escalated, because both sides wanted to demonstrate their toughness and masculinity and were unwilling to react in a conciliatory way. It would be useful to include questions in longitudinal surveys about immediate situational influences on violent events.

5. INTERACTION EFFECTS

Happily, many scholars paid heed to the emphasis in chapter 1 on studying interaction effects, especially between biological and psychosocial variables. Interaction effects tend to have been neglected in the past, because of the difficulty of studying them. However, they can have important theoretical and practical implications, especially when they concern protective factors.

It would be useful to develop a typology of interaction effects with names for the different types. This is easiest with dichotomous data and two-way interaction effects. Table 1 (based on Farrington, 1994) shows all possible types of interaction effects in a hypothetical 2 x 2 x 2 table with the following constraints:

- 100 violent and 300 nonviolent persons;
- 50 low income, low heart rate;
- 50 low income, high heart rate;
- 50 high income, low heart rate;
- 250 high income, high heart rate.

This assumes that both risk factors and violence have a prevalence of 25 per cent and that the risk factors overlap (as is usual). Moderate levels of income and heart rate are included in the high category.

Case 1 shows one main effect and no interaction effect. Case 2 shows two additive main effects and again no interaction effect. (The likelihood ratio chi-squared from a logit analysis was used to test for interaction effects: see Farrington, 1994.) Case 3 shows the first significant interaction effect, labelled an expected amplifying effect: heart rate has a greater effect on violence among low income persons than among high income persons. (Among low income persons, 60 per cent of low HR versus 2 per cent of high HR were violent; among high income persons, 40 per cent of low HR versus 20 per cent of high HR were violent.) Case 4 is the reverse of this.

Table 1. Types of interaction effects: percentage violent

Type N	Low income		High income	
	Low HR 50	High HR 50	Low HR 50	High HR 250
1. One main effect	50	16	50	17
2. Additive main effects	62	30	38	14
3. Expected amplifying	60	2	40	20
4. Reverse amplifying	38	20	62	16
5. Amplifying - protective	84	20	16	16
6. Risk - protective	20	20	80	16
7. Supressing - protective	50	50	50	10
8. Expected crossover	98	2	2	20
9. Reverse crossover	2	20	98	16

Notes: Based on Farrington (1994), Table 18.2.
Assumes 100 Violent and 300 Nonviolent.
HR = Heart Rate.

Case 5 is labelled an amplifying-protective effect: heart rate has a great effect on violence among low income persons and no effect among high income persons. High income acts as a protective factor against the risk variable of heart rate. Case 6 is the reverse, labelled a risk-protective effect: heart rate has a great effect on violence among high income persons and no effect among low income persons. Perversely, the risk factor of low income acts as a protective factor against the risk variable of heart rate.

Case 7 shows another type of protective effect, termed a suppressing-protective effect: the combination of high income and high heart rate produces a low prevalence of violence. Heart rate has a no effect on violence among low income persons, but the high percentage violent in both heart rate categories means that this is not a protective effect. The reverse effect (a low percentage violent only among those with low income and high heart rate) is not possible with the present hypothesized numbers. If it were observed, it might be termed a suppressing-risk effect, since the suppression effect would occur in the presence of the risk factor of low income.

Case 8 shows an expected crossover effect: heart rate has a large positive effect on violence among low income persons and a smaller negative effect among high income persons. Case 9 shows a reverse crossover effect: heart rate has a large positive effect on violence among high income persons and a smaller negative effect among low income persons: odds ratios 257 and 0.08 (-12) respectively. It would be useful for researchers to search for these types of interaction effects and to use these types of labels.

A key issue is on what theoretical basis interaction effects might be predicted. For example, is it expected that the effect of a biological variable would be greater in an unfavorable social background (cases 3, 5, 8) or in a favorable social background (cases 4, 6, 9)? These are prospective predictions. Retrospective predictions might be that violent people from favorable backgrounds would tend to have a biological risk factor, or conversely that nonviolent people from unfavorable backgrounds would tend to have a biological protective factor. Most biosocial interaction effects discovered so far seem to be of the expected amplifying type. More efforts are needed to discover protective and crossover effects.

6. TYPES OF INDIVIDUALS AND ANALYTIC METHODS

It seems necessary to take account of different types of individuals in developing and testing biosocial theories. It cannot be assumed that all individuals necessarily follow the same developmental pathways to violence or that risk and protective factors have similar effects on all people. Terrie Moffitt's (1993) distinction between adolescence-limited and life-course-persistent offenders was clearly influential at the Institute, as were the personality theories of Hans Eysenck and Robert Cloninger.

A key issue arising in studying biological and psychosocial variables is how far all the risk factors are concentrated in a small minority of multiple-problem individuals. A related issue is how far all the significant results might disappear if that small minority were deleted from the sample (see e.g. Magnusson & Bergman, 1988). An implication of this is that it is necessary to carry out studies focussing on types of individuals as well as studies focussing on variables.

The investigation of the independent, additive, interactive, and sequential effects of biological and psychosocial variables on violence poses many analytic challenges. Regression methods can be used to study independent and additive effects, but more techniques designed specifically to investigate interaction effects are needed. Very few tests of sequential models have been carried out; more complex structural equation modeling techniques may be useful for this purpose.

As Michael Rutter pointed out, the most important biological and psychosocial risk and protective factors may vary considerably according to different questions addressed. For example, factors influencing differences between subjects (e.g. why one person is more violent than another) may not be the same as factors influences changes within subjects (e.g. why one person is more violent in some times and places than in other times and places). Long-term developmental factors might be expected to influence between-subject differences, whilst short-term situational factors might be expected to influence within-subject changes. Rutter was particularly concerned with factors explaining aggregate differences between countries or aggregate changes over time, which may be different again from those explaining differences between individuals or changes within individuals.

7. THE WAY FORWARD

Ideally, leading biological and psychosocial researchers should hold a series of meetings to seek agreement on a general model of influences on violence, priorities in testing specific theories derivable from it, key variables to be measured, common instrumentation to be used, key design features, and common analytic methods. A coordinated program of research should be planned to test key hypotheses using samples drawn from different places and countries, ideally with central and local funding. The emphasis in this program would be on multidisciplinary collaboration and on training a new generation of truly biosocial researchers.

There is a place for retrospective case-control studies, especially in investigating characteristics of rare cases such as murderers (see e.g. the paper by Jacqueline Stoddard). There is also a place for experimental research in which biological or psychosocial variables are systematically manipulated. However, the major method of choice to study biosocial influences on violence is likely to continue to be the prospective longitudinal survey. Logically, to study developmental sequences and causes, longitudinal research is essential. Unfortunately, few previous longitudinal surveys have measured a wide range of biological and psychosocial variables, although much can be learned from further data collection and reanalyses of these studies, as this volume shows. Consequently, a new generation of longitudinal studies is needed, including ambitious biological and psychosocial measures in representative community samples of at least several hundreds, and jointly directed by biological and psychosocial researchers. Such studies should build on the firm foundations of knowledge about biosocial bases of violence provided by this volume. They could lead to a great increase in knowledge about this troubling social problem.

REFERENCES

Farrington, D. P. (1994) Interactions between individual and contextual factors in the development of offending. In R. K. Silbereisen & E. Todt (Eds.) *Adolescence in context: The interplay of family, school, peers and work in adjustment* (pp. 366–389). New York: Springer-Verlag.

Farrington, D. P., Berkowitz, L. & West, D. J. (1982) Differences between individual and group fights. *British Journal of Social Psychology*, 21, 323–33.

Farrington, D. P., Loeber, R., Stouthamer-Loeber, M. S., van Kammen, W. & Schmidt, L. (1996) Self-reported delinquency and a combined delinquency seriousness scale based on boys, mothers and teachers: Concurrent and predictive validity for African Americans and Caucasians. *Criminology*, 34, 493–517.

Huizinga, D. (1991) Assessing violent behavior with self-reports. In J. S. Milner (Ed.) *Neuropsychology of aggression* (pp. 47–66). Boston: Kluwer.

Magnusson, D. & Bergman, L. R. (1988) Individual and variable-based approaches to longitudinal research on early risk factors. In M. Rutter (Ed.) *Studies of psychosocial risk* (pp. 45–61). Cambridge: Cambridge University Press.

Moffitt, T. E. (1993) Adolescence-limited and life-course-persistent antisocial behavior: A developmental taxonomy. *Psychological Review*, 100, 674–701.

18

PREFRONTAL DYSFUNCTION IN MURDERERS LACKING PSYCHOSOCIAL DEFICITS

Jacqueline Stoddard,[1] Adrian Raine,[2] and Susan Bihrle,[2] and Monte Buchsbaum[3]

[1]School of Medicine
Department of Preventive Medicine
Institute for Health Promotion and Disease Prevention Research
[2]Department of Psychology
University of Southern California
Los Angeles, California 90089-1061
[3]Department of Psychiatry
Mt. Sinai Medical School
New York, New York

INTRODUCTION

Very few biological studies of violence have taken psychosocial influences into account as moderators of biology-violence relationships (Brennan et al, in press). Some studies have, however, shown that links between psychophysiological functioning (resting heart rate and the conditioned electrodermal response) and antisocial behavior is strongest in antisocials who come from benign social backgrounds (high social class, intact homes). Conversely, such links are minimized or even reversed in antisocials with psychosocial deficits (low SES, broken homes).

In a previous PET study, we have demonstrated that severely violent offenders (murderers) have reduced functioning of the prefrontal cortex relative to age and sex matched controls (Raine et al. 94; Raine et al. in press). This study describes a psychosocial deprivation scale on this same population in order to test the hypothesis that murderers lacking psychosocial deficits have the strongest degree of prefrontal dysfunction.

METHOD

Subjects

Forty one subjects who had been tried for murder or attempted murder in the state of California and 41 age and sex matched controls acted as subjects. All murderers pled either not guilty by reason of insanity (NGRI) or incompetent to stand trial (IST). Full details are provided in Raine et al. (in press).

Biosocial Bases of Violence, edited by Raine *et al.*
Plenum Press, New York, 1997

Measures

Two raters blind to glucose values independently rated each murderer for the degree to which evidence was presented for psychosocial deprivation, using a 5-point Likert scale (0=none, 1=minimal, 2=partial, 3=substantial 4=extreme). Twenty six murderers were identified as having had evidence of psychosocial deprivation (ratings 2 - 4 "deprived") and twelve subjects had no or minimal evidence of psychosocial deprivation (rating 0–1). Ratings of psychosocial deprivation were based on the presence of the following processes: 1) physical or sexual abuse 2) neglect 3) extreme poverty 4) foster home placement 5) having a criminal parent 6) severe family conflict and or 7) a broken home. Subjects for whom there was evidence of psychosocial deprivation (rated as 2 or greater) were classified as psychosocially deprived. Sources used to determine psychosocial deprivation included the following: 1) national or local newspaper articles, 2) criminal history transcripts 3) interviews with prosecuting or defending attorneys, 4) preliminary hearing transcripts, 5) medical records, and 6) psychological reports. Good interrater reliability was demonstrated for the psychosocial deprivation scale (r=.88). Any inconsistencies in scoring between the raters was resolved, and a consensus rating was used in all subsequent analysis.

Brain Imaging Procedures

Full details of general PET scanning procedures and quantification may be found in Raine et al, 1994. Briefly, brain regions were activated by a continuous performance challenge task and then identified using two measures: 1) a lateral measure of the surface cortical regions and 2) a measure of the medial cortical and subcortical structures.

RESULTS

Significant differences between the sub-groups are depicted in Figure 1. Using the lateral measure, non-deprived murderers had significantly lower values of right (p<.04) and left (p<.005) hemisphere glucose than the group of normals, and also had significantly lower left hemisphere values than the deprived murderers (p<.025). Using the medial measure, murderers with the deprivation had significantly (p<.05) lower right and left hemisphere relative glucose values than normals, but not the other murderer sub-group. Controls and deprived murderers did not differ significantly on any measure.

DISCUSSION

Only murderers without clear psychosocial deprivation could be characterized by lower prefrontal glucose metabolism compared to controls. Murderers with deprivation compared to those without the deprivation differed significantly on only one of the four glucose measurements. Despite the nonsignificance between murderer subgroups on the other three glucose measurements, a trend in the expected direction was observed, such that murderers without the psychosocial deprivation had the lowest mean glucose values, followed by murderers with the deprivation and controls.

Why do murderers lacking psychosocial deficits have prefrontal dysfunction, while this dysfunction is absent in deprived murderers? We would argue that, among criminals

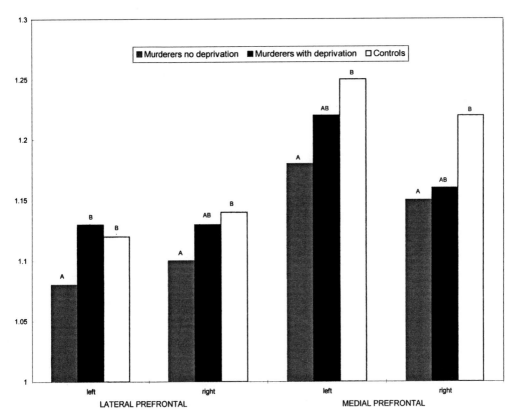

Note: Noncommon letters represent significant differences between groups within each glucose measurement

Figure 1.

from deprived social backgrounds, there would be a high proportion of crimes committed for predominantly social/psychosocial reasons and, therefore, a low remaining proportion of offenses which would be related to some biological brain abnormality. By the same token, criminals without psychosocially deprived circumstances might be expected to commit fewer crimes for social reasons, leaving a higher remaining proportion of offenses related to some brain abnormality.

There were at least three important limitations to this provisional study. First, these results are generalizable only to other murderers pleading NGRI: a select, but important population. Second, the psychosocial deprivation scale is a relatively crude and indirect index of early, adverse environmental circumstances. Nevertheless, significant differences between the three groups were observed, good interrater reliability was demonstrated and a number of important constructs were represented. Third, the small sample size limited the power of statistical analysis.

Despite these limitations, this is the first brain imaging study to demonstrate that psychosocial factors moderate brain-violence relationships. The evidence suggests that murderers who plead NGRI, and who come from a relatively normal social background, may be more likely than murderers coming from less privileged backgrounds to have select prefrontal deficits. Future research should investigate whether this pattern holds among other violent or non-incarcerated populations.

REFERENCES

Brennan, P.A & Raine, A. (In press) Recent biosocial advances in antisocial behavior research. Clinical Psychology Review.

Raine, A., Buchsbaum, M.S., & LaCasse, L. (In press) Brain abnormalities in murderers indicated by positron emission tomography. *Biological Psychiatry.*

Raine, A., Buchsbaum, M.S., Stanley, J., Lottenberg, S., Abel, & Stoddard, J. (1994). Selective reductions in prefrontal glucose metabolism in murderers assessed with positron emission tomography. *Society of Biological Psychiatry, 36,* 365–373.

19

TEMPERAMENTAL AND FAMILIAL PREDICTORS OF CRIMINAL CONVICTION

Bill Henry,[1] Avshalom Caspi,[2] Terrie Moffitt,[2] and Phil Silva[3]

[1]Department of Psychology
Colby College
Waterville, Maine 04901
[2]University of Wisconsin
Madison, Wisconsin
[3]University of Otago
Otago, New Zealand

Recent research from the Dunedin (New Zealand) Health and Development Study has identified several childhood risk factors associated with antisocial behavior. For example, Henry, Moffitt, Robins, Earls, and Silva (1993) identified eight family characteristics (e.g., number of changes in primary caregiver experienced by the child, number of residence changes occurring in the child's life, parental authoritarianism) which were associated with childhood antisocial behavior. In the same data set, Caspi, Henry, McGee, Moffitt and Silva (1995) identified a temperamental dimension among 3 and 5 year old subjects, labelled Lack of Control, which was found to be associated with teacher and parent reports of externalizing behavior problems assessed between the ages of 9 and 15.

The current study used the eight family characteristics identified by Henry et al. (1993) and the Lack of Control dimension identified by Caspi et al. (1995) to address two questions: 1.) What is the joint contribution of these variables to antisocial behavior? and 2.) do the relations between these variables and antisocial behavior differ as a function of the nature of the antisocial behavior?

METHOD

Data were drawn from the Dunedin Multidisciplinary Health and Development Study, a longitudinal investigation of 1037 children born between April 1, 1972, and March 31, 1973, in Dunedin, New Zealand. The cohort's history has been described by Silva (1990). Only males were used in the analyses reported here (n = 536.) At age 18, three groups were defined on the basis of court records. First, subjects with no record of conviction were classified as "non-convicted" (n = 404). Second, subjects who had a re-

Biosocial Bases of Violence, edited by Raine *et al.*
Plenum Press, New York, 1997

cord of conviction, but who were not convicted for violent offenses, were classified as "non-violent convicted" (n = 50). Finally, subjects who were convicted of violent offenses were classified as "violent convicted" (n = 21). (See Henry et al. (1993) for a complete description of the family characteristics, and Caspi et al. (1995) for a complete description of the Lack of Control variable.)

RESULTS

In the first stage of data analysis, multivariate analysis of variance (MANOVA) on the family and temperament variables revealed an overall significant effect for conviction group (F(8,308) = 3.22, p < .001). Four variables significantly discriminated the three conviction groups: Lack of Control at age 3, Lack of Control at age 5, number of residence changes, and number of parent changes (all ps < .01). The violent convicted group exhibited relatively high levels of Lack of Control, while the non-violent convicted group showed little difference from the non-convicted group. In contrast, both the violent and non-violent convicted groups showed similar levels of familial instability, and both groups differed from the non-convicted group (see Table 1).

Crosstabulations indicated that single caregiver at age 13 discriminated between the three conviction groups (chi-square(2) = 8.25, p < .05). Twenty-eight percent of the boys

Table 1. Means and standard deviations of three conviction groups on the continuous predictor variables

	Group		
Variable	No conviction	Non-violent conviction	Violent conviction
Lack of control (age 3)[a,b]	1.21	1.24	2.33
	(2.10)	(2.51)	(2.30)
Lack of control (age 5)[c,d]	0.85	0.93	2.31
	(1.50)	(1.24)	(2.80)
SES at birth	3.47	3.92	3.83
	(1.38)	(1.27)	(1.15)
Authoritarianism	34.20	34.94	34.95
	(7.87)	(7.95)	(5.86)
Mother's reading score	101.56	99.00	98.52
	(9.71)	(12.21)	(8.24)
Family Relations Index	19.33	18.24	18.37
	(3.47)	(4.47)	(3.66)
Number of residence changes[c]	1.74	2.50	3.11
	(2.17)	(2.76)	(2.75)
Number of parent changes[c]	0.16	0.58	0.62
	(0.52)	(1.00)	(0.88)

[a]Combined conviction groups different from non-conviction group, p = .056.
[b]Violent conviction group different from non-violent conviction group, p =.052.
[c]Combined conviction groups significantly different from non-conviction group, p < .05.
[d]Violent conviction group significantly different from non-violent conviction group, p < .05.
From Henry, B., Caspi, A., Moffitt, T., & Silva, P. (In press). Temperamental and familial predictors of violent and non-violent criminal convictions: From age 3 to age 18. Developmental Psychology. Copyright 1996 by the American Psychological Association, Inc.

with violent convictions, and 17% of the boys with non-violent convictions, had single parents, compared to 9% of boys who were not convicted.

In the second stage of data analysis, family and temperament variables were entered into logistic regression models to predict conviction outcomes.

- *Non-violent conviction versus no conviction.* Membership in the non-violent conviction group was significantly predicted by number of parent changes and by the interaction between Lack of Control and single parent status (odds ratios (OR) = 1.94 and 1.73, respectively).
- *Violent conviction versus no conviction.* Membership in the violent conviction group was significantly predicted by Lack of Control and number of residence changes (OR = 1.52 and 1.33, respectively), and by the interaction between Lack of Control and number of parent changes (OR = 1.43).
- *Violent versus non-violent conviction.* Among those subjects who had been convicted, membership in the violent conviction group was significantly predicted by one variable only: Lack of Control (OR = 1.75).

DISCUSSION

To account for these results, we differentiate between social regulation and self-regulation. Social regulation refers to those family structures and processes that act to socialize the child. Self-regulation reflects those individual difference characteristics that influence the child's capacity to regulate their own emotional experience and behavior. We argue that deficits in self-regulation will be most strongly related to antisocial outcomes when they occur in the context of relatively weak social regulation. Failure of social regulation essentially "frees" the child to engage in antisocial behavior; thus, family characteristics are associated with both types of conviction in these analyses. However, within the context of this failure of social regulation, it is the individual difference characteristics of the child — self-regulation — that determine the manner in which this antisocial behavior will be expressed. For example, within a family characterized by single parent status, high number of residence changes, or high number of parent changes, the capacity of the parent to monitor, discipline, and control the child may be greatly impaired. In this context, the child's inability to self-regulate will be more likely to result in explosive, undercontrolled behavior, increasing the likelihood of the eventual conviction for violent offenses.

REFERENCES

Caspi, A., Henry, B., McGee, R., Moffitt, T., & Silva, P. (1995). Temperamental origins of child and adolescent behavior problems: From age 3 to age 15. Child Development, 66, 55–68.

Henry, B., Moffitt, T., Robins, L., Earls, F. & Silva, P. (1993). Early familial predictors of child and adolescent antisocial behavior: Who are the mothers of delinquents? Criminal Behavior and Mental Health, 3, 97–118.

Silva, P. (1990). The Dunedin Multidisciplinary Health and Development Study: A 15 year longitudinal study. Paediatric and Perinatal Epidemiology, 4, 96–127.

20

SOCIAL, PSYCHOLOGICAL, AND NEUROPSYCHOLOGICAL CORRELATES OF CONDUCT DISORDER IN CHILDREN AND ADOLESCENTS

Jean Toupin, Michele Dery, Robert Pauze, Laurier Fortin, and Henri Mercier

University of Sherbrooke
Sherbrooke, Quebec, Canada

INTRODUCTION

An important number of studies have obtained associations between social and familial factors and antisocial behaviors. Some researchers have distinguished between adversity factors, such as life stresses, monoparentality and parent's mental health, and proximal factors such as parental education practices and family interactions. Studies have demonstrated that these factors remain significant even when socioeconomic status (SES) is statistically controlled (Frick, 1994).

Another category of studies have suggested a relationship between antisocial behavior and neuropsychological performance (Gorenstein, 1990). The aspects of cognitive functioning that have been convincingly associated with delinquency are verbal skills and executive functions. Surprisingly, however, Frost, Moffitt & McGee (1989) failed to find a significant association between Conduct Disorder (CD) and cognitive impairment in subjects without a comorbid diagnosis of Attention Deficit and Hyperactivity Disorder (ADHD). As a consequence, the objective of this study is to examine the relative association of social, familial, comorbid and neuropsychological factors to a DSM-III-R diagnosis of Conduct Disorder, while statistically controlling for socioeconomic status.

METHOD

Subjects were children 7–12 years old and adolescents 13–17 years old. The sample of children included 62 cases with a DSM-III-R diagnosis of Conduct Disorder and 36 controls from regular schools. In the sample of adolescents there were 68 cases and 41 controls. Cases were recruited from residential treatment programs, day services and school based programs for disordered children in order to get different severity of conduct

Biosocial Bases of Violence, edited by Raine *et al.*
Plenum Press, New York, 1997

problems. Each sample included a small proportion of girls, but there were no significant differences between cases and controls on age and gender. A wide range of measures were used to assess familial and parental characteristics (stressful events, socioeconomic status, psychopathology in parent, use of punishment), and child characteristics (comorbid diagnoses, severity of symptomatology, social competence and neuropsychological performance).

The statistical analysis involved a logistic regression in each category of variable (social and familial variables, psychological profile, neuropsychological performance) with socioeconomic status first forced in the regression. Variables contributing significantly to the classification of subjects beyond socioeconomic status in each category were retained for the final logistic regression. The interaction effects were also considered.

RESULTS

As it can be seen in table 1, the statistical analysis conducted on the sample of children shows that the use of punishment by parents, the number of ADHD symptoms according to parents and the score obtained in performing the Rey Complex Figure Test were significant over and above SES. They provide the best prediction equation for CD status in children. This regression equation including SES can correctly classify 94% of the subjects in the groups. No interactions were found to contribute beyond this model.

The logistic regression analysis on the sample of adolescents indicated that the significant variables in the equation were stressful events, punishment by parents, oppositional symptoms, ADHD symptoms and language skills (see table 1). Together with SES these variables provide a correct classification of 88% of the subjects of the sample.

Table 1. Summary of hierarchical regression analysis for variables predicting conduct disorder in children (n = 98) and adolescents (n = 104)

Variables	*B*	*SE* B	Odds ratios[a]
Children			
Step 1			
Socioeconomic status	48	25	1.6**
Step 2			
Punishment by parents	-0.69	2.5e+07	2.0**
ADHD symptoms (according to parents)	0.36		1.4**
Rey Complex Figure Test	-0.20		1.2**
Adolescents			
Step 1			
Socioeconomic status	-33	27	14
Step 2			
Language skills		0.63	
Punishment by parents	-0.40		1.5**
ADHD symptoms (according to parents)	0.62		1.9**
Oppositional symptoms (according to adolescents)	1.82		6.1**

*p < .05.
**p < .01

[a]The ratio of the likelihood that an event will occur in one group as compared to the likelihood that it will occur in the other group.

Again no interaction effects were detected when entered in the final step of the regression procedure.

DISCUSSION

The results of this study are consistent with Terrie Moffitt's contention that CD may be the result of a developmental chain of events initiated by low neuropsychological functioning in certain domains and exacerbated by a pathogenic environment (Moffitt, 1993). However, because this study is cross-sectional this hypothesis could not be tested. Our results do suggest that certain cognitive skills and parental punitive practices are key issues in CD for both children and adolescents. These aspects should be considered priorities when thinking about treatment for these youth.

Our results also indicate that this association of cognitive abilities with CD in both children and adolescents is independent of ADHD symptoms. This would suggest that youth with CD need not to show ADHD symptoms in order to perform below average on certain neuropsychological tests. We need to acknowledge however that since the youngsters in this study were in treatment, these findings may not apply to CD in the community. We also need to mention that they apply to a sample that is mostly composed of males.

REFERENCES

Frick, P.J. (1994). Family dysfunction and the disruptive behavior disorders: A review of recent empirical findings. In T.H. Ollendick & R.J. Prinz (Eds.) *Advances in Clinical Child Psychology* (Vol. 17: pp 203–226) New York: Plenum Press.

Frost, L.A., Moffitt, T.E., & McGee, R. (1989). Neuropsychological function and psychopathology in an unselected cohort of young adolescents. *Journal of Abnormal Psychology, 98,* 307–313.

Gorenstein, E.E. (1990). Neuropsychology of juvenile delinquency. *Forensic Reports, 3,* 15–48.

Moffitt, T.E. (1993). Adolescence-limited and life-course-persistent antisocial behavior : a developmental taxonomy. *Psychological Review, 100,* 674–701.

21

A BIOSOCIAL EXPLORATION OF THE PERSONALITY DIMENSIONS THAT PREDISPOSE TO CRIMINALITY

L. Arseneault, B. Boulerice, R. E. Tremblay, and J. -F. Saucier

Research Unit on Children's Psycho-Social Maladjustment
University of Montréal
Montréal, Québec
H1N 3V2 Canada

INTRODUCTION

Personality theorists have used a dimensional approach based on behavioral variations in large scale populations in order to describe and explain antisocial behaviors. The model proposed by Cloninger in 1986 is based on the interaction between three genetically controlled inheritable personality traits. In Cloninger's view, antisocial individuals would be those with a low level of harm avoidance and reward dependence associated with a high level of novelty seeking. Behaviors would be the result of an interaction between specific biogenetic predispositions and environmental influences.

Some studies have shown that the interaction between biological factors like birth complications or neuropsychological deficits and early adverse environment predispose to violent crimes (Raine & Mednick, 1989; Raine, Brennan, & Mednick, 1994) and physical aggression (Moffitt, 1990). Fetal brain damage resulting from traumatic events during delivery may increase the risk of antisocial personality development, especially for those who were raised in disadvantaged environment (Kandel & Mednick, 1991). The present study is an attempt to verify if the interaction between birth complications and a stressful environment during childhood could predict the stability of the personality dimensions that predispose to antisocial behaviors according to Cloninger's personality model.

METHOD

The subjects came from the Montréal Longitudinal and Experimental Study. The boys (N = 567) came from low-socioeconomic areas of Montréal, Québec and have been followed since age 6 (Tremblay et al., 1994). Only French speaking boys whose mother

Biosocial Bases of Violence, edited by Raine *et al.*
Plenum Press, New York, 1997

Table 1. Results of the regression analyses predicting cloninger's personality
dimensions that predispose to criminality

Novelty seeking dimension

Step	Predictors	F	R^2	Beta	t
1	S.I.	9.77**	0.065	0.014	1.65
	M.I.			-0.002	-0.15
	U.I.			-0.022	-0.31
2	E.S.			0.022	6.03**

Harm avoidance dimension

Step	Predictors	F	R^2	Beta	t
1	S.I.	4.09**	0.035	0.015	1.65
	M.I.			-0.031	-1.41
	U.I.			-0.023	1.03
2	E.S.			0.023	3.86**
3	M.I. x E.S.			0.007	1.78

Reward dependence dimension

Step	Predictors	F	R^2	Beta	t
1	S.I.	4.34**	0.037	-0.025	-1.58
	M.I.			-0.015	-0.12
	U.I.			0.023	1.98*
2	E.S.			-0.023	-3.69**
3	S.I. x E.S.			0.007	2.07*

Logistic regression analysis predicting the antisocial profile

Step	Predictors	B	SE	Wald	df	p level
1	S.I.	-0.0031	0.0643	0.0023	1	0.96
	M.I.	0.1549	0.1217	1.6196	1	0.20
	U.I.	-0.0188	0.0970	0.0374	1	0.85
2	E.S.	-0.0392	0.1034	0.1439	1	0.70
3	M.I. x E.S.	-0.0583	0.0289	4.0642	1	0.04

* p < 0.05
** p < 0.001
(N = 567)
S.I.: surgical intervention indicator   U.I.: urgent intervention indicator
M.I.: minor intervention indicator   E.S.: environmental stress

tongue was French and whose mother and father were born in Canada were included in the sample in order to create a homogeneous group.

Teachers were asked to complete the Social Behavior Questionnaire (Tremblay et al., 1991) in the spring of each school year, when the boys were aged 6, 10, 11 and 12. Composite scores following an optimal scaling were used. These scores are indices of the stability of each dimension. Information on birth complications were collected by consulting medical records. Each complication was given a score according to the results of an optimal scaling analysis. Three factors emerged from a principal component analysis. A high score on the first factor indicates a situation that might necessitate surgical intervention without trying less invasive interventions like episiotomy. A low score on the second factor indicates a delivery without any complications, but a high score represents a situation where the foetus' condition was stabilised with minor interventions. A high score on the third factor indicates a delivery that might necessitate an urgent intervention. A scale of environmental stress was developed using familial status, mother's educational level and occupational prestige when the boys were aged 6. The age of the mother at the time of the first child's birth was also a variable included in this scale. A score of 0, 1 or 2 was given on each variable. The 30th percentile was used as the low end cut-off point for the mother's educational level, occupational prestige and age at the time of the first child's birth while the 70th percentile was used as the high end cut-off point. The score for familial status was given depending on whether a separation occurred between the parents and if it happened before or after the boy reached age 11.

RESULTS

Results from regression analyses indicated that the interaction between the minor intervention indicator (factor 2) and the environmental stress had a marginal impact on stable harm avoidance (p = 0.08). This interaction effect suggests that the environmental stress increases the risk of having a stable high level of harm avoidance among those who had a high level of birth complications only. Surprisingly, the impact of this interaction goes in the opposite direction of Cloninger's theory. The interaction between the surgical intervention indicator (factor 1) and the environmental stress significantly predicted reward dependence (p = 0.04). The level of reward dependence is lower for subjects who had a high level of environmental stress among those who had a low level of birth complications. This interaction effect indicates that the environmental stress induces more variation on the reward dependence dimension only for those who had a low level of birth complications. Although significant, the impact of this interaction was very small. The total variance explained by the predictors was low ($R^2 = 0.037$) and the effect of each variable was weak. The interaction between a stressful environment and birth complications did not predict novelty seeking.

CONCLUSION

Our results suggest that a low level of birth complications associated with a high level of environmental stress affect Cloninger's reward dependence dimension which represents a propensity to overreact to reinforcements. According to Tremblay and his colleagues (1994), having a low level of reward dependence in childhood would increase the risk of developing delinquency during adolescence. In fact, subjects who showed a profile corresponding to the antisocial personality at age 6 (low level of harm avoidance and reward dependence associated with a high level of novelty seeking) were 4 times more at risk of developing delinquent behaviors compared to those who were similar on the harm avoidance and novelty seeking dimensions, but had a high level of reward dependence. Also, it is interesting to note that reward dependence includes aspects of Eysenck's psychotism dimension which he qualified as the closest representation of the antisocial personality (Eysenck, 1970). Those who are high on psychotism are described as tough-minded and socially detached, two characteristics also found in low reward dependent individuals. The birth complications effect in the interaction term was in the opposite direction of what was expected. This result may suggest that the environmental stress level has no impact on subjects who had a high level of birth complications. It may also be that a high level of reward dependence is facilitated by birth complications.

The exploratory statistical technique used to quantify the birth complications data set might explain some of the unexpected results. The interest of using this technique was not to search for specific delivery situations like those we did end up with, but rather to avoid doubtful interpretations of other scoring procedures. Future studies should focus on predicting the antisocial profile resulting from the interaction between the three personality dimensions. They also should consider the impact of mediator variables. Disinhibition measures, IQ or neuropsychological scores might help understand the pathways from birth to antisocial personality.

REFERENCES

Cloninger, C. R. (1986). A unified biosocial theory of personality and its role in the development of anxiety states. *Psychiatric Developments*, *44*, 573–588.

Eysenck, H.J. (1970). *The structure of human personality*. London: Methuen.

Kandel, E. & Mednick, S.A. (1991). Perinatal complications predict violent offending. *Criminology*, *29*(3), 519–529.

Moffitt, T.E. (1990). Juvenile delinquency and attention deficit disorder: Developmental trajectories from age 3 to age 15. *Child Development, 61*, 893–910

Raine, A. & Mednick, S.A. (1989). Biosocial longitudinal research into antisocial behavior. *Revue Épidémiologique de Santé Publique, 37*, 515–524.

Raine, A., Brennan, P., & Mednick, S.A. (1994). Birth complications combined with early maternal rejection at age 1 year predispose to violent crime at age 18 years. *Archives of General Psychiatry, 51*, 984–988.

Tremblay, R.E., Pihl, R.O., Vitaro, F., & Dobkin, P.L. (1994). Predicting early onset of male antisocial behavior from preschool behavior: A test of two personality theories. *Archives of General Psychiatry, 51*, 732–739.

Tremblay, R.E., Loeber, R., Gagnon, C., Charlebois, P., Larivée, S., & LeBlanc, M. (1991). Disruptive boys with stable and unstable high fighting behavior patterns during junior elementary school. *Journal of Abnormal Child Psychology, 19*, 285–300.

22

REDUCED HEART RATE LEVELS IN AGGRESSIVE CHILDREN

Traci Bice Pitts

3368 Belford Road
Reno, Nevada 89509

1. INTRODUCTION

Heart rate is one of the best studied and well replicated psychophysiological corre-lates of antisocial behavior in children. In a recent review (Raine, 1993), 14 studies of an-tisocial children all found lower resting heart rates in antisocial youth relative to same age peers. Despite this strong empirical link, there remain a number of important theoretical and methodological issues which need to be addressed. First, this literature does not ex-plore psychophysiological arousal during tasks which aggressively challenge these youth. This is important because a trait theory of fearlessness would suggest that aggressive youths heart rates would remain low even during an aggression challenge situation, while stimulation seeking theory (Quay, 1965) would predict that aggressive youths heart rates would increase back to normal during an aggressively challenging situation.

Second, an important theoretical question concerns whether low resting heart rate is a general predisposition to all antisocial behavior or whether it is specific to a subtype of antisocials. Finally, potentially important confounds to heart rate concern height and weight. Most studies have not controlled for the effects of these variables on heart rate.

This study addressed these issues by measuring heart rate levels of proactive and re-actively aggressive children in both a resting and challenge state, while taking into ac-count the effects of height and weight. The key hypotheses were as follows: First, aggressive children, in general, will have lower heart rates relative to nonaggressive chil-dren in a resting state. Second, the heart rate levels in aggressives will remain low during aggressively challenging situations (fearlessness theory). Third, a subtype of aggressive youths, specifically, "reactive only" aggressive youths, will respond to challenge with in-creased heart rate levels (stimulation-seeking theory).

2. METHODS

103 male subjects were recruited from 3rd to 6th grade classrooms in a low SES school district near Los Angeles. Dodge and Coie's (1987) teacher rating checklists were

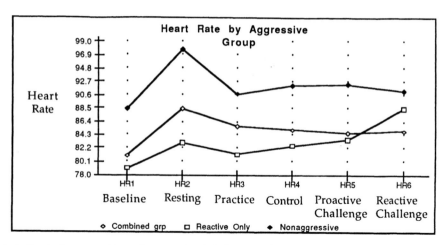

Figure 1. Heart rate measured in resting, control and challenge conditions for proactive-reactive, reactive-only and control subjects.

used to rate subjects as nonaggressive or in one of three aggressive groups (proactive only, reactive only, combined proactive-reactive). Teacher ratings were collected on 395 boys with top and bottom quartile scores categorizing subjects into groups. Data were collected from 95 boys: 38 from the combined proactive-reactive group, 19 from reactive only subjects and 38 nonaggressive subjects (randomly chosen from the large pool of nonaggressive subjects). Subjects with scores consistent with the theoretical subtype of 'proactive only' were not found in this large sample, suggesting that this subtype, if existent, is quite rare. The ethnic mix of the sample was consistent with the school population (44 Caucasian, 18 African American, 36 Hispanic and 7 Asian).

2.1. Measures

2.1.1 Articulated Thoughts in Simulated Situations (ATSS). ATSS (Davison, Robins, & Johnson, 1983) is a think-aloud cognitive assessment in which subjects listen to realistic audiorecorded vignettes and, at predetermined points, subjects were asked to verbalize what they were thinking. Four ATSS tapes were produced, a practice tape, a control tape, a proactively challenging tape and a reactively challenging tape.

2.1.2 Heart Rate (HR). HR was measured through a small pulse meter attached to the index finger of the subject's nondominant hand. HR was assessed before and after an initial waiting period (HR1, HR2) and then during each of the four ATSS tapes, at points predetermined to be most provocative (HR3 - HR6).

3. RESULTS

Chi-square analyses comparing the aggressive and nonaggressive groups found no significant differences on height (p<.84), weight (p<.44), grade (p<.17) or race (p<.06). A Multivariate analysis of Covariance (MANCOVA) conducted to assess ethnic group differences in heart rate (with height and weight as covariates) found no effect for race (p<.30).

3.1. Heart Rate

HR differences were tested through MANCOVA with height and weight as covariates. A significant main effect for group was found (F(6,85)=2.61), p<.02. A preplanned contrast comparing mean heart rate levels of the two aggressive groups to the nonaggressive group found significant between group differences on 5 of 6 heart rate readings: HR1 F(1,90)=6.4, p<.01;HR2 F(1,90)=14.4, p<.000;HR3 F(1,90)=6.32, p<.014;HR4 F(1,90)=8.6, p<.004;HR5 F(1,90)=9.0, p<.003;HR6 F(1,90)=2.1; p<.148. Consistently, the aggressive subjects' heart rates were significantly lower than the nonaggressive subjects. The one occasion which did not produce differences in heart rate between the groups appeared due to a rise in the heart rate levels of the reactive only group during the reactive challenge. A within-group t-test comparing baseline (HR1) to reactive challenge condition (HR6) found that the reactive-only group indeed exhibited a significant increase in HR during this challenge (p<.05), though they never reached HR levels as high as nonaggressive subjects.

4. DISCUSSION

These results suggest that aggressive boys differ from nonaggressive boys on HR in a predictable fashion. Aggressive groups maintained lower HR levels throughout the experiment relative to nonaggressive peers. This study also examined the potentially important effect of an aggression challenge on heart rate levels and found that the heart rate levels of aggressive groups remained low during provocation with the one predicted exception of the reactive only subjects showing an increase in heart rate during the reactively challenging task. However, even during this challenge, the mean heart rate levels of the reactive only group never reached levels higher than the nonaggressive subjects.

Taken as a whole, the findings of lower heart rate levels for aggressives in resting and challenge states generally supports a fearlessness theory of aggression. However, there is a subtype of aggressive youth, termed reactive-only in Dodge's typology, who do respond to challenge with increased heart rates. These are boys who react aggressively to challenge, but who do not proactively seek out trouble. Thus, there is some support for the sensation-seeking model for this subtype. Overall, it is likely that various subtypes of aggressive youths respond aggressively through different cognitive and physiological processes.

In sum, aggression relates to low heart rate levels during a resting and aggressively challenging state. Confounds of height and weight do not explain the heart rate differences between aggressive and nonaggressives, and there is evidence to suggest a differential heart rate response for specific subtypes of aggressive youth.

5. REFERENCES

Davison, G.C., Robins, C. & Johnson, M.K. (1983). Articulated thoughts during simulated situations: A paradigm for studying cognition in emotion and behavior. *Cognitive Therapy and Research*, 7(1), 17–40.

Dodge, K.A. and Coie, J.D. (1987). Social-information-processing factors in reactive and proactive aggression in children's peer groups. *Journal of Personality and Social Psychology*, 53(6), 1146–1158.

Quay, H.C. (1965). Psychopathic personality as pathological stimulation-seeking. *American Journal of Psychiatry*, 122, 180–183.

Raine, A. (1993). *The psychopathology of crime*. San Diego, CA: Academic Press.

23

HEART RATE AND PSYCHOSOCIAL CORRELATES OF ANTISOCIAL BEHAVIOR IN HIGH-RISK ADOLESCENTS

Friedrich Lösel and Doris Bender

Institute of Psychology
University of Erlangen-Nürnberg
Bismarckstr. 1, 91054 Erlangen
Germany

1. INTRODUCTION

Children and adolescents with antisocial behavior show a relatively low resting heart rate (Raine, 1993). This is explained through theories predicting autonomic under-arousal, passive emotional withdrawal, increased vagal tone, and reduced fear of punishment in antisocial individuals (Quay, 1993; Raine, 1993). Consistencies between cardiovascular, electrodermal, and cortical response systems (Raine, Venables, & Williams, 1990) and symmetrical findings for inhibited children (Kagan, 1989) support such psychophysiologic concepts. Although measurement problems, developmental changes, and alternative theoretical explanations must be taken into account, results on the relationship between heart rate (HR) and antisocial behavior (ASB) are rather consistent and substantial in effect size (Raine, 1993). Lower HR in childhood and adolescence is even a long-term predictor of crime and violence (Farrington, 1987, this volume; Raine et al., 1990, Wadsworth, 1976). In contrast, high autonomic arousal has a protective effect against adult criminality (Raine et al., 1995).

The HR-ASB relationship is particularly pronounced in younger, unselected and noninstitutionalized samples (Raine, 1993). Confounds with social risks, life styles and effects of incarceration may also be responsible for the inconsistent results on HR correlates of adult criminality and psychopathy (Hare, 1978). However, the influence of social factors on the HR-ASB relationship is not yet clear. Recent research, for example, also demonstrates correlations within a low SES and even a disruptive sample (Kindlon et al., 1995). Against this background, the present paper studies relations between resting HR and ASB in adolescents from multiproblem milieus. After ascertaining whether there is a HR effect in such a high-risk sample, we analyze social factors that may influence this relation. In particular, we investigate (low) supervision/control in education as an environmental risk factor and alcohol consumption as a personal risk factor for ASB.

The Biosocial Bases of Violence, edited by Raine *et al.*
Plenum Press, New York, 1997

2. METHOD

The present data are from a subsample of the Bielefeld-Erlangen Study on Resilience (e.g., Lösel & Bliesener, 1994). This study investigated 100 adolescents from 27 residential welfare institutions (group homes) in a 2-year follow-up. These individuals come from disturbed family backgrounds and are at enhanced risk for ASB. The initial sampling used case conferences and objective measures to select severely antisocial as well as relatively well adapted (resilient) adolescents. A risk index contained variables like parental divorce, unemployment, alcohol abuse, and criminality, as well as subjective factors like experiences of parental violence or being unloved. The mean of the risk index in our sample was located at the 90th percentile of a normal school sample.

The present study focuses on 37 adolescents from our male subsample who were still in institutions 2 years later and then attended a medical examination. Their mean age was 17.41 years ($SD = 1.12$). Pulse rate/resting heart rate was measured about 20 minutes after the beginning of a standard neurological and psychiatric examination and was scored in beats per minute. The subject was seated quietly. ASB was measured by an own 27 item self-report scale with proven reliability and validity. The amount of supervision/control in education (CON) was assessed 2 years before through educators' reports to a specially adapted family climate scale (Schneewind et al., 1985). It contains items about obligatory norms in the institution, supervision of adolescent's behavior, and sanctioning of deviance. Alcohol consumption (ALC) was assessed within the medical interview and scored as "regular" if juveniles drank at least 1–2 beers daily.

3. RESULTS

Elementary statistics on the variables are shown in Table 1. Although the correlation between HR and ASB took the expected negative direction, it was not significant. A closer inspection of the bivariate distribution suggested that this may have been due to a nonlinear tendency. We trichotomized the group approximately at the 25th and 75th percentile. The high group (HR $\geq$ 90; $n = 10$) showed homogeneously low ASB scores ($M = 7.28$, $SD = 3.10$). Results for the medium group ($70 < HR < 90$; $n = 19$) were $M = 11.11$ ($SD = 5.43$) compared with $M = 8.80$ ($SD = 7.68$) for the low group (HR $\leq$ 70, $n = 8$). The mean and variance of ASB in the high HR group differed significantly from the medium group ($t(27) = 2.42, p < .05$; $F(18,9) = 4.05, p = .05$) and also from both other groups combined ($t(31) = 2.06, p < .05$; $F(26,9) = 5.26, p < .05$). This was not the case for the low HR group ($t(25) = .89$, ns; $F(18,7) = .72$, ns; and $t(35) = .44$, ns; $F(28,7) = 1.50$, ns). Thus, high HR had a protective function, whereas the other pole (low HR) did not show a symmetric risk effect.

Educational supervision/control correlated significantly with ASB measured 2 years later. Adolescents from institutions with more supervision were less antisocial. Furthermore, they showed a higher HR. Whereas ALC was not related significantly to HR, it correlated positively with ASB. In a hierarchical regression analysis, ALC had a significant effect on ASB (Beta = .34, $t = 2.17, p < .05$), whereas the independent effect of CON was weaker (Beta = -.30, $t = 1.84, p = .07$) and that of HR not significant. The predictors explained a substantial amount of variance in ASB ($R^2 = .23, F(2,34) = 3.32, p < .05$).

As the results on HR suggested more complex relations, we computed two further ANOVAs with HR (dichotomized at the 75th percentile) and CON (median split) or ALC as independent variables. In the analysis with CON and HR there was a weak main effect

Table 1. Means, Standard deviations, and intercorrelations of antisocial behavior, heart rate, supervision/control in education, and regular alcohol consumption

Variable	M	SD	ASB	HR	CON
Antisocial behavior (ASB)	9.57	5.61	—		
Heart rate (HR)	81.81	16.48	−.12	—	
Control in education (CON)	11.78	1.54	−.35*	.33*	—
Regular alcohol consumption (ALC)	73[a]	—	.38*	.01	−.14

[a] Percentage. * $p \leq .05$, two-tailed significance.

of CON, $F(1,33) = 3.48$, $p = .07$, but there was no significant effect of HR and for the interaction CON x HR. In contrast, the model with ALC and HR as predictors demonstrated not only the main effect of ALC, $F(1,33) = 6.51$, $p < .05$, but also a weak main effect of HR, $F(1,33) = 2.71$, $p = .10$, and a substantial interaction effect for HR x ALC, $F(1,33) = 5.41$, $p < .05$. Whereas alcohol drinkers with low to medium HR showed the highest antisociality ($M = 12.31$, $n = 20$), they scored low when HR was high ($M = 6.68$, $n = 7$). The respective results for the two other groups were $M = 8.67$ (no ALC/high HR, $n = 3$) and $M = 5.05$ (no ALC/low to medium HR, $n = 7$).

4. DISCUSSION

The present study investigates the HR-ASB relationship within a multiply deprived, high-risk, and institutionalized sample. Under these conditions, a HR effect may well be eclipsed by social influences. A HR effect could also be lowered because the subjects were already familiar with a psychological test setting and HR was not measured at the beginning of the medical examination (Raine, 1993). In addition, we have performed no sophisticated analyses of HR variability (Mezzacapa et al., in press). The biological variable was measured only once within a standard medical examination in different settings (no well-controlled laboratory procedure), and it was not corrected for body mass, smoking, exercise, and so forth. However, we did not find significant relations with potential moderators like body build or regular smoking. Despite this simple measurement, most results agree with theoretical expectations. Within our group from multiproblem milieus, the relation of HR to ASB is weaker than in more normal samples. However, there is still a relation in the expected direction that primarily depends on a protective effect of high HR. This protective effect of high HR in at-risk samples corresponds to Raine et al. (1995), and a similar tendency can also be seen in other data (Farrington, this volume; Kindlon et al., 1995). Our results suggest that we should not just expect symmetric risk versus protective effects of the different poles of a variable. They fit Rutter's (1985) view that a protective effect is not a general characteristic of a variable but a mechanism in the context of given risks. On the other hand, low to medium HR in combination with ALC seems to represent an additive risk effect for ASB. The heterogeneity of ASB in the medium- and low-HR groups and the relatively low ASB in their nondrinking subgroups need further clarification. Moderators like anxiety or comorbidity problems may be relevant here. There are hints from our medical examination that several "atypical" subjects show other neurological or psychiatric disorders. However, the small size of these subgroups must also be taken into account.

Whereas the relation between ALC and ASB can be viewed as part of a broader problem behavior syndrome, the correlation of CON (measured 2 years before) with ASB and HR is more interesting from a causal perspective. Together with positive emotionality, a norm-oriented educational style is a strong protective factor in our resilience study (Lösel, 1994). This is consonant with the positive impact of a responsive *and* controlling education in families or at school (Baumrind, 1989; Olweus, 1993). Intensive supervision of high-risk juveniles may stimulate their cardiovascular response system and thus increase their sensitivity for social stimuli like punishments or delayed rewards. Perhaps, HR contributes to a mediating function between a protective educational style and low ASB. More HR measurements, a longer time interval, and control of anxiety (Mezzacappa et al., in press) are needed to test such causal links. Nonetheless, our data underline the importance of integrated biopsychosocial approaches in the study of antisociality.

REFERENCES

Baumrind, D. (1989). Rearing competent children. In W. Damon (Ed.), *Child development today and tomorrow* (pp. 349–378). San Francisco: Jossey-Bass.

Farrington, D.P. (1987). Implications of biological findings for criminal research. In S.A. Mednick, T.E. Moffitt, & S.A. Stack (Eds.), *The causes of crime: New biological approaches* (pp. 42–64). New York: Cambridge University Press.

Hare, R.D. (1978). Electrodermal and cardiovascular correlates of psychopathy. In R.D. Hare & D. Schalling (Eds.), *Psychopathic behaviour* (pp. 107–144). London: Wiley.

Kagan, J. (1989). Temperamental contributions to social behavior. *American Psychologist, 44*, 668–674.

Kindlon, D.J., Tremblay, R.E., Mezzacappa, E., Earls, F., Laurent, D., & Schaal, B. (1995). Longitudinal patterns of heart rate and fighting behavior in 9- through 12-year-old boys. *Journal of the American Academy of Child and Adolescent Psychiatry, 34*, 371–377.

Lösel, F. (1994). Protective effects of social resources in adolescents at high risk for antisocial behavior. In E.G.M. Weitekamp & H.-J. Kerner (Eds.), *Cross-national longitudinal research on human development and criminal behavior* (pp. 281–301). Dordrecht, NL: Kluwer.

Lösel, F., & Bliesener, T. (1994). Some high-risk adolescents do not develop conduct problems: A study of protective factors. *International Journal of Behavioral Development, 17*, 753–777.

Mezzacappa, E., Tremblay, R.E., Kindlon, D., Saul, J.P., Arsenault, L., Seguin, J., Pihl, R.O., & Earls, F. (in press). Anxiety, antisocial behavior and heart rate regulation in adolescent males. *Journal of Child Psychology and Psychiatry.*

Olweus, D. (1993). *Bullying at school.* Oxford: Blackwell.

Quay, H.C. (1993). The psychobiology of undersocialized aggressive conduct disorder: A theoretical perspective. *Development and Psychopathology, 5*, 165–180.

Raine, A. (1993). *The psychopathology of crime: Criminal behavior as a clinical disorder.* San Diego: Academic Press.

Raine, A., Venables, P.H., & Williams, M. (1990). Relationships between central and autonomic measures of arousal at age 15 years and criminality at 24 years. *Archives of General Psychiatry, 47*, 1003–1107.

Raine, A., Venables, P.H., & Williams, M. (1995). High autonomic arousal and electrodermal orienting at age 15 years as protective factors against criminal behavior at age 29 years. *American Journal of Psychiatry, 152*, 1595–1600.

Rutter, M. (1985). Resilience in the face of adversity. Protective factors and resistance to psychiatric disorder. *British Journal of Psychiatry, 147*, 598–611.

Schneewind, K.A., Beckmann, M., & Hecht-Jackl, A. (1985). *Das Familienklima-Testsystem.* [Family Climate Test]. München: Institut für Psychologie der Universität.

Wadsworth, M.E.J. (1976). Delinquency, pulse rate and early emotional deprivation. *British Journal of Criminology, 16*, 245–256.

24

VIOLENT CRIME PATHS IN INCARCERATED JUVENILES

Psychological, Environmental, and Biological Factors

Hans Steiner,[1] Sharon E. Williams,[2] Lisa Benton-Hardy,[2] Mindy Kohler,[3] and Elaine Duxbury[3]

[1]Department of Child Psychiatry
Stanford University
Stanford, California 94305
[2]Division of Child Psychiatry and Child Development
Stanford University School of Medicine
Stanford, California 94305
[3]Research Division, California Youth Authority
Sacramento, California

INTRODUCTION

Environmental, biological and psychological factors have all been shown to be influential in the genesis and perpetration of violent crime, but rarely are they simultaneously tested within the same subject pool of incarcerated violent offenders to ascertain their relative contribution. Raine, Brennan, & Mednick (1994) showed that the presence of biological factors such as birth complications combined with rejection from the mother interacted to predict violent crimes by age 18. In regard to psychological factors, Feldman and Weinberger (1994) studied adolescent boys, and found that the internal psychological variable "restraint" mediated the influence of parenting to produce either prosocial or antisocial outcome in mid-adolescence.

In the current study, environmental, biological and psychological factors were investigated in the same population of violent juvenile delinquents. It was postulated that of the three variables, psychological factors and in particular restraint would best predict the perpetration of violent crime. This was postulated for two reasons. First, this model is interactionistic and developmental. Multiple factors are believed to be causally involved. However, as internal psychological structures develop, they come to represent the sum of an individual's constitutional and life experiences and serve as a kind of internal map of where an individual has been and where they are likely to go in the future. Second, the variable of restraint has been shown to predict recidivism in juvenile delinquents over a 3.5 and ten year life span (Steiner, Garcia & Huckaby, 1994; Tinklenberg, Steiner, Huck-

Biosocial Bases of Violence, edited by Raine *et al.*
Plenum Press, New York, 1997

aby & Tinklenberg, in print). Such findings also seem applicable to normal adolescent boys in which restraint again was found to be more influential than family qualities in predicting 5 year prospective outcome (Feldman & Weinberger, 1994).

METHOD

Subjects

Subjects were 60 incarcerated males in the Sex Offenders Program at the California Youth Authority (CYA), the state prison system for adolescents. Subjects varied in terms of the type of sex crime committed and were consistent with the overall CYA population in terms of number of previous crimes (mean = 2.6). Mean age of the subjects was 18 years, 11 months (range 15 years, 7 months to 22 years, 5 months). Ethnic backgrounds were as follows: 34% Caucasian, 36% African-American, 28% Hispanic and 2% Other.

Instruments

Psychological factors were evaluated using the Weinberger Adjustment Inventory (WAI; Weinberger & Schwartz, 1990), a 84-item questionnaire which assesses the way in which a subject reacts to conflict and stressful situations, yielding scores for restraint and distress. Environmental factors, which were defined by problems with family support, social support, education, and housing, were derived from official case histories and resulted in an overall stressor score. Independent ratings of this stressor score by two blind raters correlated .70. Biological factors were defined by pulse rate (PR) which was recorded twice with a 2-minute rest period using a Micronta digitized blood pressure cuff. First and second pulse rates correlated .94 and were averaged for analysis. Body mass index (BMI), combining height and weight, was also recorded.Additionally, offenses against persons and property, drug offenses, and technical offenses were recorded using official CYA crime records by a blind and independent rater .

Procedure

Questionnaire data were collected on the living unit at the CYA in group format. Questions were read out loud. Following group administration, all responses were quality checked and re-administered individually as needed. Biological data were collected individually following questionnaire administration.

RESULTS

As seen in Table 1, regressions were used to assess significant associations in relation to the type of committing offense. Five variables and one interaction were selected a priori to be entered into the equation due to sample size (10 subjects/variable). Significance was found only for crimes against persons. A hierarchical regression in which the order of entry was psychological, environmental and then biological factors and their interaction with stressful events (consistent with what we would expect in this age group) revealed the following: marginally significant results in step one with psychological variables only (adjusted R square = .06); significant results when environmental factors were

Table 1. Dependent Variable - past offenses agaimst persons

Step	Variable	Simple R	Beta	Adj. R^2	Overall F	p
		Hierachical regression				
1				0.06	2.7	0.07
	Distress	0.18	0.03			
	Restraint	−0.25	−0.31			
2				0.14	4.1	0.01
	Distress	0.18	0.02			
	Restraint	−0.25	−0.33			
	Environment	0.26	0.3			
3				0.09	2	0.08
	Distress	00.018	−0.04			
	Restraint	−0.25	−0.31			
	Environment	−00.26	0.03			
	BMI	−0.02	−0.07			
	PR	0.2	−0.08			
	Environment X PR	−0.32	0.3			

bold: p<.05; italic: p=0.06

added (adjusted R squared = .11). Adding BMI and PR and the interaction of environment and PR revealed marginally significant results with a small drop in adjusted R squared (.09). No other main effects or interactions were significant.

DISCUSSION

Results of the present study support a relationship between environmental and psychological factors and the perpetuation of violent crime against persons in adolescents. High levels of environmental stressors are associated with higher crimes against persons and the inverse is true for levels of restraint. Similar results were reported in earlier research which found that incarcerated juvenile males who reported stressful life events and met criteria for PTSD also had low levels of restraint (Steiner, Garcia & Matthews, in print). This current pattern also replicates findings from previous studies (Steiner & Huckaby, 1993; Steiner, Garcia, & Huckaby,1994; Tinklenberg, Steiner, Huckaby & Tinklenberg, in print) which assigns prospective validity to restraint in terms of criminal recidivism. Our findings indicate that the crime activity of youth at this mid-adolescent age are influenced by environmental adversity, and the psychological factor of restaint, which in our model represents the outcome of interaction between biological and environmental factors operative during many years of growing up. Together, they appear to be more sensitive measures for violent crimes commited prior to subject's assessment rather than pulse rate, BMI or the interaction of pulse with stressful events. This result supports our concept of restraint as a proximal variable for explaining criminal activity in adolescents, in line with the findings of Feldman and Weinberger (1994).

Pulse rate did not differentiate crime histories in this sample which could be due to its relative lack of sensitivity in this skewed population although PR values collected were within the ranges of what has been described in the literature (mean=68, SD=15, range=42–105). In comparison to simultaneously measured psychological structures, a single channel of arousal such as pulse rate can be expected to have less predictive power compared to a variable representing multiple interactive predispositions and life influences.

The results of this study support the role of higher personality functions in the perpetuation of violent crime. It should be noted, however, that there were limitations to the study including small sample size, lack of control group, and the retrospective nature of the assessment of criminal activity and environmental stressors. Future studies should assess data prospectively.

REFERENCES

Feldman, S.S. & Weinberger, D.A. (1994). Self-restraint as a mediator of family influences on boys' delinquent behavior: a longitudinal data analysis. *Child Development, 65,* 195–211.

Raine, A., Brennan, P., & Mednick, S.A. (1994). Birth complications combined with early maternal rejection at age 1 year predispose to violent crime at age 18 years. *Archives of General Psychiatry, 51,* 984–988.

Steiner, H., Garcia, G.I., & Matthews, Z. (in print). Posttraumatic stress disorder in incarcerated juvenile delinquents. *Journal of the American Academy of Child and Adolescent Psychiatry.*

Steiner H., Garcia G.I., & Huckaby W. (1994). Recidivism along personality dimensions. *Scientific Proceedings of the Annual Meeting, American Academy of Child and Adolescent Psychiatry, 9,* 52.

Tinklenberg, J., Steiner, H., Huckaby, W., & Tinklenberg, J. (in print).Criminal recidivism predicted from narratives of violent juvenile delinquents, *Child Psychiatry and Human Development.*

Weinberger, D.A., Schwartz, (1990). Distress and restraint as superordinate dimensions of self-reported adjustment: a topological perspective. *Journal of Personality, 58,* 381–417.

Widom, C.S. (1989). The cycle of violence. *Science, 244,* 160–166.

25

SEROTONERGIC FUNCTIONING IN PARTNER-ABUSIVE MEN[*]

Alan Rosenbaum, Susan S. Abend, Paul J. Gearan, and Kenneth E. Fletcher

University of Massachusetts Medical School
Department of Psychiatry
55 Lake Ave. N.
Worcester, Massachusetts 01655

Interpersonal aggression is one of the most significant problems confronting contemporary society. People are concerned for their personal safety not only in the streets, but in their homes, schools, and places of employment, as well. Currently, the political rhetoric is focused on punishment based deterrence; building more prisons and enforcing sentences. Despite mounting evidence that exposure to violence while growing up is a major contributor to violent crime, the preventive impact of addressing intrafamilial aggression is rarely mentioned as a potential antidote to societal violence. Irrespective of its contribution to violence in society, relationship aggression is a serious problem in its own right, affecting approximately a third of the married/dating population. Reducing such aggression requires prevention and treatment, and developing effective interventions is predicated on an understanding of the etiology and dynamics of the problem.

Aggression is a complex behavior that defies simplistic explanation. Univariate models, such as frustration-aggression, may explain some instances of aggression, but fail when invoked as global explanations for all manifestations of the behavior. Even intra-individually, aggression occurs for different reasons, at different times. Human behavior, in general, can only be explained as resulting from an interaction of psychological, socio-cultural, and biological factors. Research and theorizing regarding the etiology of relationship aggression has focused almost exclusively on social, cultural, political, and to a lesser extent, psychological influences, and has all but ignored biological factors. Since all perceptions are filtered through the brain, and all voluntary behaviors are initiated and controlled by the brain, it would seem impossible to explain behavior, even partner directed aggression, without attention to physiology.

In our own research, we have been examining the influence of physiological factors on aggression in adult, intimate, heterosexual relationships. Thus far it has been demonstrated that a history of significant head injury increases the probability of relationship ag-

[*] This research was supported by NIMH Grant #MH44812 to the first author.

gression (Rosenbaum & Hoge, 1989; Rosenbaum et al., 1994) and that batterers can be distinguished from non-batterers on the basis of their performance on several neuropsychological tests reflective of frontal lobe dysfunction (Cohen et al., unpublished). The mechanism by which head injury affects aggression, however, is unclear. Serotonin (5-HT) deficit has been linked with various forms of aggressive behavior, especially impulsive aggression (Coccaro, 1992). There is also a study suggesting that head injury may produce serotonergic deficiencies (van Woerkom et al., 1977). The foregoing suggested the importance of examining serotonergic activity among partner abusive men in order to evaluate the relationships between head injury, serotonergic functioning, and relationship aggression.

METHOD

Thirty-six partner abusive men (PA) were compared to 38 non-abusive men (NA) on levels of serotonergic activity as indexed by prolactin response to fenfluramine challenge. The PA subjects were recruited from referrals to a batterers treatment program; NA subjects were recruited via newspaper advertisement. Subjects were screened out if they were currently taking any medication, had a history of cardiac disease (including arrhythmia and ischemia), hypertension, pituitary disease, visual field defect, hypogonadism, had abused drugs or alcohol within the past six months, or had been diagnosed with psychosis. They also had to be currently involved in some level of heterosexual relationship. Participants were paid $200 ($US). The fen challenge methodology was utilized to provide a summary measure of pre- and post-synaptic serotonergic activity without the risks and side effects inherent in measuring levels of CSF 5-HIAA (Coccaro & Kavoussi, 1994). A double-blind, placebo control design was employed. Each subject maintained a low monoamine diet for three days prior to each day of participation. Subjects completed the protocol twice, with a two week inter-trial interval. The order of placebo-fenfluramine was counterbalanced. A standard 60 mg dose was administered orally at one hour post placement of the intravenous catheter, which was used to draw hourly blood samples. Assessment of head injury was made by a physician blind to the aggression history of the subject, which was ascertained in a structured interview conducted by a psychologist blind to head injury status. Additional data were gathered via structured interview and a written questionnaire battery.

RESULTS

Baseline prolactin (PRL) levels at hour 2 were subtracted from PRL levels at hours 4, 5, and 6, for both placebo and fenfluramine conditions. The change scores for placebo were then subtracted from the change scores under the fenfluramine condition for each subject. The resultant scores represent the prolactin response to the fen challenge and were analyzed using a repeated measures MANOVA. The results indicated a significant Head injury x Abuse x Time interaction with PA subjects showing a blunted prolactin response as compared to NA subjects, only in the absence of a significant head injury [Quadratic univariate $t(1)= 2.43$ $p<.02$]. PA and NA subjects did not differ in PRL response over hours 4, 5, and 6. A follow-up repeated measures MANOVA comparing only non head-injured PA and NA men confirmed that PA men showed significant blunting of the PRL response as compared to the NA controls [Wilks lambda=.85, F $(2,41)=3.49$, $p<.04$].

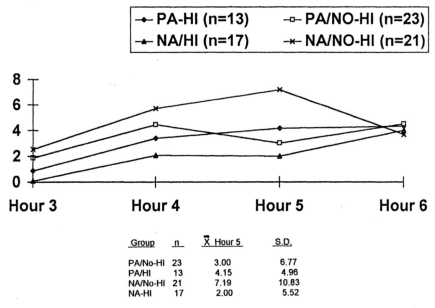

Figure 1. Mean Prolactin levels (ng/ml) following d,l-fenfluramine (60 mg) administration at hour 1.

Although as a group head-injured subjects did not differ from non head-injured subjects over the same interval, among NA subjects, there was a trend for head injured subjects to have lower peak prolactin levels than non-head injured subjects [t for unequal variances (df=30.92)= 1.92, p=.065]. Further, there was a significant difference in variances between the two groups [according to Levene's test of equality of variances, p=.023], providing a reasonable explanation for the lack of significance.

DISCUSSION

The results suggest that serotonergic deficits may be a factor for some partner abusive men and also that some types of head injury, as van Woerkom noted, may negatively impact the serotonergic system. Absent a significant head injury, PA subjects can be distinguished from NA subjects based on their reduced prolactin response to a fenfluramine challenge. The finding of a trend for a head injury effect among NA subjects suggests that head injury may be one mechanism by which prolactin response is blunted. As expected, however, the results confirm that head injury is not the only factor determining serotonergic functioning, nor do serotonergic deficits necessarily lead to domestic aggression. Men abuse their partners , just as people behave aggressively, for a multitude of reasons. Batterers are a heterogeneous group and it is foolish to expect that a unitary cause will be discovered. These results are important in that they demonstrate that a biological factor, serotonergic functioning, may be implicated in this form of aggression, as it has already been demonstrated to be a factor in some types of impulsive aggression. It suggests the value in pursuing biological determinants of aggressive behavior, and of taking this class of variables into account in developing models of relationship aggression. It further suggests the possibility that pharmacologic agents which impact on the serotonin system (e.g.,

SSRIs), may be useful adjuncts to more traditional, psychoeducational interventions for some partner abusers.

REFERENCES

Coccaro, E.F. (1992) Impulsive aggression and central serotonergic system function in humans: An example of a dimensional brain-behavior relationship. *Int. Clin. Psychopharmacology*, 7, 3–12.

Coccaro, E.F. & Kavoussi, R.J. (1994). The neuropsychopharmacologic challenge in biological psychiatry. *Clinical Chemistry*, 40, 319–327.

Cohen, R.A., Rosenbaum, A., Fletcher, K.E., Kane, R.L., Warnken, W.J., Benjamin, S., & Hoge, S.K., Neuropsuchological correlates of domestic violence. Unpublished manuscript.

Rosenbaum, A. & Hoge, S.K. (1989). Head injury and marital aggression. *Am. J. of Psychiat.*, 146, 1048–1051.

Rosenbaum, A., Hoge, S.K., Adelman, S.A., Warnken, W.J., Fletcher, K.E., & Kane, R.L. (1994). Head injury in partner-abusive men. *J. of Consulting & Clinical Psychology*, 62, 6, 1187–1193.

van Woerkom, T.C.A.M., Teelken, A.W. & Minderhound, J.M. (1977). Difference in neurotransmitter metabolism in frontotemporal lobe contusion and diffuse cerebral contusion. *Lancet* i, 812–813.

26

NEUROTRANSMITTER-NEUROENDOCRINE RESPONSES TO AGGRESSION

Personality Influences

Gilberto Gerra

Centro Studi Farmacotossicodipendenze
Az. USL, Via Guasti S. Cecilia 3
43100 Parma, Italy

1. INTRODUCTION

The biological correlates of aggressiveness have been extensively studied in humans. Decreased serotonin (5-HT) activity has been found to be involved in the pathogenesis of aggressiveness (Brown and Linnoila, 1990) and also changes in noradrenergic system function correlate with pathological aggressiveness (Coccaro et al.,1991). Contrasting data are available concerning the role of sexual steroids, prolactin and hypothalamus-pituitary adrenal axis in the aggressive behavior (Olweus et al., 1988; Constantino, 1995; Scerbo and Kolko, 1994).

These studies have always investigated aggressiveness occurring in personality disorders, antisocial or suicidal behavior, murderer, arsonists and sexual offenders: we therefore decided to investigate the biochemical variables that might interfere with the quality and the quantity of aggressiveness in normal individuals.

2. METHOD

Thirty healthy male subjects, aged 18–19 years (M = 18.7 years, SD = 0.6 years), entered the study after giving written informed consent. Axis I and II disorders, especially personality disorders, were excluded utilizing SCID and SIDP, according to DSM III-R and DSM IV criteria.

Characteristics and quantification of aggressiveness (defined as direct, as indirect or verbal, as irritability, negativism, resentment, suspiciousness) were analyzed by the Buss-Durkee Hostility Inventory (BDHI), by the clinical impression of the psychiatrist, by a visual analog scale and a semi-structured interview with the school teachers (a modified version of the interview utilized by Bank) (Bank et al., 1993).

The 30 selected subjects were then divided into two groups, group A (15 cases) with low-normal aggressiveness levels, group B (15 cases) with high-normal aggressiveness. The subjects included in group B (higher aggressiveness) scored over a median split on the measures and over BDHI standard level for normal subjects in the Italian version. Aggressiveness was experimentally induced by the free-operant procedure (Cherek, 1981): the subjects had the free choice of working or fighting against another research subject, depicted as participating in the same study, in another room, with many provocative events during the 30 minutes task. Testosterone (TE), Prolactin (PRL) , Growth Hormone (GH), Cortisol (CORT), Norepinephrine (NE) and Epinephrine (EPI) concentrations in plasma were measured before, during (20 min.) and after the task. Data were analyzed statistically by ANOVA, ANOVA for repeated measures and by Pearson's coefficient of correlation.

3. RESULTS

During the free operant methodology session group B gained more money than group A (p<0.05) and subtracted more than group A(p<0.05) this being considered evidence of higher aggressiveness of group B than group A subjects.

After the free-operant methodology session, heart rate increased significantly in both groups, but more significantly in the B group (p<0.001). The same was true for systolic blood pressure (p=0.001).

After the stimulus, NE concentrations were significantly increased in group B but not in group A (p-0.0004). EPI levels did not change after the stimulus in either group. GH values rose significantly after the stimulus in both groups. ANOVA for repeated measures showed a significantly higher increased of GH in group B (p=0.01). PRL levels decreased in group A and increased in group B, neither changes being significant. CORT concentrations rose significantly after the stimulus in group B and not in group A. ANOVA for repeated measures revealed a significant effect of diagnosis (p=0.04). TE basal values were significantly higher in group B (p<0.05) and rose in group A but not in group B: the difference was not significant.

4. DISCUSSION

Our data confirm the results obtained by us in a previous study (Gerra et al., in press) that showed that in psychologically healthy male subjects unstimulated aggressiveness is greater in subjects with higher basal Te concentrations, even though either the hormonal or the psychological parameters are still included in a range of normality. Since GH and Cort secretions are stimulated by NE (Müller et al., 1977) the greater rises in the two hormones after aggressiveness might suggest that NE modulates the quantity and quality of aggressiveness, through its stimulatory effect on Te secretion, influencing basal Te levels (Mayerhofer et al., 1993); on the other hand it is also possible that aggressiveness, as an aspecific distress condition, is the cause and not the consequence of increased NE (Wyatt et al., 1971).

Our data, including higher cortisol responses in aggressive subjects during aggressive task, apparently contrast with the findings of low cortisol levels in aggressive subjects (Scerbo and Kolko, 1994; Bergman and Brismar, 1994), but the subjects included in our study were not affected by personality disorders characterized by aggressiveness; they showed high aggressiveness in the range of normal behavior. Aggressiveness and fighting task could be perceived as a more stressful condition by normal subjects, respect to psychiatric patients, who could present an "habituation" to aggressive behavior with lower awareness of negative emotions: the more intense psychological involvement of normal subjects with higher aggressiveness (group B) could explain the activation of the hypothalamus-pituitary-adrenal axis.

Testosterone secretion, that is NE-dependent, was already higher before the task in the more aggressive group, which could be a pre-existing condition, not correlated to fighting task or dominant behavior.

Therefore, we tentatively suggest that normal subjects with high-normal aggressiveness might have a potentially hyperactive central noradrenergic system, responsible for the downstream NE-dependent hormonal stimulation and for the behavioral characteristics of the individual.

5. REFERENCES

Bank, L., Duncan, T., Patterson, G.R. and Reid, J. (1993), J. Pers. 61, 693–709
Bergman, B. and Brismar, B. (1994), Alcohol Clin. Exp. Res. 18, 311–316.
Blanchard, D.C., Veniegas, R., Elloran, I. and Blanchard, R.J. (1993), J. Stud. Alcohol Suppl.11, 9–19.
Brown, G.L. and Linnoila, M.I. (1990), J. Clin. Psychiat. 51, 31–43.
Cherek, D.R. (1981), Psychopharmacology 75, 339–345.
Coccaro, E.F., Lawrence, T., Trestman, R.I., Gabriel, S., Klar, H.M. and Siever, L.J. (1991), Psychiatry Res.., 43, 587–599.
Constantino, J.N. (1995), J. Am. Acad. Child. Adolesc. Psychiatry 34, 535–536.
Gerra, G., Avanzini, P., Zaimovic, A., Fertonani, G., Caccavari, R., Delsignore, R., Gardini, F., Talarico, T., Lecchini, R., Maestri, D. and Brambilla, F. (1996) Behavioral Brain Research., in press.
Mayerhofer, A., Bartke, A., Began, T. (1993) Biol. Reprod. 48, 883–888.
anterior pituitary function". Academic Press, New York.
Müller, E.E., Nistic˜, G. and Scapagnini, U. (1977) "Neurotransmitter and anterior pituitary function". Academic Press, New York.
Olweus, D., Mattsson, A., Schalling, D. and Low, H. (1988), Psychosom. Med. 50, 261–272.
Scerbo, A.S. and Kolko, D.J. (1994), J. Am. Acad. Child. Adolesc. Psychiatry 33, 1174–1184.
Wyatt, R.J., Portnoy, B.D., Kupfer, J., Snyder, F., Engelman, K. (1971), Archives of General Psychiatry 24, 65–71.

27

TRYPTOPHAN DEPLETION AND BEHAVIORAL DISINHIBITION IN MEN AT RISK FOR ALCOHOLISM AND ANTISOCIAL BEHAVIOR

David LeMarquand, Robert O. Pihl, Simon N. Young, Richard E. Tremblay,
Roberta M. Palmour, and Chawki Benkelfat

Departments of Psychology and Psychiatry, McGill University
l'École de Psychoéducation, Université de Montréal
Montréal, Québec
Canada

1. INTRODUCTION

The present studies tested the hypothesis that experimentally altering central nervous system serotonin (5-HT) synthesis through tryptophan depletion (T-) would increase behavioral disinhibition in susceptible individuals. The oral administration of a mixture of amino acids devoid of tryptophan (trp), the amino acid precursor of 5-HT, is a safe and effective method of lowering brain 5-HT synthesis, and presumably 5-HT function. Previous clinical studies have reported a negative relationship between brain serotonergic functioning and impulsivity. Impulsive fire-setters, particularly those with a family history of alcoholism, as well as impulsive violent offenders, have lower levels of cerebrospinal fluid (CSF) 5-hydroxyindoleacetic acid (5-HIAA), a metabolite of 5-HT, compared to nonimpulsive violent offenders and healthy controls (Linnoila, De Jong & Virkkunen, 1989; Virkkunen et al., 1994). Two groups, selected on the basis of an increased risk for alcoholism or antisocial behavior, respectively, were tested: young men with multigenerational family histories of alcoholism, and adolescent males with past histories of aggressive, disruptive behavior.

2. METHOD

Participants in study 1 were recruited from the community, screened for current and prior psychopathology, and interviewed to confirm their family history status. Family history positive young men had a multigenerational family history (MFH) of paternal male alcoholism; family history negative men had no family history of alcoholism in two previous generations (FH-). The adolescent men in study 2 were recruited from a large sam-

Biosocial Bases of Violence, edited by Raine *et al.*
Plenum Press, New York, 1997

ple of boys from low socioeconomic backgrounds, followed longitudinally since age 6. Rated by their teachers at ages 6, 10, 11, and 12 on the Social Behavior Questionnaire (Tremblay, Vitaro, Gagnon, Piché & Royer, 1992), those high on the aggression and anxiety subscales were classified as aggressive-anxious (Ag-Anx); those high in anxiety and low in aggression comprised the anxious (Anx) group, and those with moderate levels of aggression and anxiety served as controls (C). Both studies employed double-blind, placebo-controlled designs. Participants ingested an amino acid mixture devoid of (T-) or including (B) trp, the amino acid precursor of 5-HT (for a full description, see Benkelfat et al., 1994). This mixture reliably decreases plasma trp levels five hours following ingestion, and significantly decreases brain 5-HT synthesis by a factor of 9 in men and 40 in women, as measured by positron emission tomography (Nishizawa et al., 1996). Study 1 employed a between-subjects design (each participant received either a T- or B mixture), while a within-subjects design (each participant received both the B and T- mixtures in separate counterbalanced sessions) was used in study 2. Blood samples were drawn before and 5 hours following amino acid consumption to determine plasma trp concentrations. Behavioral disinhibition was assessed 5 to 6 hours following amino acid consumption using a go/no-go learning task (Newman & Kosson, 1986). Participants learned by trial-and-error to press a button to numbers (presented on a computer screen) associated with reward (winning $0.10), and to not press to stimuli associated with punishment (the loss of $0.10). Commission errors (CEs), or failures to withhold responses to stimuli associated with punishment or loss of potential reward, provided a measure of behavioral disinhibition. Omission errors (OEs), failures to respond to stimuli associated with reward or the avoidance of punishment, were also recorded. Participants completed four conditions of varying feedback (see Iaboni, Douglas & Baker, 1995 for a description of the conditions).

3. RESULTS

Across studies, plasma free trp significantly increased five hours following ingestion of the B mixture, and significantly decreased after ingestion of the T- mixture (amino acid by time interaction, study 1: $F(1, 54) = 167.43, p < .0001$; study 2: $F(1, 23) = 191.40, p < .0001$). In study 1, T- increased commission errors (averaged across go/no-go conditions) in the MFH group only. In study 2, the Ag-Anx group made more commission errors (averaged across amino acid status and go/no-go conditions) compared to controls, and marginally more than the Anx group. There were no main effects or interactions involving T- on commission errors in study 2. There were no significant effects on OEs in either study. Table 1 presents the OEs and CEs by amino acid status group for each study.

4. DISCUSSION

Tryptophan depletion increased behavioral disinhibition in young men at risk for alcoholism to a level comparable to that of disruptive adolescent males following the balanced amino acid mixture, suggesting that behavioral disinhibition displayed by the latter group may be a consequence of low baseline central serotonergic functioning. The lack of a T- effect on disinhibition in the aggressive, disruptive male adolescents may have been a consequence of a floor effect on central serotonergic functioning. The results of the first study suggest that serotonergic functioning in men at risk for alcoholism is susceptible to perturbation by T-, and that impairments in this system have important effects on clini-

Table 1. Omission and commission errors ($M \pm SE$) by Amino acid status and group for studies 1 and 2

Amino acid status, group	OEs	CEs
Study 1		
B, FH- (n=18)	3.33 (0.64)	6.53 (1.23)
B, MFH (n=11)	3.18 (0.82)	6.84 (1.58)
T-, FH- (n=16)	3.5 (0.68)	6.48 (1.31)
T-, MFH (n=13)	4.12 (0.75)	11.87 (1.45)*
Study 2		
B, Ag-Anx (n=14)	3.11 (0.56)	11.77 (0.82)**
B, Anx (n=11)	3.77 (0.63)	7.66 (0.93)
B, C (n=6)	1.75 (0.85)	5.38 (1.26)
T-, Ag-Anx	4.25 (0.56)	12.3 (0.82)**
T-, Anx	3.05 (0.63)	6.64 (0.93)
T-, C	3.42 (0.85)	5.04 (1.26)

Note. OEs indicates omission errors; CEs, commission errors; B, balanced amino acid mixture; T-, tryptophan depleted amino acid mixture; MFH, young men with a multigenerational family history of paternal alcoholism; FH-, those with no family history of alcoholism; Ag-Anx, aggressive anxious adolescent males; Anx, anxious adolescent males; C, nonaggressive, nonanxious adolescent males.
*Amino acid by group interaction on square root transformed CEs, $F(1, 54) = 4.84$, $p < .05$.
**Group main effect on CEs (averaging across amino acid status), $F(2, 28) = 4.07$, $p < .05$.

cally significant behaviors such as impulsivity. Clearly, further study of disinhibition following T- in individuals at risk for future psychopathology is warranted.

5. REFERENCES

Benkelfat, C., Ellenbogen, M., Dean, P., Palmour, R., & Young, S. N. (1994). Mood-lowering effect of tryptophan depletion: Enhanced susceptibility in young men at genetic risk for major affective disorders. *Archives of General Psychiatry, 51,* 687–697.

Iaboni, F., Douglas, V. I., & Baker, A. G. (1995). Effects of reward and response costs on inhibition in ADHD children. *Journal of Abnormal Psychology, 104,* 232–240.

Linnoila, M., De Jong, J., & Virkkunen, M. (1989). Family history of alcoholism in violent offenders and impulsive fire setters. *Archives of General Psychiatry, 46,* 613–616.

Newman, J. P., & Kosson, D. S. (1986). Passive avoidance learning in psychopathic and nonpsychopathic offenders. *Journal of Abnormal Psychology, 95,* 252–256.

Nishizawa, S., Benkelfat, C., Young, S. N., Leyton, M., Mzengeza, S., de Montigny, C., Blier, P., & Diksic, M. (1996). *Differences Between Rates of Serotonin Synthesis in the Brains of Human Males and Females.* Manuscript submitted for publication.

Tremblay, R. E., Vitaro, F., Gagnon, C., Piché, C., & Royer, N. (1992). A prosocial scale for the Preschool Behavior Questionnaire: Concurrent and predictive correlates. *International Journal of Behavioral Development, 15,* 227–245.

Virkkunen, M., Rawlings, R., Tokola, R., Poland, R. E., Guidotti, A., Nemeroff, C., Bissette, G., Kalogeras, K., Karonen, S. -L., & Linnoila, M. (1994). CSF biochemistries, glucose metabolism, and diurnal activity rhythms in alcoholic, violent offenders, fire setters, and healthy volunteers. *Archives of General Psychiatry, 51,* 20–27.

28

AGGRESSION IN PHYSICALLY ABUSED CHILDREN

The Interactive Role of Emotion Regulation

Angela Scarpa

University of Georgia
Department of Psychology
Athens, Georgia 30602-3013

1. INTRODUCTION

While child physical abuse is related to a heightened risk of aggression, the majority of abused children do *not* become aggressive (Widom, 1989). Factors which put some physically abused children at increased risk for aggressive behavior remain to be clarified. A transactional, biopsychosocial model of development suggests that aggressive behavior problems arise when intrinsic child vulnerabilities interact with a nonoptimal environment (Sanson, Smart, Prior, & Oberklaid, 1993). Internalizing problems, such as anxiety and depression, have been found to be markedly higher in youth with vs. without conduct disorder (Hinshaw, Lahey, & Hart, 1993). This suggests that vulnerability to emotional distress (i.e., emotion dysregulation) may be one intrinsic child vulnerability related to later aggression.

Emotion regulation is defined as the ability, when experiencing a strong emotion, to self-soothe (by reducing autonomic arousal), focus attention, and organize behavior in the service of an external goal (Katz & Gottman, 1995). As part of this definition, it should be noted that emotion regulation is part of the broader concept of self-regulation and also involves regulation of physiological arousal in distressing situations. In this sense, emotion dysregulation can be measured in several ways, including examination of level of emotional distress as well as physiological reactivity in stressful situations. In the current preliminary study, emotion dysregulation in children was measured both by high levels of internalizing behavior and by increased cortisol reactivity over the course of a stressor task. Cortisol is a hormone released as part of our sympathetic response to stress and is used here as a measure of the ability of individuals to modulate their internal physiological state when distressed (see Stansbury & Gunnar, 1994 for a review).

This preliminary study examined the influence of emotion regulation of distress on aggression in physically abused children. In line with the transactional model of development, it

Biosocial Bases of Violence, edited by Raine *et al.*
Plenum Press, New York, 1997

was hypothesized that the experience of physical abuse in children will interact with emotion dysregulation to heighten the risk of their involvement in aggression. In other words, this study was conducted to test the hypothesis that the relationship between abuse and aggression will be strongest in the presence of emotion dysregulation, using data from two consecutive summer treatment programs for 7 to 15 year-old children with disruptive behavior disorders. It was also hypothesized that emotion regulation might also be related to beneficial gains in the treatment of aggressive behavior, such that children who improve in treatment might be those who are better at regulating their emotional distress.

2. METHOD

In the first program, parent report of internalizing behavior from the Child Behavior Checklist (CBCL; Achenbach & Edelbrock, 1983) was used to measure poor emotion regulation, and aggressive behavior was rated independently by parents using the CBCL, teachers using the Teacher Report Form (TRF; Edelbrock & Achenbach, 1984), and clinic staff using the Overt Aggression Scale (OAS; Yudofsky et al., 1986) for 52 children (45 boys, 7 girls; 38 African American, 14 Caucasian). In the second program, salivary cortisol reactivity after a provocation task was measured as a physiological index of poor emotion regulation for 19 children (17 boys, 2 girls; 14 African American, 5 Caucasian). Children were divided into reactors (i.e., cortisol levels increased after the task) and nonreactors (i.e., cortisol levels remained the same or decreased after the task). Aggressive behavior was rated by clinic staff. History of abuse was assessed if medical records indicated that "a caregiver or responsible adult inflicted physical injury upon a child by other than accidental means" (Barnett et al., 1993). Data from the second program was further analyzed to assess improvement in aggression over the 7-week treatment program. Of the 19 children assessed, 10 were further selected based on the criterion that their staff aggression ratings fell above the median in the first week of the treatment program. These 10 were then further subdivided into those who improved (N=7) or did not improve (N=3), based on a median split of aggressive behavior ratings during the last week of the program.

3. RESULTS

According to all three sources in the first program, the highest reports of aggressive behavior occurred in children who had been both abused and rated with high internalizing behavior ($t(48)=7.17$, $p<.000$ for parents; $t(30)=2.15$, $p<.02$ for teachers; $t(45)=1.44$, $p<.08$ for staff). In the second program, a trend was found showing the highest rates of aggression reported in children who were both abused and classified as cortisol reactors ($t(15)=1.45$, $p<.084$). This replicated the same pattern of aggression scores as found in the first program sample. Table 1 presents the means and standard deviations for these findings.

Furthermore, relative to children who did not improve over the course of treatment, children who did improve were more likely to be classified as cortisol nonreactors ($Chi(1)=3.90$, $p<.049$). That is, 100% of children who improved over the course of the treatment program were classified as cortisol nonreactors, whereas 43% of those who did not improve were so classified.

Table 1. Mean (SD in parentheses) aggressive behavior ratings as a function of physical abuse history and level of emotion regulation ability for programs 1 and 2

	Program 1			
	Abused/low internalizing	Nonabused/low internalizing	Nonabused/high internalizing	Abused/high internalizing
Parent (CBCL)	57.14 (7.95)	65.42 (7.95)	79.00 (8.07)	82.18 (6.35)
Teacher (TRF)	69.40 (9.18)	70.50 (9.74)	73.91 (11.61)	84.25 (9.22)
Staff (OAS)	0.63 (.77)	0.66 (.66)	0.97 (.86)	1.15 (.58)
	Program 2			
	Abused/cortisol nonreactor	Nonabused/cortisol nonreactor	Nonabused/cortisol reactor	Abused/cortisol reactor
Staff (OAS)	3.33 (.26)	3.45 (.88)	3.68 (1.26)	4.42 (.69)

4. CONCLUSIONS

In both samples, a similar pattern emerged in which disruptive children who have *co-occurring* physical abuse history and emotion dysregulation showed the highest levels of aggressive behavior. This pattern of results is consistent with the transactional model of aggression suggesting that aggressive behavior problems arise when child vulnerabilities are coupled with an adverse family environment. Thus, vulnerability to emotional distress may be seen as one factor related to the increased risk of aggression in some physically abused children. On the other hand, children who are better at regulating their emotional distress may be protected from such adverse environmental experiences. The results also indicated that emotion regulation ability might allow for beneficial gains of treatment, suggesting that treatment programs might target such specific needs as appropriate.

REFERENCES

Achenbach, T.M., & Edelbrock, C. (1983). *Manual for the Child Behavior Checklist and Revised Child Behavior Profile*. Queen City, VT: Queen City Publishers.

Barnett, D., Manly, J.T., & Cicchetti, D. (1993). Defining child maltreatment: The interface between policy and research. In Eds. D. Cicchetti and S.L. Toth, *Child abuse, child development, and osical policy* (pp. 7–74). Northwood, NJ: Ablex Publishing.

Edelbrock, C., & Achenbach, T.M. (1984). The teacher version of the Child Behavior Profile I. Boys Aged 6–11. *Journal of Consulting and Clinical Psychology, 52*, 207–217.

Hinshaw, S.P., Lahey, B.B., & Hart, E.L. (1993). Issues of taxonomy and comorbidity in the development of conduct disorder. *Developmental Psychopathology, 5*, 31–49.

Katz, L.F., & Gottman, J.M. (1995). Vagal tone protects children from marital conflict. *Developmental Psychopathology, 7*, 83–92.

Sanson, A.V., Smart, D., Prior, M., & Oberklaid, F. (1993). Precursors of hyperactivity and aggression. *Journal of the American Academy of Child and Adolescent Psychiatry, 32*, 1207–1216.

Stansbury, K., & Gunnar, M.R. (1994). Adrenocortical activity and emotion regulation. *Monographs of the Society for Research in Child Development, 59*, 108–134.

Widom, C.S. (1989). Does violence beget violence? A critical examination of the literature. *Psychological Bulletin, 106*, 3–28.

Yudofsky, S.C., Silver, J.M., Jackson, W., Endicott, J., & Williams, D. (1986). The Overt Aggression Scale for the objective rating of verbal and physical aggression. *American Journal of Psychiatry, 143*, 35–39.

29

CHILDREN'S AGGRESSION AND DSM-III-R SYMPTOMS PREDICTED BY PARENT PSYCHOPATHOLOGY, PARENTING PRACTICES, CORTISOL, AND SES

Keith McBurnett,[1] Linda J. Pfiffner,[1] Lisa Capasso,[1] Benjamin B. Lahey,[2] and Rolf Loeber[3]

[1]University of California, Irvine
Child Development Center
4621 Teller Suite 108
Newport Beach, California 92660
[2]University of Chicago
Chicago, Illinois
[3]University of Pittsburgh
Pittsburgh, Pennsylvania

1. INTRODUCTION

A handful of studies over the past 10 years have examined whether cortisol (a hormone which in humans is associated with stress and catabolic processes) is related to behavioral measures of aggression or conduct problems. In reviewing this small literature, McBurnett and Pfiffner (McBurnett & Pfiffner, in press) reached the tentative conclusion that low cortisol may be associated only with the most severe, chronic, and aggressive subtype of disruptive behavior. The basis for this hypothesis was that three studies found no significant association between cortisol and diagnostic status when the experimental groups were based on a diagnosis of a disruptive behavior disorder (Kruesi, Schmidt, Donnelly, Euthymia & others, 1989) or diagnosis of Conduct Disorder (CD), (Scerbo & Kolko, 1994) (Targum, Clarkson, Magac-Harris, Marshall & Skwerer, 1990), and that significant findings only emerged from designs in which diagnostic groups were more finely subtyped. In a sample of adults, cortisol was abnormally low only in prisoners who had both Antisocial Personality Disorder and a history of habitual violence (Virkkunen, 1985). When that sample was reanalyzed using childhood characteristics as a grouping variable, only prisoners with a developmental history of Undersocialized Aggressive CD were found to have abnormally low cortisol. This suggested that low cortisol was to be found only among groups who were chronically violent and psychopathic. In a clinic sample of

Biosocial Bases of Violence, edited by Raine *et al.*
Plenum Press, New York, 1997

boys, low cortisol was not found to be characteristic of the diagnosis of CD, but cortisol was found to be significantly low in boys who had CD and no anxiety disorder when compared to boys with CD who also had an anxiety disorder (McBurnett et al., 1991). Analyses of follow-up data on these same subjects suggested that cortisol was low in boys with DSM-IV Childhood Onset CD when compared to those with Adolescent Onset CD (McBurnett & Pfiffner, in press) and that low cortisol was associated only with overt, aggressive CD symptoms and not with covert CD symptoms (McBurnett, Lahey, Capasso & Loeber, 1996). These findings are consistent with the interpretation that low cortisol is characteristic only of those boys with CD who have a chronic, aggressive, low anxiety (psychopathic?) form of the disorder, similar to the interpretation of the Virkkunen (1985) study. We were interested in finding out whether this biological variable (low cortisol) was associated with chronic CD independently of known psychosocial correlates of CD, or whether the biological and the psychosocial correlates accounted for essentially the same variance in the development of chronic CD.

2. METHOD

The design compared four predictors of chronic aggression and covert antisocial behavior in 42 school-age boys referred to the Georgia Children's Center, a psychological assessment clinic. The data was part of the first four years of the Developmental Trends Study (DTS), a longitudinal study of delinquency in which clinic-referred boys are assessed at annual intervals (Lahey et al., 1990). The methodology for cortisol studies is described in greater detail elsewhere (McBurnett et al., 1991) (McBurnett et al., 1996). Regression models were tested for three criterion variables: *Aggressive CD* and *Covert CD* were composed of the total number of aggressive and non-aggressive CD symptoms, respectively, that were identified in separate DISC interviews of child subjects, their parents, and their teachers repeated at annual intervals over four years; and *Peer-Nominated Aggression* was the total number of classroom peer nominations for "Fights Most" and "Meanest" in sociometric exercises conducted in the first and second years of DTS assessments. Data for two of the four predictor variables, *Parent Psychopathology* and *Parent-Child Relations,* came from parent interviews conducted in the first year of DTS assessment. Parent Psychopathology was scored as the total number of positive diagnoses in both parents from: Antisocial Personality Disorder (301.70), Adult Antisocial Behavior (V71.01), Alcohol Dependency (303.90), Alcohol Abuse (305.00), any Psychoactive Substance Dependency, and any Psychoactive Substance Abuse. Parent-Child Relations was scored from parents' responses to questions about their communication with their child, discussions of child's past activities and future plans, and degree of relatedness and positive regard for their child. The value for *Cortisol* was the average concentration of cortisol from a single saliva sample taken in the second year assessment and a similar sample taken in the fourth year assessment. Finally, *SES* was computed from parental educational attainment and job status using Hollingshead's method (Hollingshead, 1975).

3. RESULTS

Table 1 presents the results of the regression analyses. For each of the individual predictors, the columns listed under "Model Parameters" represent how the predictors operate in the model: R-square (amount of variance-accounted-for), Beta (standardized mul-

Table 1. Multiple regression models predicting aggressive CD, covert CD,
and peer-nominated aggression

Predictors	Model parameters				Unique parameters		
	r^2	Beta	t/F	p	r^2	F	p
Aggressive CD symptoms							
Parent psychopathology	.074	−.009	−.81	NS	0	—	NS
Parent-child relations	.209	.374	3.47	.001	.112	12	.001
Cortisol	.178	−.368	-3.67	.0005	.126	12.8	.001
SES	.247	−.287	−2.55	.014	.06	6.5	.05
Model	.466		12.45	.001			
Covert CD symptoms							
Parent psychopathology	.187	.23	1.99	.052	.04	3.94	.055
Parent-child relations	.325	.482	4.29	.0001	.187	18.4	.001
Cortisol	.04	−2.02	−1.94	.058	.038	3.8	NS
SES	.107	−.019	−.16	NS	0	.03	NS
Model	.421		10.4	.0001			
Peer-nominated aggressive symptoms							
Parent psychopathology	.008	.098	.74	NS	.008	.55	NS
Parent-child relations	.028	.248	1.91	.062	.051	3.6	NS
Cortisol	.16	−.467	−3.81	.0004	.202	14.5	.001
SES	0	.214	1.55	.127	.033	2.41	NS
Model	.236		4.24	.0046			

tiple regression coefficient), the relevant *t* or *F* statistic, and the probability of obtaining the statistic by chance. The columns listed under "Unique Parameters" represent the strength of the individual predictors after controlling for the variance accounted for by all the other predictors in the model. When symptoms of anxiety was used as the criterion, the regression equations showed no significant results.

4. DISCUSSION

Despite numerous limitations to this study (e. g., data gathered from different assessment years, sample size limitations on ability to interpret interactions), the findings are provocative. Low cortisol was associated only with aggressive behavior (both Aggressive CD and Peer-Nominated Aggression) and not at all with covert antisocial behavior. This finding is not surprising, because similar results had already been found in this dataset using a different design (McBurnett & Pfiffner, in press). Parent Psychopathology showed the opposite pattern: it was associated only with Covert CD and not with aggression. Poor Parent-child Relations was not associated with Peer-Nominated Aggression, but it was associated with Aggressive CD and even more strongly with Covert CD. Finally, SES was associated only with Aggressive CD. This pattern supports the very tentative hypothesis that low biological arousal and deviant/rejecting parental behavior represent distinct mechanisms having differential effects on persistent episodic aggression and on inadequate internalization of social constraints. Raine (1993) proposed that the literature on biological bases of crime supported a genetic influence for petty, nonviolent offending; whereas violent offending is associated with several types of atypical biological findings, but no strong evidence of a genetic influence. The current findings regarding cortisol's relation to aggression and parental psychopathology's relation to covert CD is consistent with Raine's conclusion.

REFERENCES

Hollingshead, A. B. (1975). *Four factor index of social status.* New Haven, CT: Yale University.

Kruesi, M. J., Schmidt, M. E., Donnelly, M., Euthymia, D., & others. (1989). Urinary free cortisol output and disruptive behavior in children. *Journal of the American Academy of Child & Adolescent Psychiatry, 28*(3), 441–443.

Lahey, B. B., Loeber, R., Stouthamer-Loeber, M., Christ, M. A. G., Green, S. M., Russo, M. F., Frick, P. J., & Dulcan, M. (1990). Comparison of DSM-III and DSM-III-R diagnoses for prepubertal children: Changes in prevalence and validity. *Journal of the American Academy of Child and Adolescent Psychiatry, 29*, 620–626.

McBurnett, K., Lahey, B. B., Capasso, L., & Loeber, R. (1996). Aggressive symptoms and salivary cortisol in clinic-referred boys with Conduct Disorder. *Annals of the New York Academy of Sciences, 794*(Sept. 20: Understanding Aggressive Behavior in Children), 169–179.

McBurnett, K., Lahey, B. B., Frick, P. F., Risch, S. C., Loeber, R., Hart, E. L., Christ, M. A. G., & Hanson, K. S. (1991). Anxiety, inhibition, and conduct disorder in children: II. Relation to salivary cortisol. *Journal of the American Academy of Child and Adolescent Psychiatry, 30*, 192–196.

McBurnett, K., & Pfiffner, L. J. (in press). Estimating developmental risk for psychopathy using subtypes, comorbidities, and biological correlates of conduct disorder. In R. D. Hare, D. Cooke, & A. Forth (Eds.), *Psychopathy: Theory, Research, and Implications for Society* .

Raine, A. (1993). The psychopathology of crime. San Diego: Academic Press.

Scerbo, A. S., & Kolko, D. J. (1994). Salivary testosterone and cortisol in disruptive children: Relationship to aggressive, hyperactive and internalizing behaviors. *Journal of the American Academy of Child & Adolescent Psychiatry, 3*(8), 1174–1184.

Targum, S. D., Clarkson, L. L., Magac-Harris, K., Marshall, L. E., & Skwerer, R. G. (1990). Measurement of cortisol and lymphocyte subpopulations in depressed and conduct-disordered adolescents. *Journal of Affective Disorders, 18*(91–96).

Virkkunen, M. (1985). Urinary free cortisol secretion in habitually violent offenders. *Acta Psychiatrica Scandinavica, 72*(1), 40–44.

NAME INDEX

SUBJECT INDEX